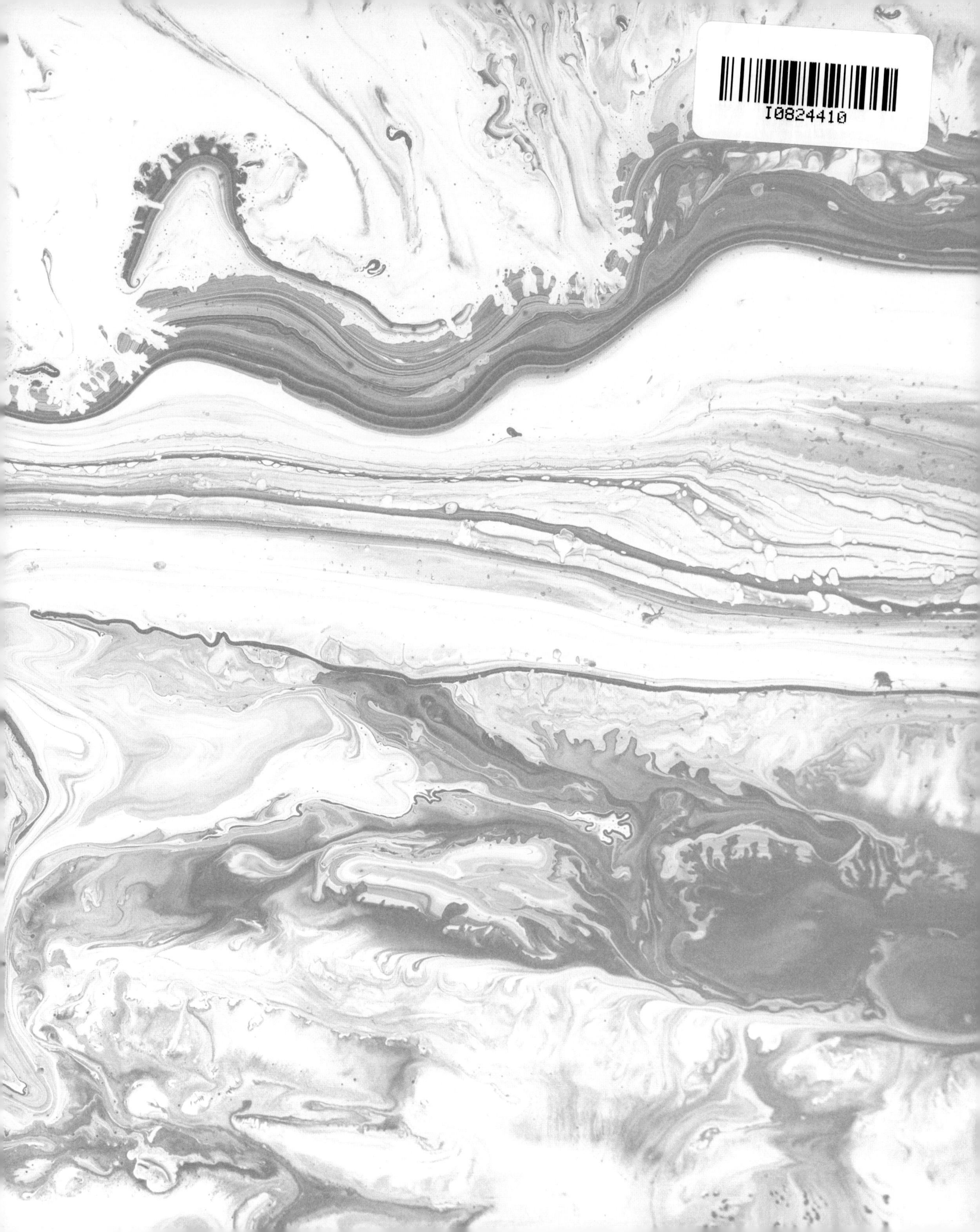
I0824410

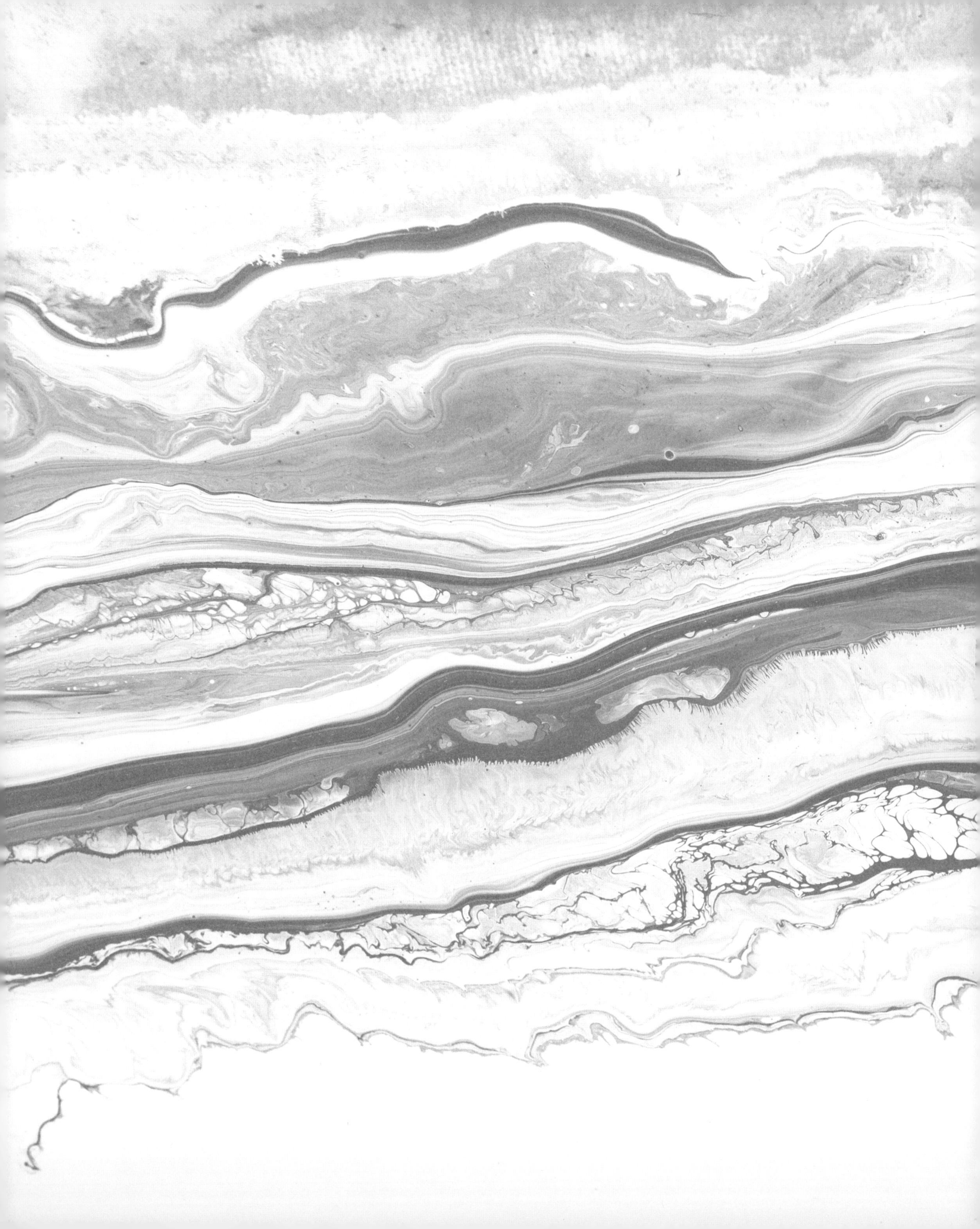

ACRYLIC PAINTING
MEDIUMS & METHODS

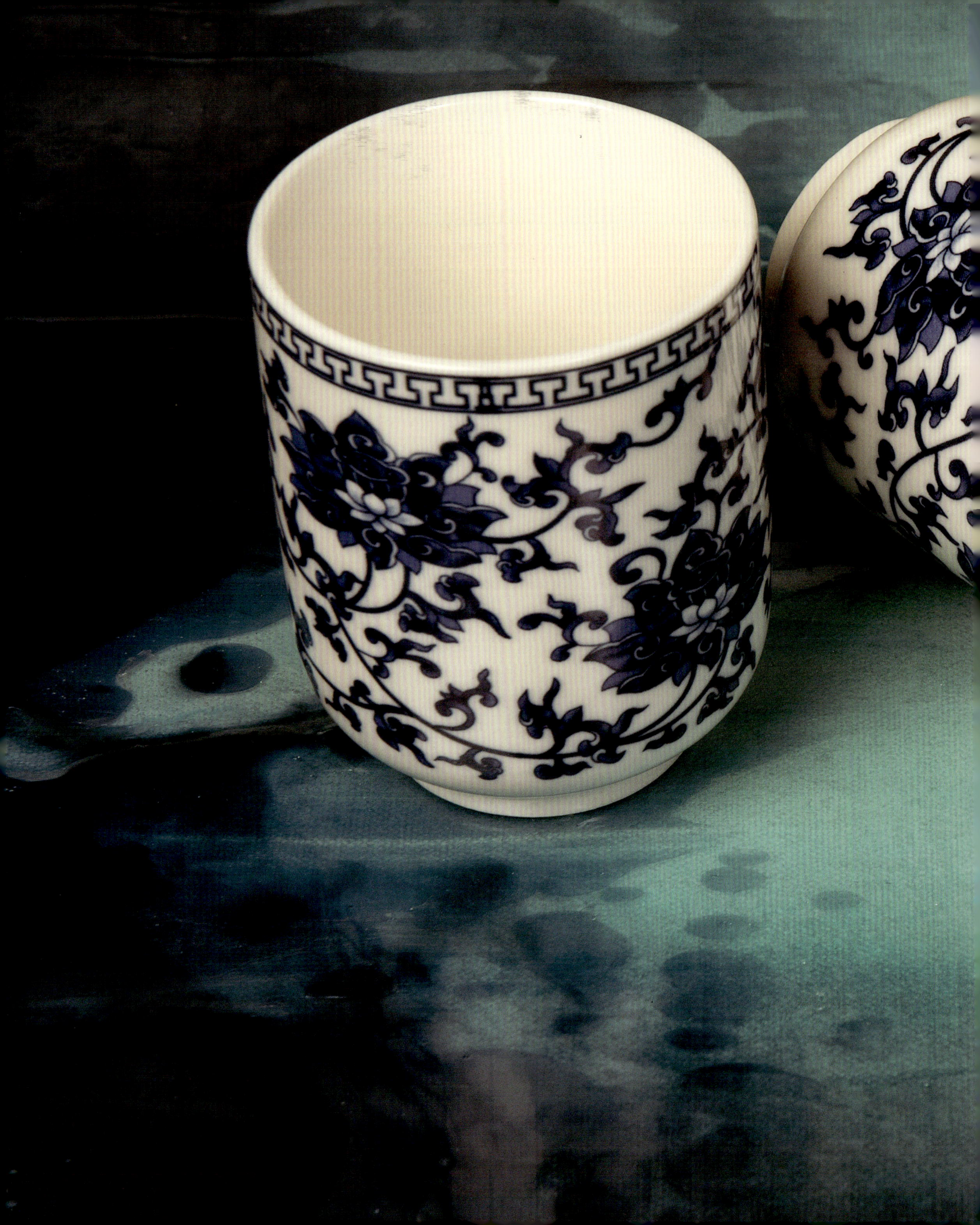

RHÉNI TAUCHID

ACRYLIC PAINTING MEDIUMS & METHODS

A Contemporary Guide to Materials, Techniques, and Applications

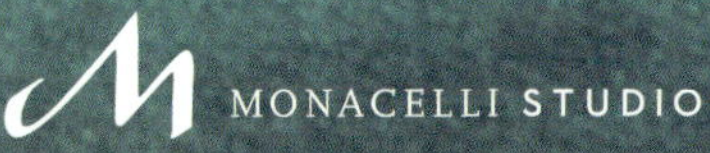

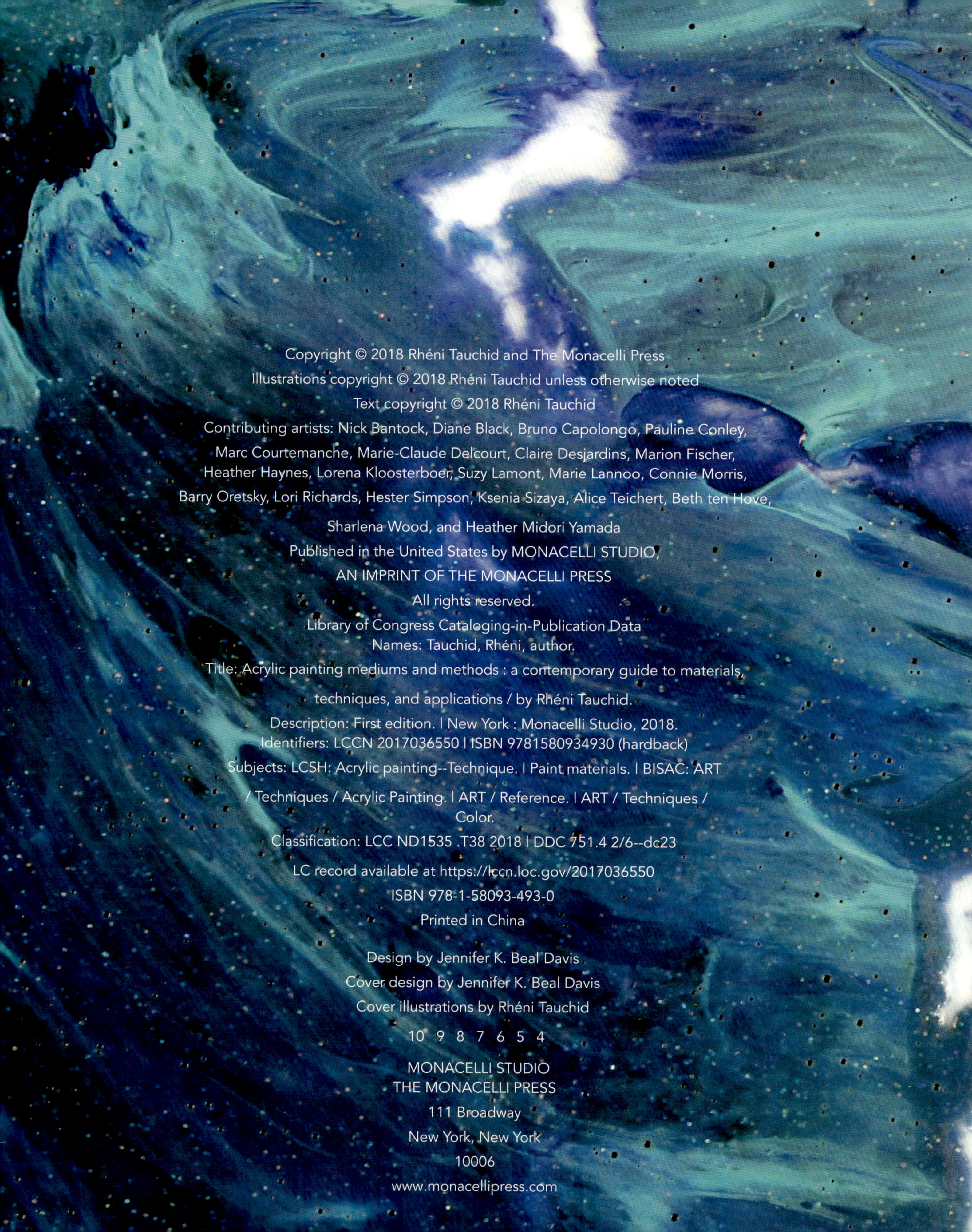

Contributing artists: Nick Bantock, Diane Black, Bruno Capolongo, Pauline Conley, Marc Courtemanche, Marie-Claude Delcourt, Claire Desjardins, Marion Fischer, Heather Haynes, Lorena Kloosterboer, Suzy Lamont, Marie Lannoo, Connie Morris, Barry Oretsky, Lori Richards, Hester Simpson, Ksenia Sizaya, Alice Teichert, Beth ten Hove, Sharlena Wood, and Heather Midori Yamada

Published in the United States by MONACELLI STUDIO,
AN IMPRINT OF THE MONACELLI PRESS

Library of Congress Cataloging-in-Publication Data
Names: Tauchid, Rhéni, author.
Title: Acrylic painting mediums and methods : a contemporary guide to materials, techniques, and applications / by Rhéni Tauchid.
Description: First edition. | New York : Monacelli Studio, 2018.
Identifiers: LCCN 2017036550 | ISBN 9781580934930 (hardback)
Subjects: LCSH: Acrylic painting--Technique. | Paint materials. | BISAC: ART / Techniques / Acrylic Painting. | ART / Reference. | ART / Techniques / Color.
Classification: LCC ND1535 .T38 2018 | DDC 751.4 2/6--dc23
LC record available at https://lccn.loc.gov/2017036550

ISBN 978-1-58093-493-0

Printed in China

Design by Jennifer K. Beal Davis
Cover design by Jennifer K. Beal Davis
Cover illustrations by Rhéni Tauchid

10 9 8 7 6 5 4

MONACELLI STUDIO
THE MONACELLI PRESS
111 Broadway
New York, New York
10006
www.monacellipress.com

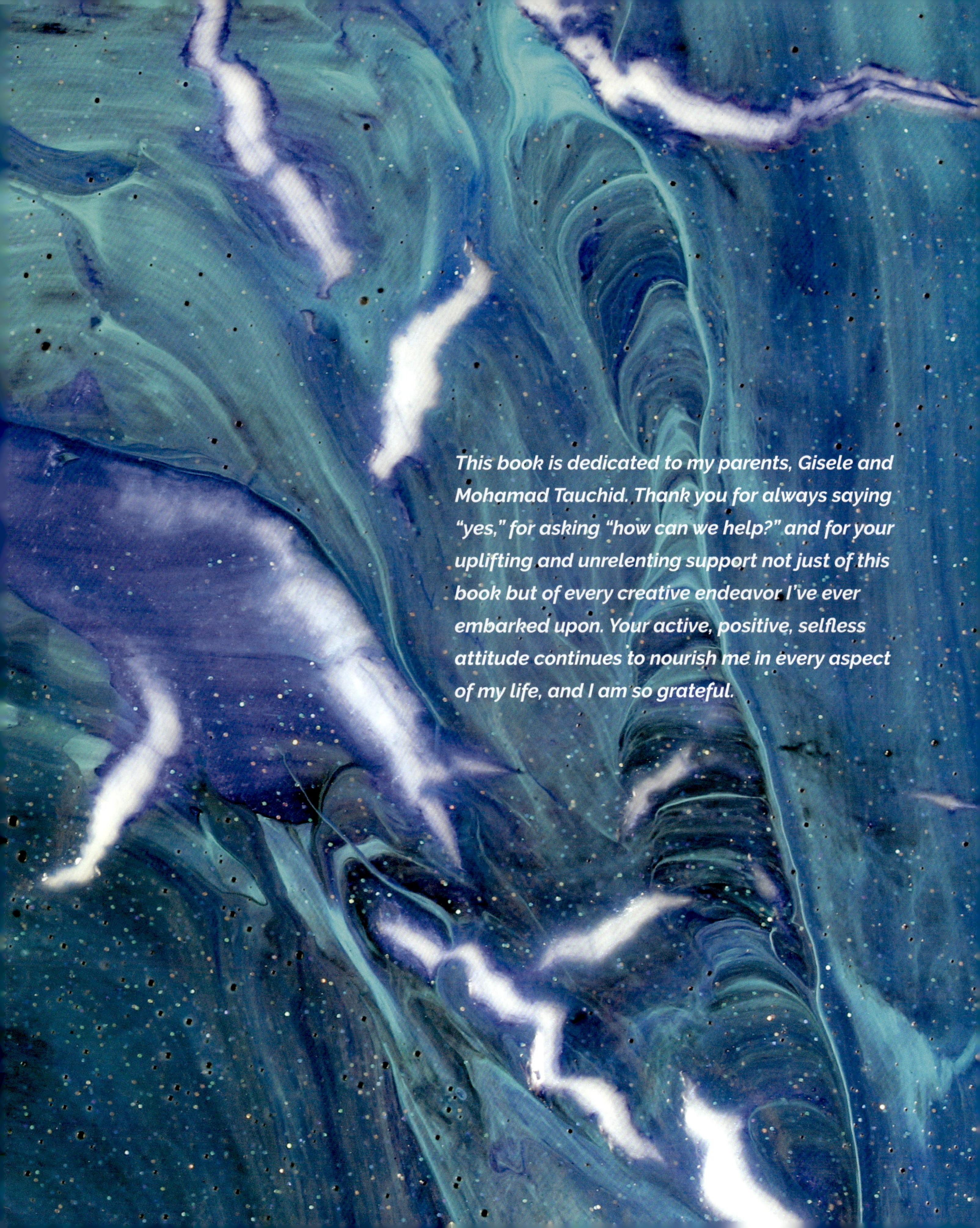

This book is dedicated to my parents, Gisele and Mohamad Tauchid. Thank you for always saying "yes," for asking "how can we help?" and for your uplifting and unrelenting support not just of this book but of every creative endeavor I've ever embarked upon. Your active, positive, selfless attitude continues to nourish me in every aspect of my life, and I am so grateful.

CONTENTS

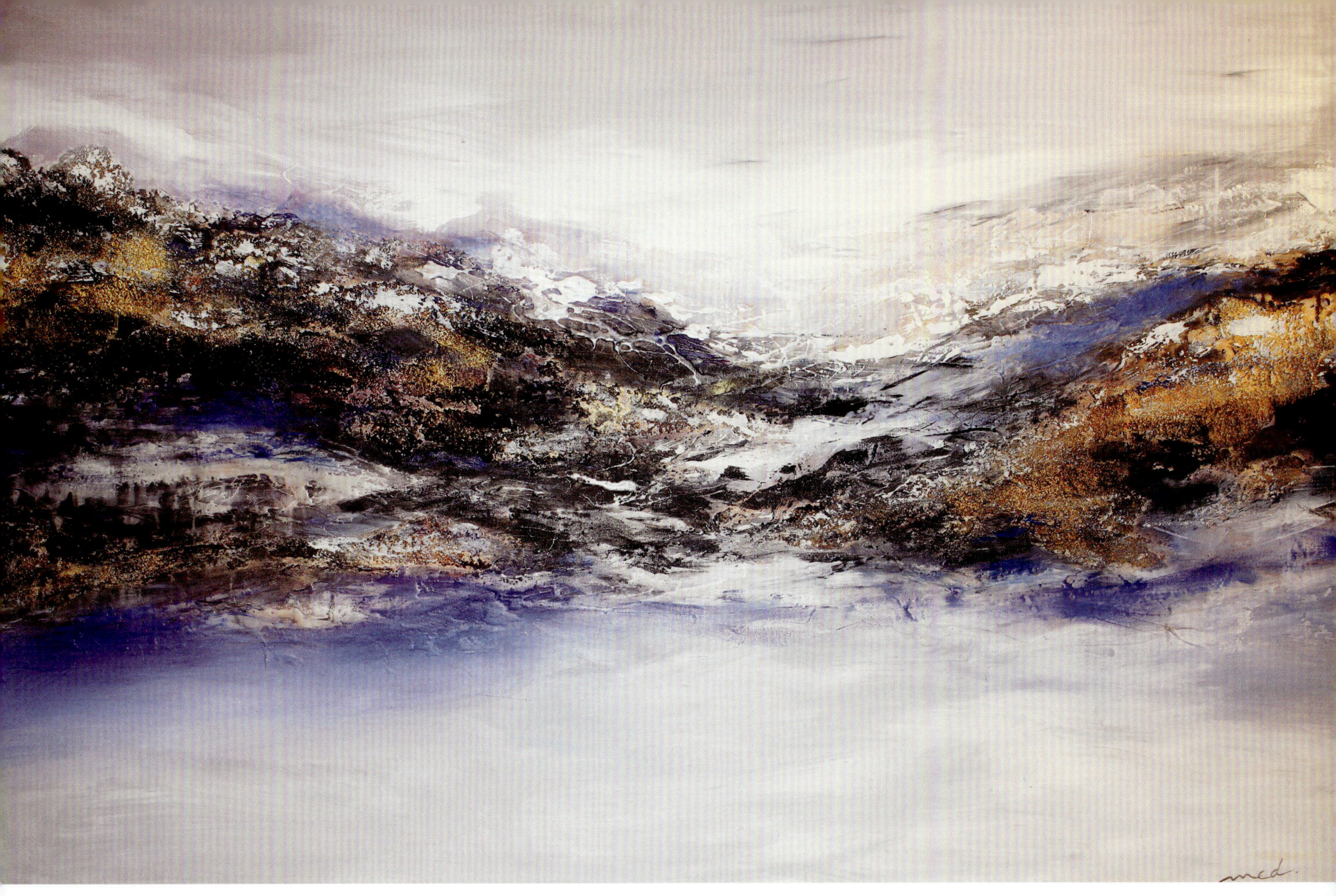

ABOVE Marie-Claude Delcourt, *Crossing Over*, 2015, acrylic on canvas, textured with acrylic mediums including gel medium, glass bead gel, and string gel, 40 x 60 inches (102 x 152 cm). Collection of the artist.

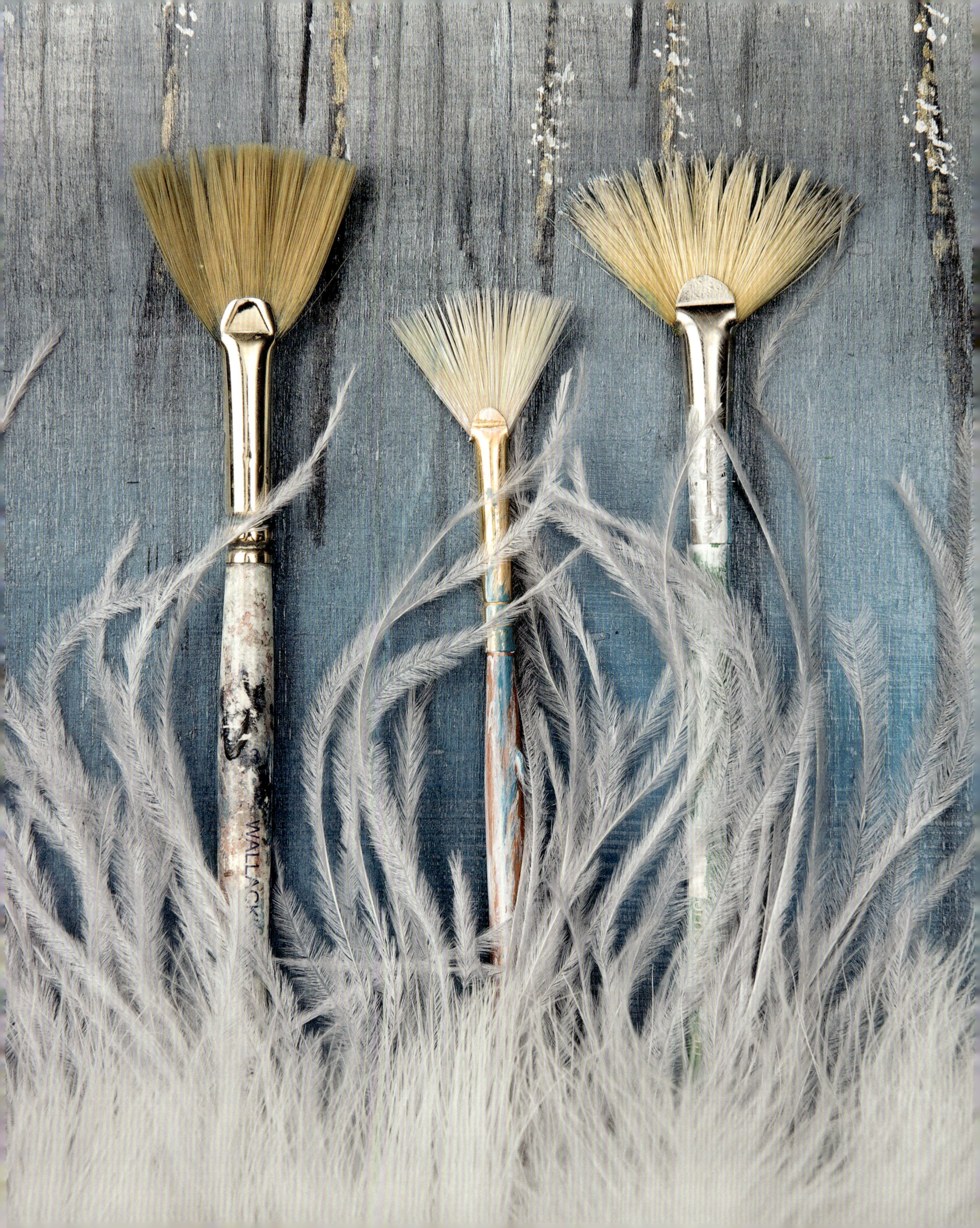
WALLACK

ACKNOWLEDGMENTS

As solitary as the act of conceiving, writing, and organizing a book of this sort is, it does take a great many people to give it substance and bring it to life. The people at Tri-Art Manufacturing and the products they create are the soul of this book. These products give life to my creative impulses. Tri-Arts' high-quality materials are the best I have worked with, and I am proud to be a part of the company. Everyone at Tri-Art, from the owner and brilliant visionary behind the company, my husband Steve Ginsberg, to the last labeler, has given their support to this project. My thanks, too, to Evan and Jess and the Art Noise staff for their support and assistance throughout the completion of this project.

Producing the photographs for this book has been a joy. My team—Jonathan Sugarman (photography), Connie Morris (photography), and Grace Paulionis (styling)—has been able to translate and capture my love of painting materials and the painting process. These photos, our collective art, elevate and celebrate my obsession with mediums to breathtaking effect.

Many thanks to my art community (local and online) for partaking in my acrylic mediums roundtable discussion and exploration. Your input was invaluable. I thank Lori Richards, Sue Langlois, Jane Colden, Heather Haynes, Beth ten Hove, Diane Black, Babi Sugarman, Mat Poirier, Kyla Munro, Suzanne Charo, Pauline Conley, Drew Harris, Marie Lannoo, Laurens Hoste, Peter John Reid, Jeanne Krabbendam, Marnelle North, Heidi Bergeron, Joslyn Wilson, Margaret Mclauchlan, and all of those on Facebook who took the time to fill out my questionnaire. I also appreciate the advice and conversations about acrylic paint use and conservation with Alison Murray and Michael O'Malley from the Art Conservation department of Queen's University (Kingston, Ontario).

This book was made possible by the hard work of my stellar team at The Monacelli Press. I am so very grateful to Victoria Craven for reaching out and working together with me to conceive of the concept for this book, and for her patience and guidance as we developed and produced its contents. Thanks to James Waller, project editor, whose calm, precise, and meticulous editorial skills have honed and polished my manuscript into a cohesive whole (while keeping my tendency to run amok in check). Thank you to Jennifer K. Beal Davis, designer, for her delicious and sophisticated execution of the the harmonious design and layout of this book.

Lastly, thank you in particular to Ma for proofreading, to Steve, Mica, and Tilda for your support and patience, and to everyone in my extended family and my closest friends, for your encouragement and support through all aspects of this project.

FOREWORD: TWO MEDIUMS

by Alice Teichert

In my world, everything has a melody, a music within. Likewise, each medium I work with is unique, with its own specific, physical attributes to explore and embrace. Over the course of creating with acrylic mediums, a visual notation slowly reveals itself to me. It realizes as a resonance much like the ripples of a language dancing through space, water, and time.

The layering aspect is vital in my painting process. The layers ultimately enable me to move deeper into a playful dance of light and textural exchanges—a space devoid of known references.

I have always nurtured the need to examine the media I work with. I also strive to become a conduit for the energy that exists beyond simple cognitive knowledge, color, texture, and flat planes. I have thus managed to develop my own connective links to the endless possibilities presented by acrylics. These links appear time and again throughout my paintings and are still the most successful ways to voice the fundamentals of my visual expression—*transparency* and *translucency*. Gloss gel medium for transparency and dry media ground for translucency: applied in successive layers, both find their own distinctive tone and character. Up to thirty fine-pigmented layers interspersed with other textures or graphic elements compose the signatures that are unique to my work.

My work relies on immediacy, simplicity, fluidity, and spontaneity. But it takes time to get to know the mediums personally and to create one's own relationship with each one. Therefore, books on acrylic painting have always been a part of my studio and my personal resources. Instead of having to go through every trial and error on my own, I receive supportive assistance that enables me to speed up and thereby release my own creative process. Such is the case with this new guide to materials, techniques, and applications. Here, Rhéni Tauchid offers a palpable sense of how mediums behave in relation to their various environments and applications. Her expertise is exceptionally well presented, a clarity essential for artists who want to explore with confidence the worlds that acrylic mediums have to offer.

A bio of Alice Teichert appears on page 265.

OPPOSITE Alice Teichert, *Nuance*, 2013, acrylic, thick pencil, and pastel crayons on canvas, 55 x 40 inches (140 x 102 cm). Private collection, Cobourg, Ontario, Canada.

This composition uses successive applications of acrylic medium products in many fine layers to create an exchange between translucency and transparency, allowing the eye to penetrate the deeper layers while intensifying the viewer's gaze into the wide open space.

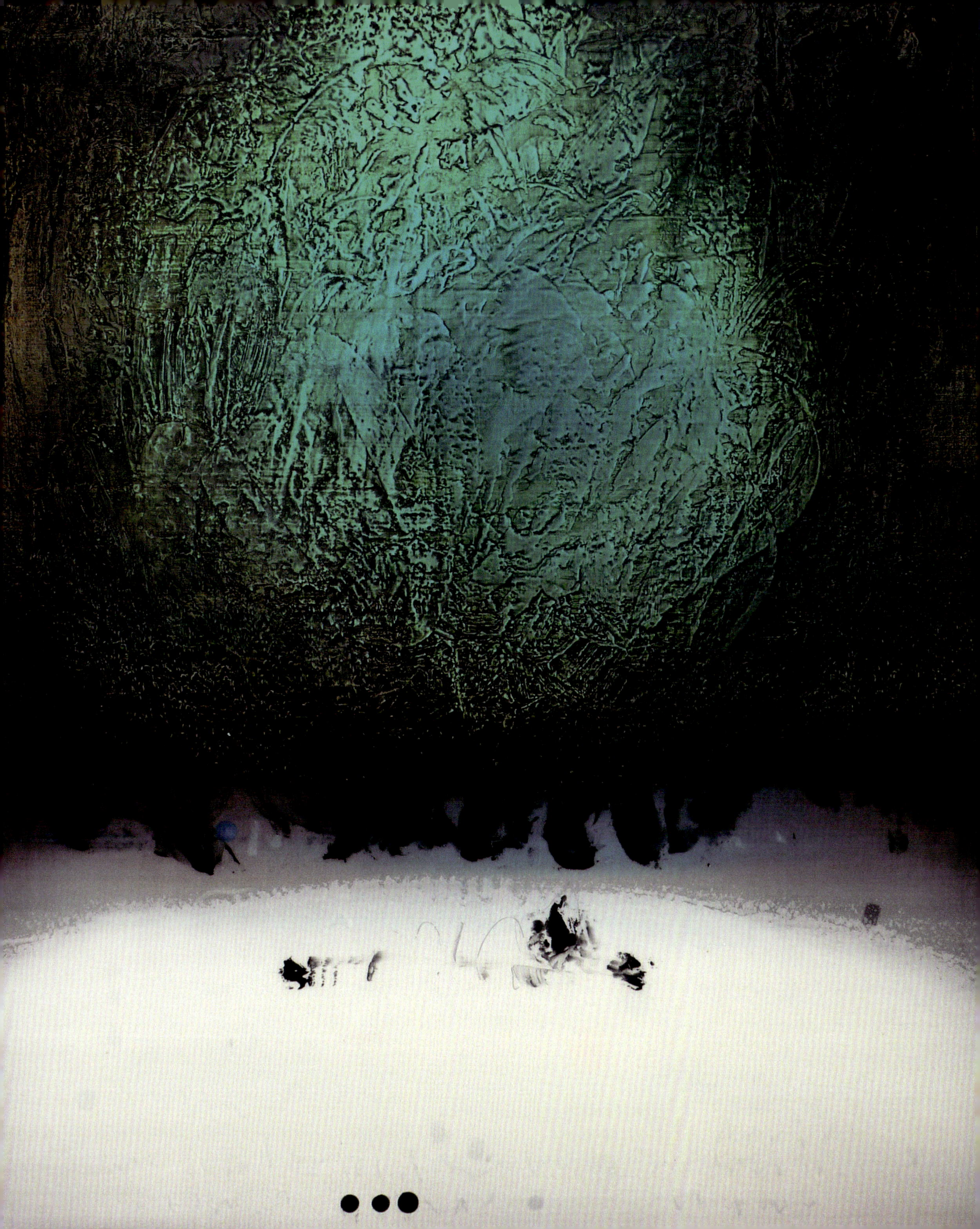

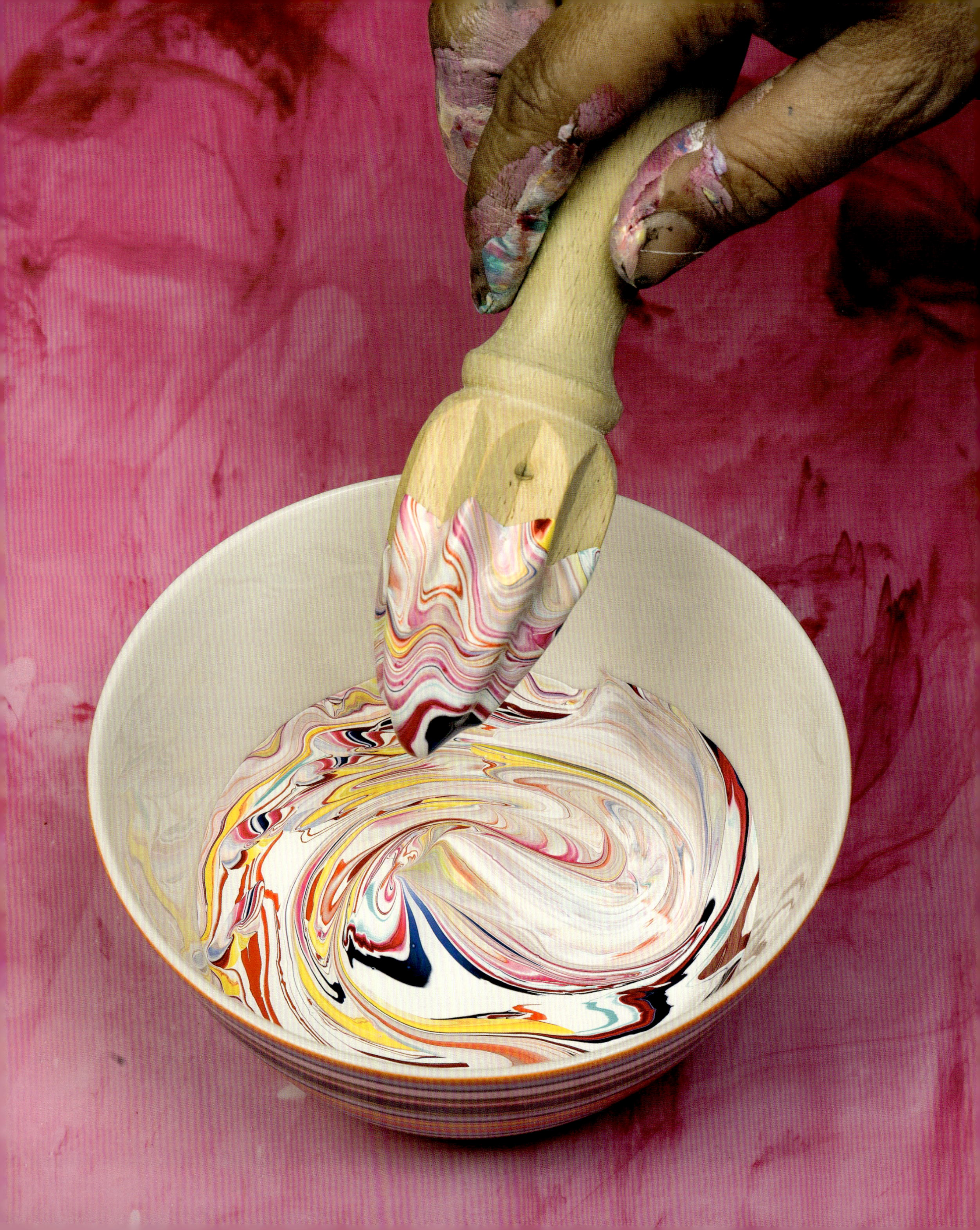

INTRODUCTION

They are called auxiliary, accessories, additives, accompaniments to acrylic paints. But in point of fact, acrylic mediums are *essential* to acrylic painting today, and this book will show you why.

An acrylic medium is a substance that is mixed or used with paint to impart a particular physical characteristic to the paint or to the painting's surface. They are the conduit through which painters can more vividly express color, form, and the poetry of surface. So while they may seem like sidekicks to the superhero that is color, mediums are the ones with the real superpowers.

What artists know about acrylic mediums they primarily know from experimentation. These materials are just beginning to be explored in art classes and schools. We are all gathering information, teaching ourselves and each other—each of us a novice. The new educators are art-supplies stores' staff, manufacturers' materials consultants, and fellow artists and craftspeople.

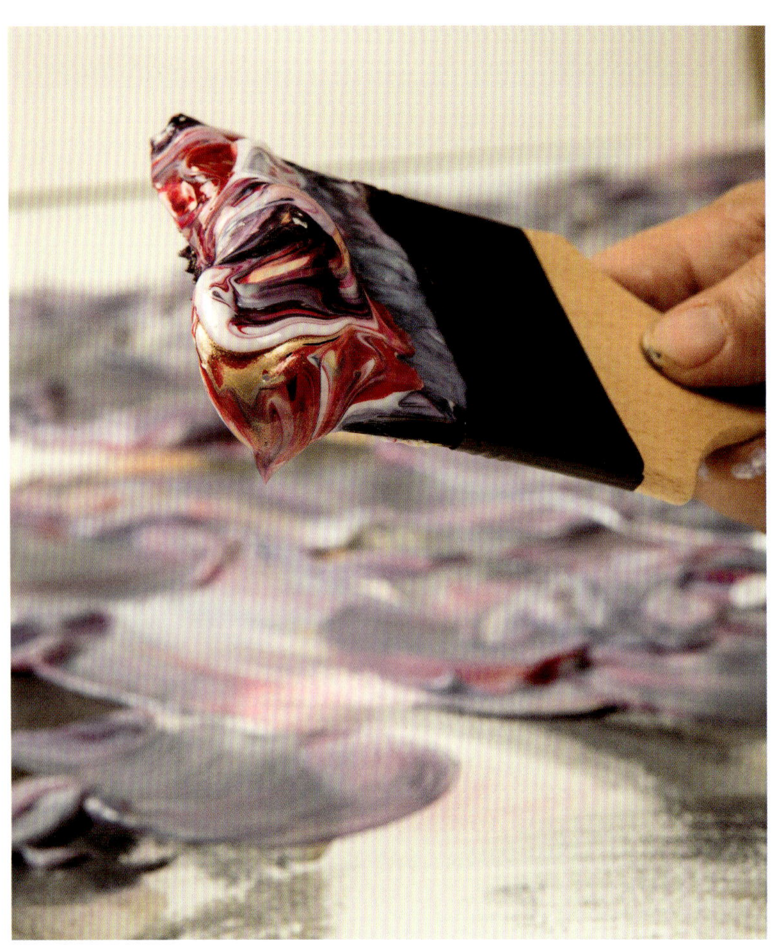

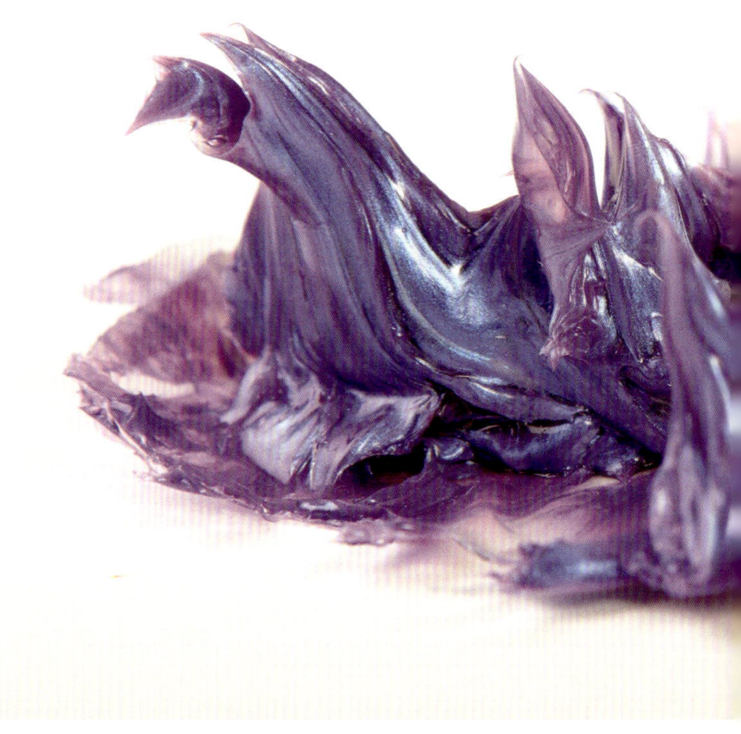

ABOVE, LEFT Brush with gel medium and loosely mixed color.

ABOVE, RIGHT Gloss gel with dioxazine violet and interference blue.

I first started painting with acrylics in the late 1970s, but it wasn't until the early '90s that I even became aware that there was such a thing as an acrylic medium. Before that, it was all just color and water to me. It took me years to uncover the complexity of this deceptively simple material. First, there was just the paint itself: it could be mixed with water, it adhered to a variety of supports, and it was easy to manipulate. But then came the confusion of differing formats: liquid acrylics, high-viscosity paints, acrylic inks! And finally the medley of mediums, exploding onto the marketplace with a staggering diversity of choices. Whoa, mind blown.

I started by messing with them, to see what would happen. Just to see if I could correctly predict the results. But what began as a tactile exploration became a necessary and often revelatory exercise. Intuitive exploration is the

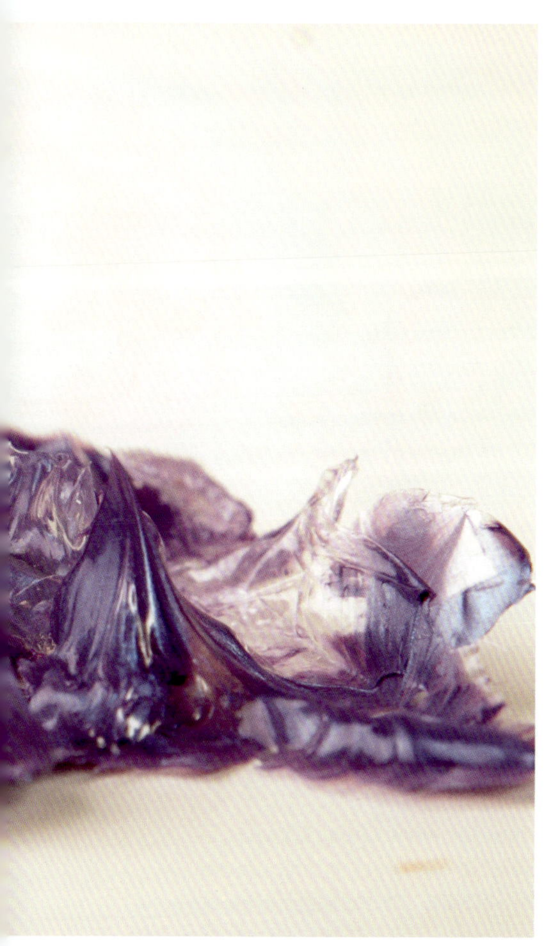

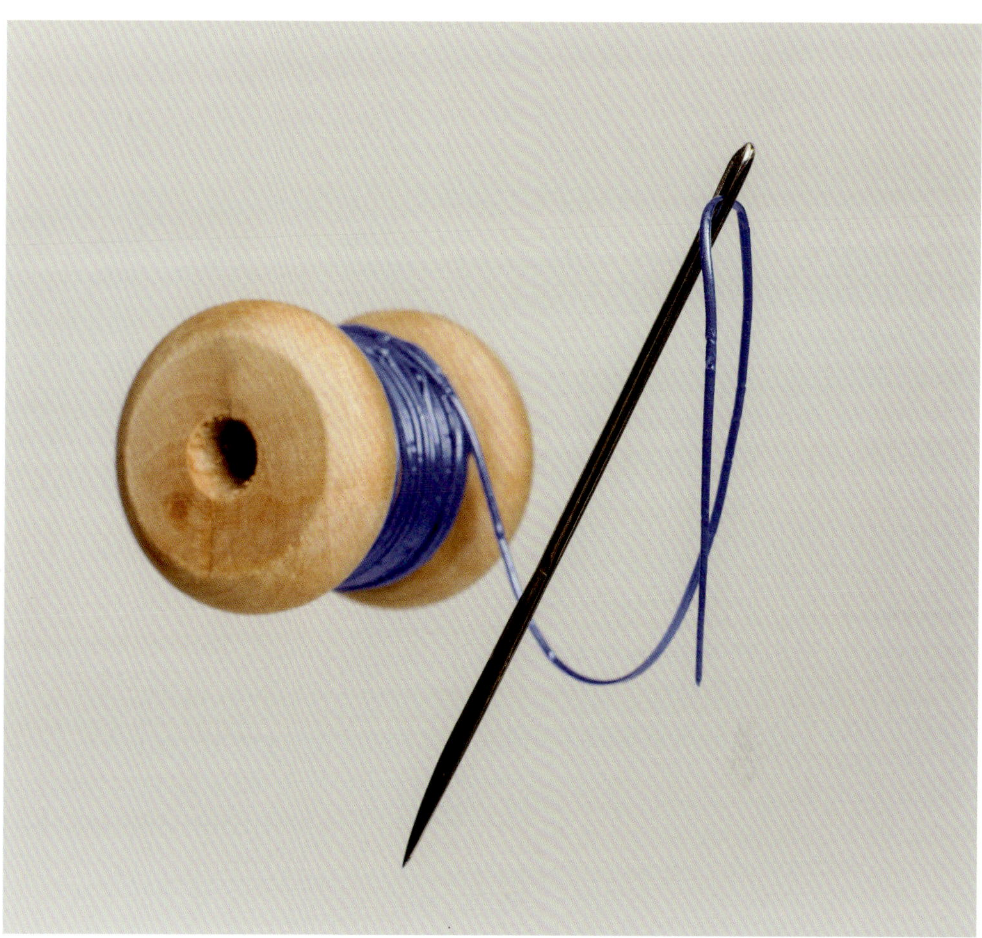

ABOVE, RIGHT Acrylics are the thread with which I create my paintings. (The photo shows gloss gel tinted blue and extruded.)

key to finding out how these materials can work as part of your creative process. Through my experimentation—and what writer Elizabeth calls "the random hurricanes of outcome"—I developed an understanding of the indispensable nature and deep potential of these materials.

It's sometimes hard to smash your habits as an artist and to wade into unknown waters, be it with a new material, subject, or approach. And the world of acrylic mediums and additives is big—continuously expanding and bewildering to navigate. There is a widespread assumption that acrylic mediums are really useful only for goopy, loose, abstract-style paintings. But this tendency to associate mediums with less formal compositions inaccurately pigeonholes these tools, which can have a transformative effect on paintings of *any* style, from the very expressive and abstract to exquisitely detailed realism.

There is no single way to explore acrylic mediums. Consider this book a jumping-off point and let inspiration, fueled by information, take you the rest of the way. Mix your mediums together in varying proportions, throw water at them, sink objects into them, expand the parameters of your usual art practice. Become familiar with the essence of the materials and understand the principles behind the methods. Through this process your paintings will gain depth, dimension, and individuality.

The photographs in this book portray mediums being used in both simple and complex ways. Many of the photos utilize my paintings as background. They are the culmination of years of experimentation with acrylic mediums, and an expression of my creative materials journey. (They also make terrific backdrops!) But, really, they're just starting points, showing you the potential of the materials and methods of application. It is up to you to interpret these tools to find your own creative voice.

The results of your experimentation may not stick, but they'll leave a residue of knowledge, and knowledge always informs practice. Mediums will enhance your artwork if you allow them to shift your perceptions and your process. By using acrylic mediums, you will have more control over the outcome of your work, gaining the potential to produce work that has finesse, subtlety, and sophistication or—if you wish—bold, lush textural expressions.

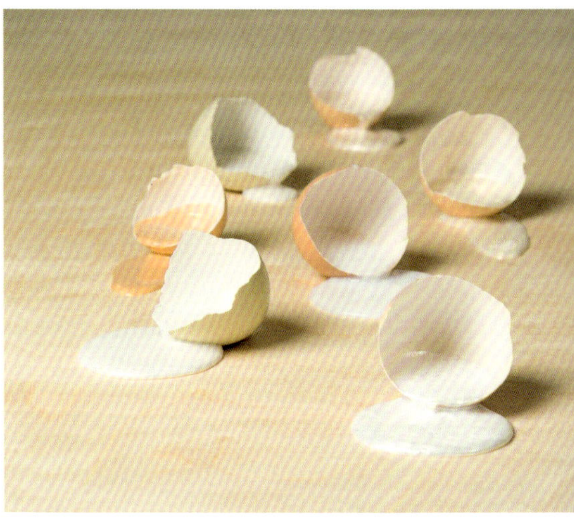

ABOVE Iridescent color mixes into various liquid mediums, the birthplace of new ideas.

OPPOSITE Lorena Kloosterboer, *Tempus ad Requiem V*, 2015, acrylic on canvas, 31 1/2 x 15 3/4 inches (80 x 40 cm). Private collection, UK.

About her work, artist Lorena Kloosterboer says, "I build up the support with many layers of gesso and wet-sand it to obtain a smooth surface, hiding the texture of the canvas. I trace my composition, then carefully build up transparent layers of color by diluting my acrylics with gloss medium and a bit of water laced with flow improver, which helps glazes run more easily. When I'm happy with an area, I protect it with a layer of pure gloss medium—a 100 percent acrylic polymer dispersion, basically an acrylic paint without color pigments that dries crystal clear—so that when I continue painting and make a mistake I can quickly wipe the top layer off with a wet sponge without damaging previous work. Once I'm finished, I cover my painting with an isolation coat of gloss medium. To protect the painting surface and change the surface luster, I varnish with a mixture of gloss and matte varnish for a satin sheen."

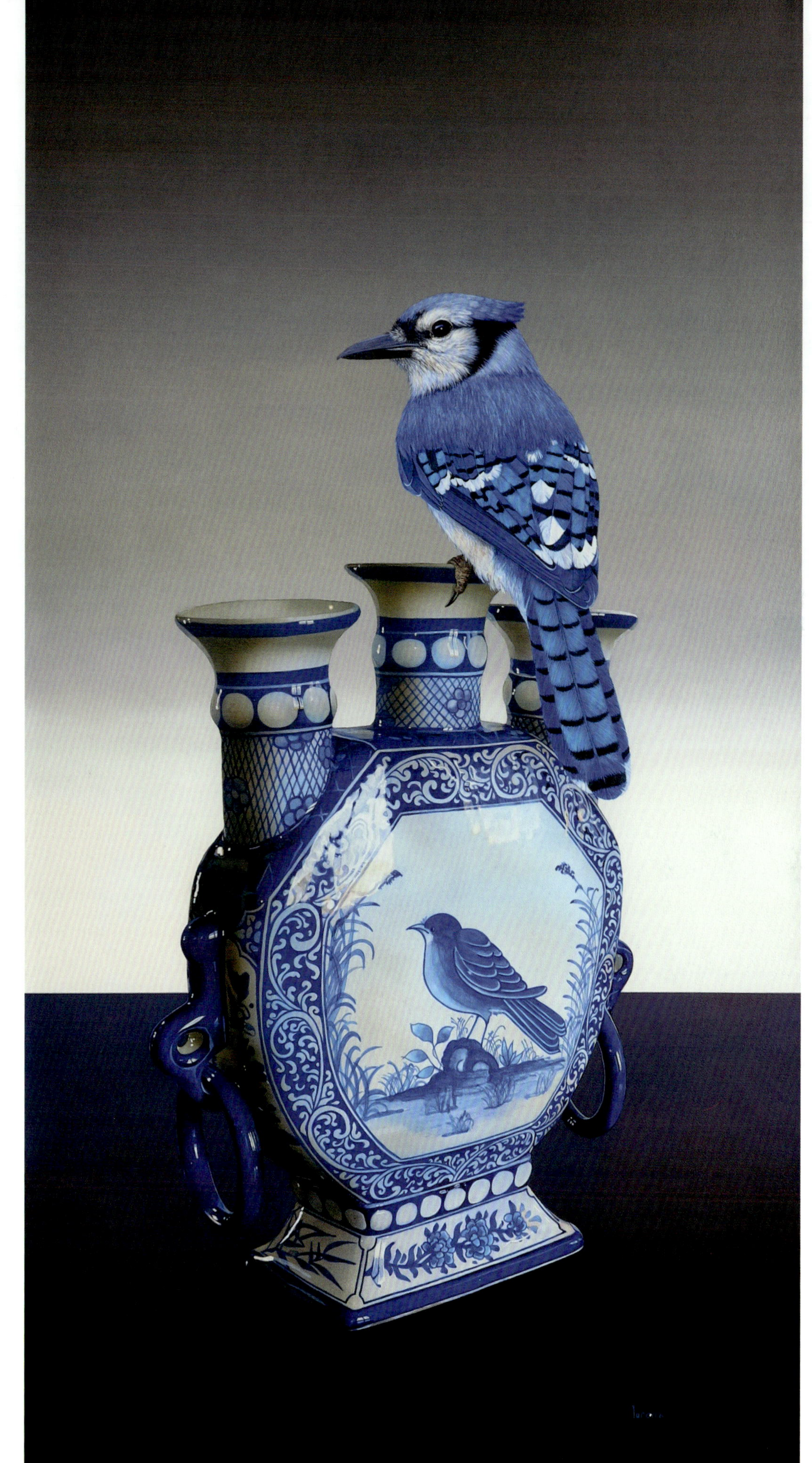

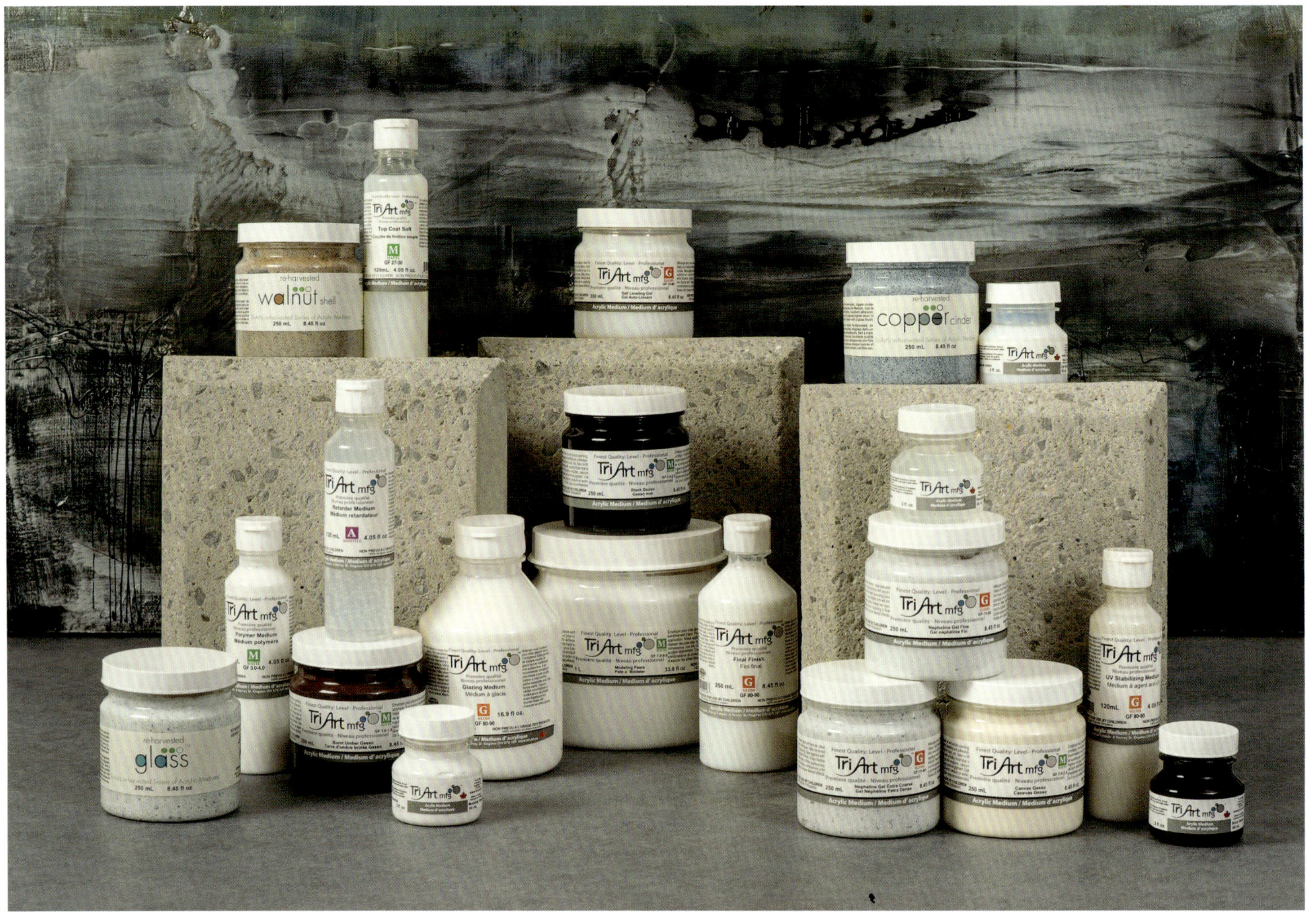

ABOVE An assortment of acrylic mediums and additives.

WHY USE MEDIUMS?

Mediums are diverse and they are legion, and only by getting to know them can you assess their necessity in your art-making process. Mediums are indispensable to some painters, while others paint successfully without mediums. Using them is a choice, not an imperative. I will stress, however, that mediums can *expand* the physical capabilities and attributes of your colors, can enable you to *express* variety and depth in your paintings, and are *economical* to use. You will do well to give them a serious look.

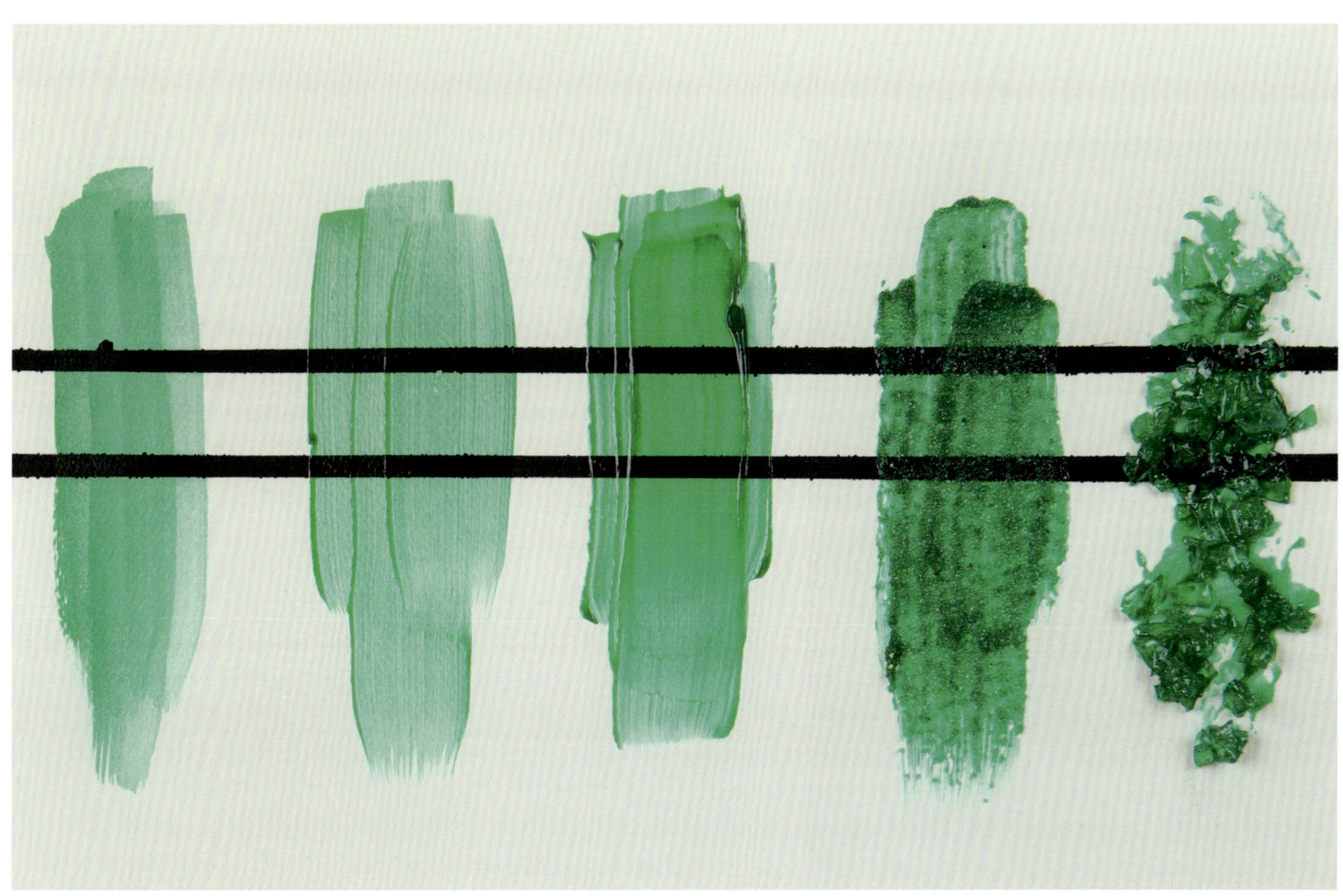

ABOVE Phthalo green diluted with (from left to right) water, liquid medium, gel medium, granular medium, and recycled plastic medium. The black line illustrates the relative coverage (transparency/opacity) of the paint mixtures.

EXPANSION

Mediums make good things better. Why settle for salt and pepper when there's a whole world of herbs and spices available to you? Mediums increase everything: size, textural capability, transparency, strength of the paint film. They are the great augmenters of color. For example, desaturating a strong color can bring out the bright undertone of a professional-grade color, imparting a lush depth to a transparent pigment.

EXPRESSION

Changing the physical appearance of acrylic paints in innumerable ways, mediums offer a staggering variety of expressive choices. They are capable of changing many aspects of painting, from adhesion and transparency to surface texture and luster. However you want your paints to look or feel, there is a medium for that.

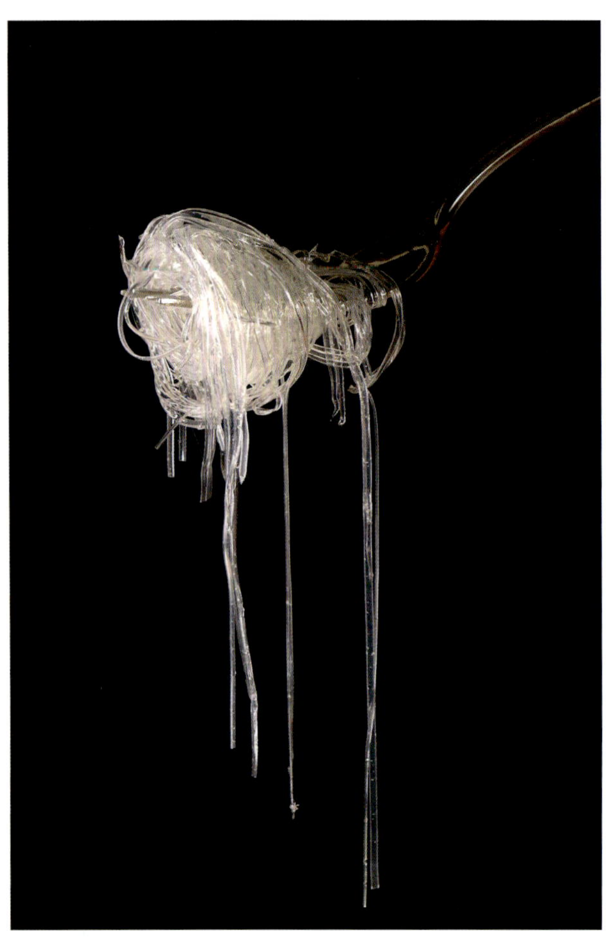

ABOVE Gloss gel "noodles" on a fork. You should "taste" all the possibilities.

ECONOMY

It makes good, economical sense to use acrylic mediums. The most expensive component of any paint material is pigment, and mediums (for the most part) don't contain any. Mediums are made up of acrylic emulsion and sometimes contain texture-building particulates, so they're inexpensive. For example, a liter (34 fl oz) of gel medium is half the price or less of the same quantity of an average professional-grade acrylic color.

If you have a limited art materials budget, you can stretch your dollar by using mediums. Because pigment loads differ greatly according to brand and the quality of the paint, it can be more economical to buy highly loaded, professional-grade colors in small sizes and extend them with acrylic mediums. (Student- or economy-grade paints will lose color saturation at lower levels of dilution, requiring more color to achieve the same result.)

Painting with mediums makes sense. They don't cost a fortune, there are so many interesting ones to pick from, and they make your color sing!

COOKING AND PAINTING

I paint as often as I can, which some weeks is not at all. I prepare food, in some capacity, every day. Cooking and enjoying food is a continuous source of inspiration that, though sometimes frustrating or irritating, is a creative outlet for me. And for me there's an analogy between cooking and my use of acrylic paints and mediums: Textures and flavors are expressed through new and varied combinations, culminating in something complex that is formed through learned and improvisational methods. My materials—my "ingredients"—are always influencing the outcome.

Like my cooking, my paintings are rarely subject-driven. They are a layered exploration of material, reaching for an end result in which the ingredients and processes coalesce into something resembling harmony.

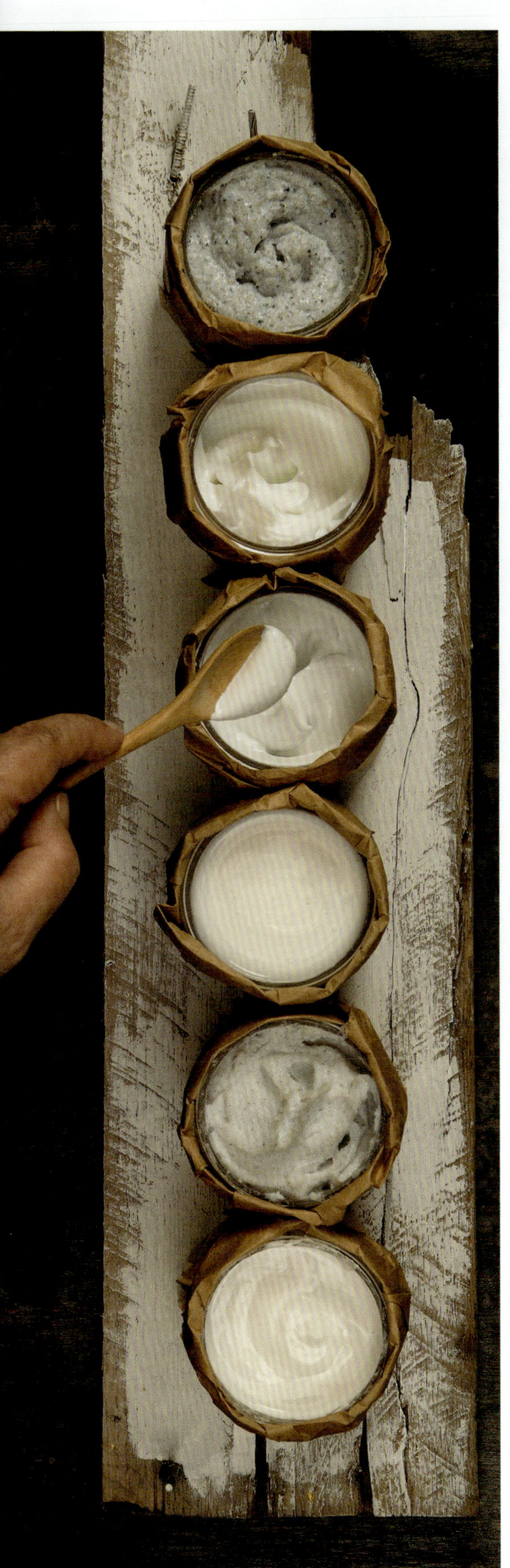

HOW THIS BOOK IS BUILT

Consider this book the speed-dating portion of your acrylic education. A whole lot of mediums are examined in detail but presented in a way that will bring you up to speed on how they work, what they work with, and how they enhance your particular style of painting.

The book has two parts: Part 1: Materials identifies, categorizes, and describes each member of the acrylic-mediums family, as well as supports and tools. Part 2: Methods explores how the mediums can be used, both individually and in combination, to produce and augment acrylic paintings.

It should be noted that the information presented in the book is not exhaustive. New mediums are being invented and manufactured on an ongoing basis. And regarding methods, every artist and craftsperson using mediums has the potential to come up with a variation on existing practices and to create new techniques unique to their process. The information presented here, however, will give you a good overview, and the knowledge you'll gain will help you navigate what is to come.

ABOUT THE PRODUCTS

Full disclosure: The description of acrylic mediums and their use in this book comes with a caveat. While I have done my best to make this a general reference on existing acrylic mediums, I primarily reference acrylic materials produced by Tri-Art Manufacturing because of my affiliation with the company and my familiarity and appreciation of their products. There are many dissimilarities between acrylic brands, only some of which will be explored here. For further information on other brands' mediums, visit their manufacturers' websites.

LEFT Dipping into acrylic mediums.

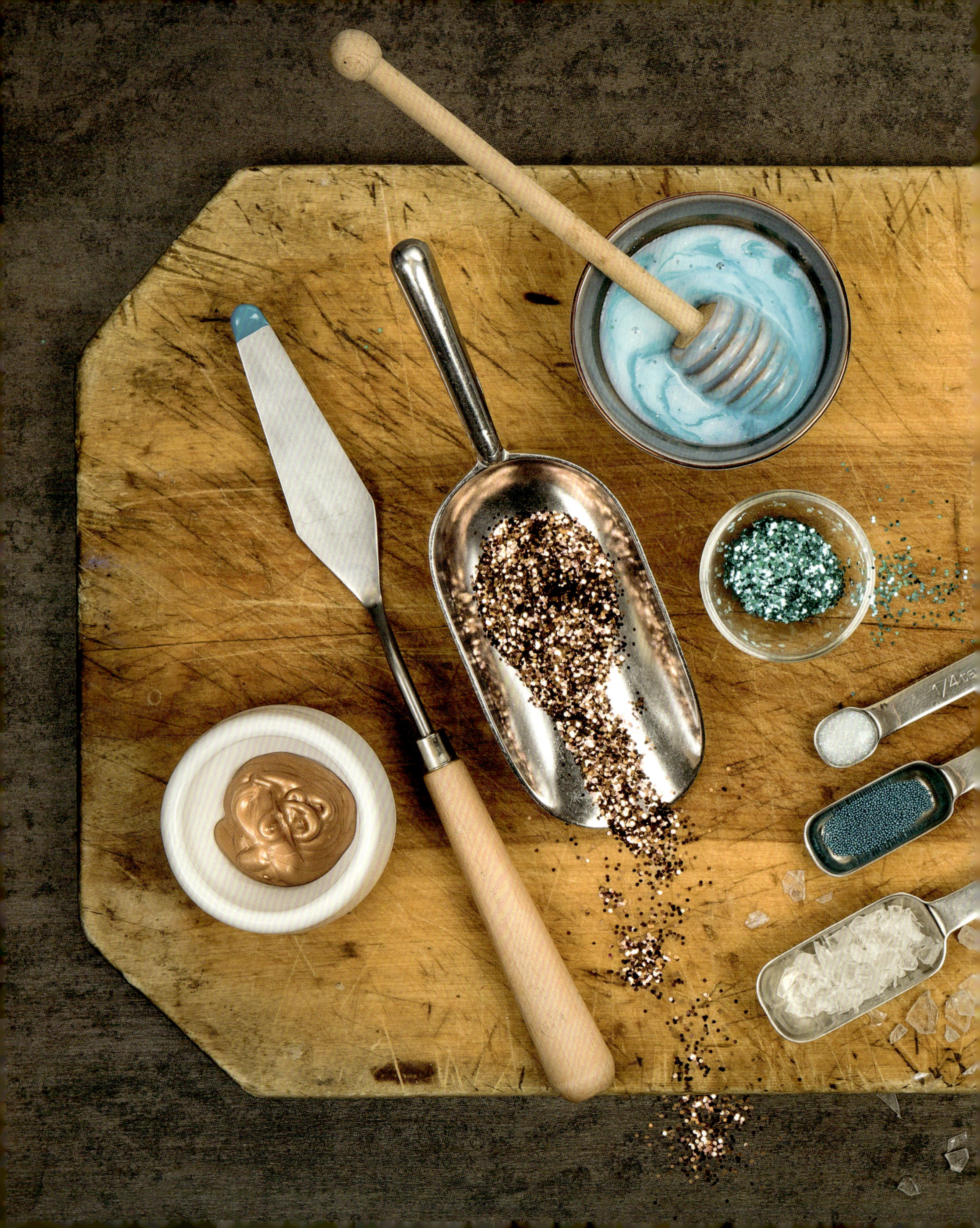

PART 1

MATERIALS

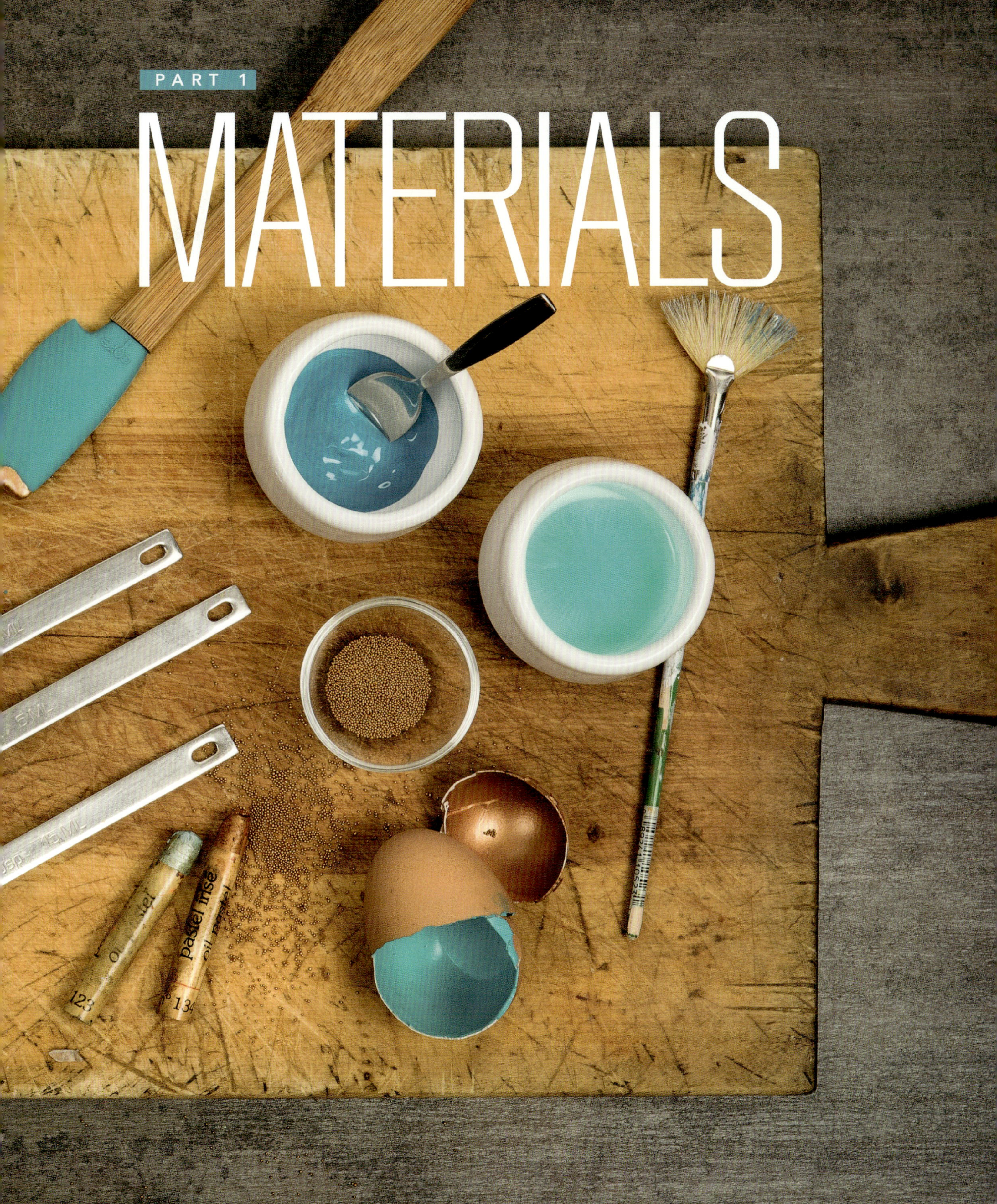

CHAPTER 1

ACRYLIC MEDIUM BASICS

FROM THE UBIQUITOUS GLOSS GEL, to polymer mediums that react to UV light, to custom mediums, acrylic mediums offer a world of visual and tactile possibilities to be explored. This book is your primer, your texture lexicon, your Rosetta Stone for decrypting the complexity of acrylic mediums from the ground to the final top coat.

Are acrylic mediums necessary? People who are new to acrylic painting, and even those who have been painting with acrylics for years, have asked me this question countless times. I'm never really sure how to answer, other than to say that for me and my painting process, they are absolutely essential.

It is, however, a fact that mediums will enhance and facilitate many aspects of acrylic use. For many artists, the primary barrier to incorporating mediums into their art practice is not aversion to mediums but rather ignorance of how they work. Acrylic mediums are a study in contrasts, ranging from matte to gloss, thin to thick, transparent to opaque. How they behave with each other and in conjunction with acrylic colors is dictated by their intrinsic properties. Understanding these basic properties will help you use them to their best potential.

Why use mediums? Because they augment and amplify both your paint and your process.

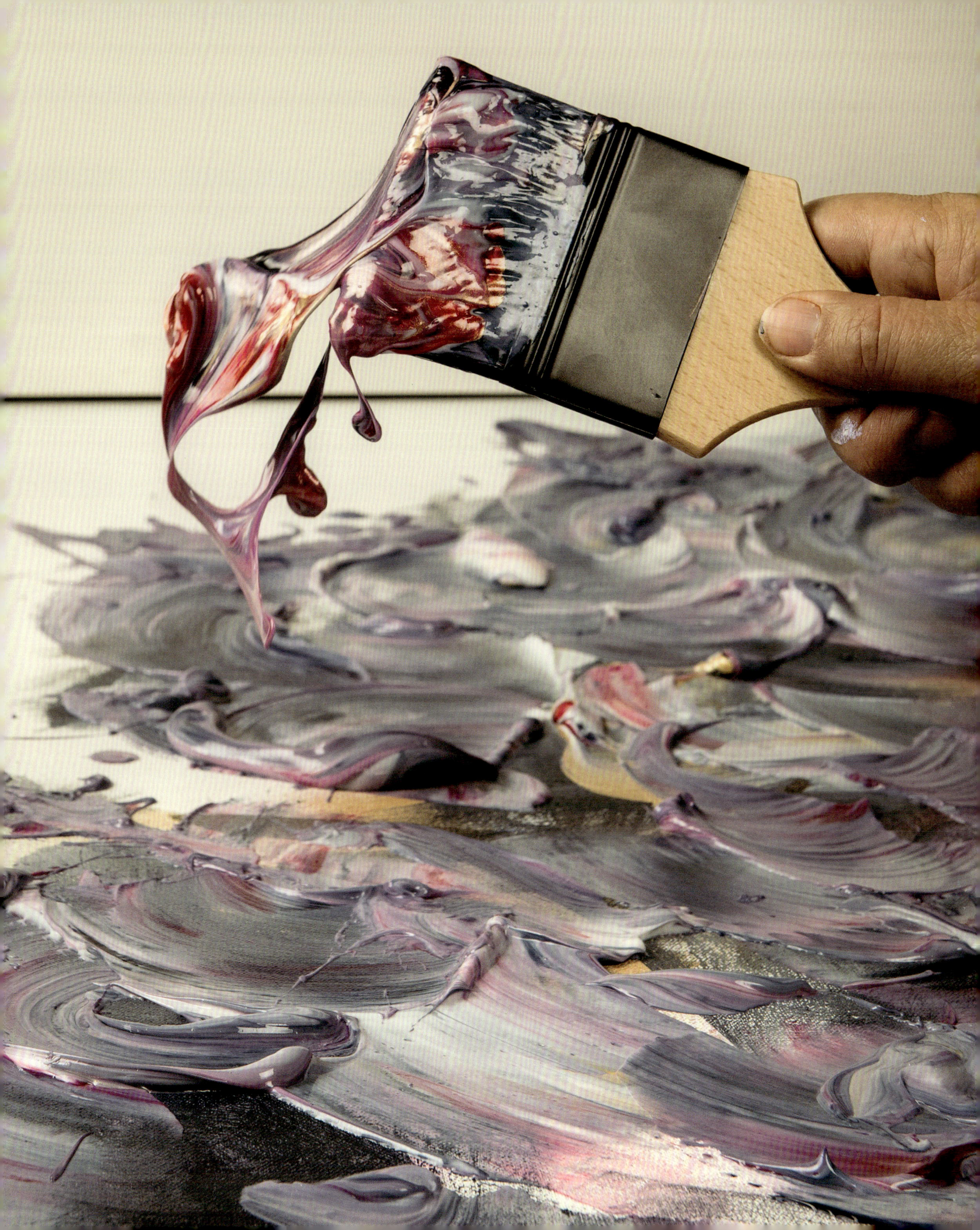

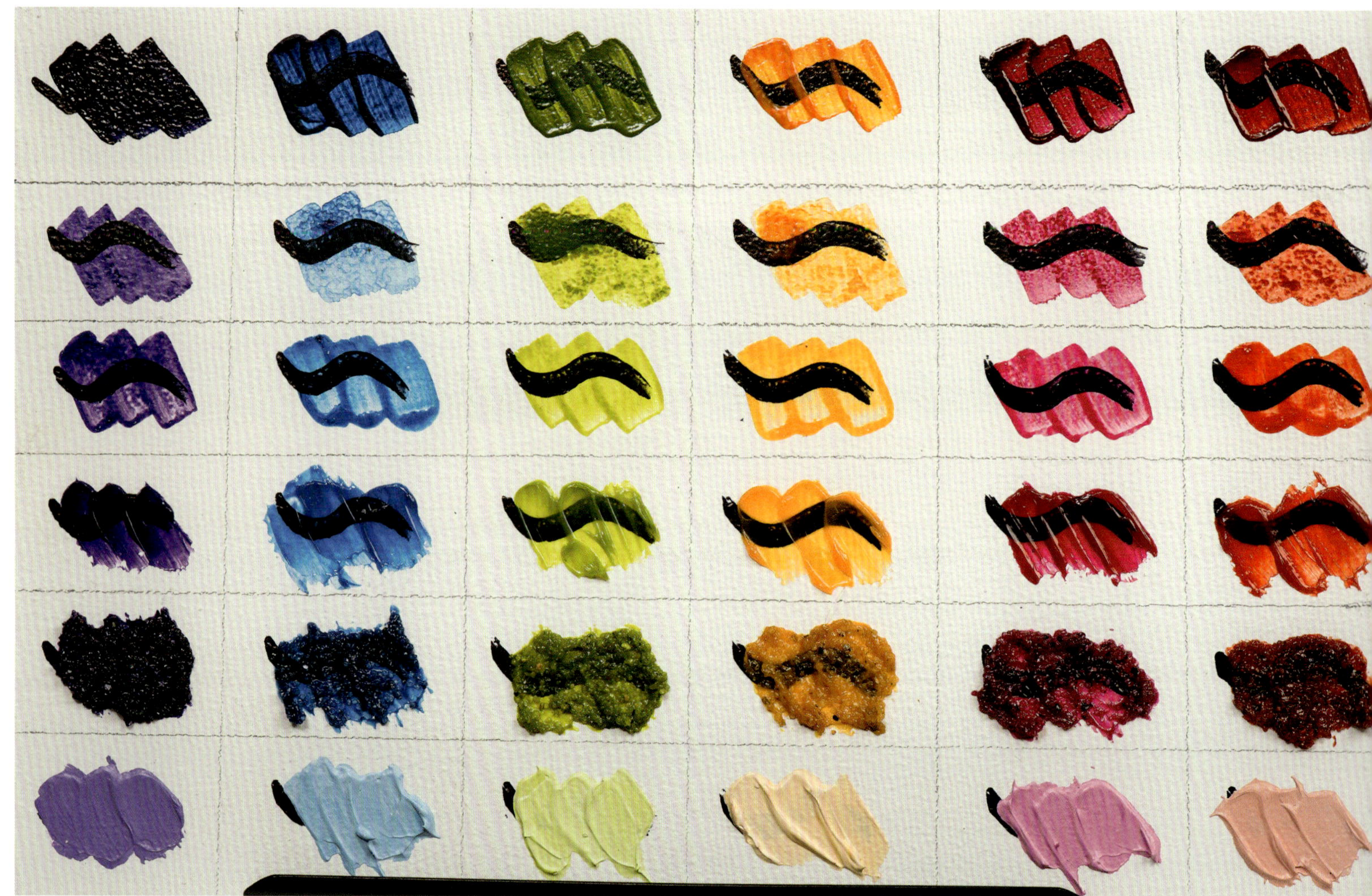

Although the composition of acrylic mediums is not all that complex (they are simpler to produce than colors), the process of figuring out how to make them behave in a given way can certainly be fraught. That's because it's about chemistry. Manipulating raw materials to manufacture paint mediums that are so nuanced and varied is a delicate and exacting science. I find the science fascinating, but I understand that most people don't really want or need to know about it. Still, it's valuable to understand, even on a layman's level, how some of this is accomplished.

ABOVE Transparent acrylic colors mixed with mediums. From left to right, the colors are dioxazine violet, phthalo blue green shade, green gold, Indian (isoindolinone) yellow, quinacridone magenta, transparent pyrrole red. From top to bottom, the colors are undiluted (top) and mixed with the following mediums (dilution: 5% color + 95% medium): water, liquid medium, gel medium, nepheline gel, modeling paste. The black squiggles show relative transparency/opacity.

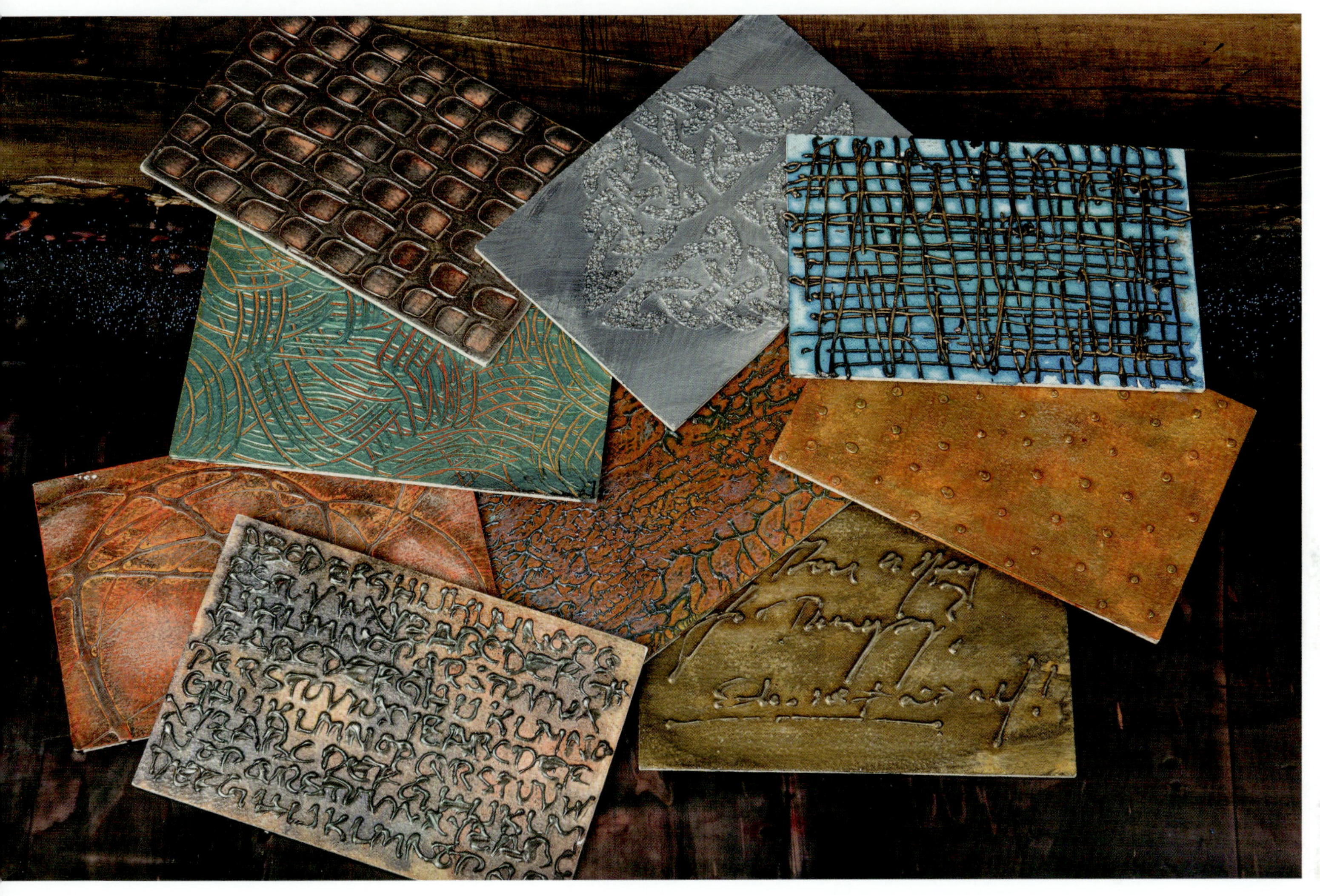

ABOVE Creating samples of textures and paint treatments can be a great source of inspiration when starting a new project.

Know your stuff. Be the engineer of your artwork, the architect of your creative output.

I can't stress enough that knowledge is power, in art-making as in all things. It's astounding to me that people use materials without having any knowledge about what they are or how they are made. Then they react with confusion and frustration when they get less than favorable results. Most problems they encounter can be explained or solved through logic—logic based on even the most cursory understanding of how those materials are made.

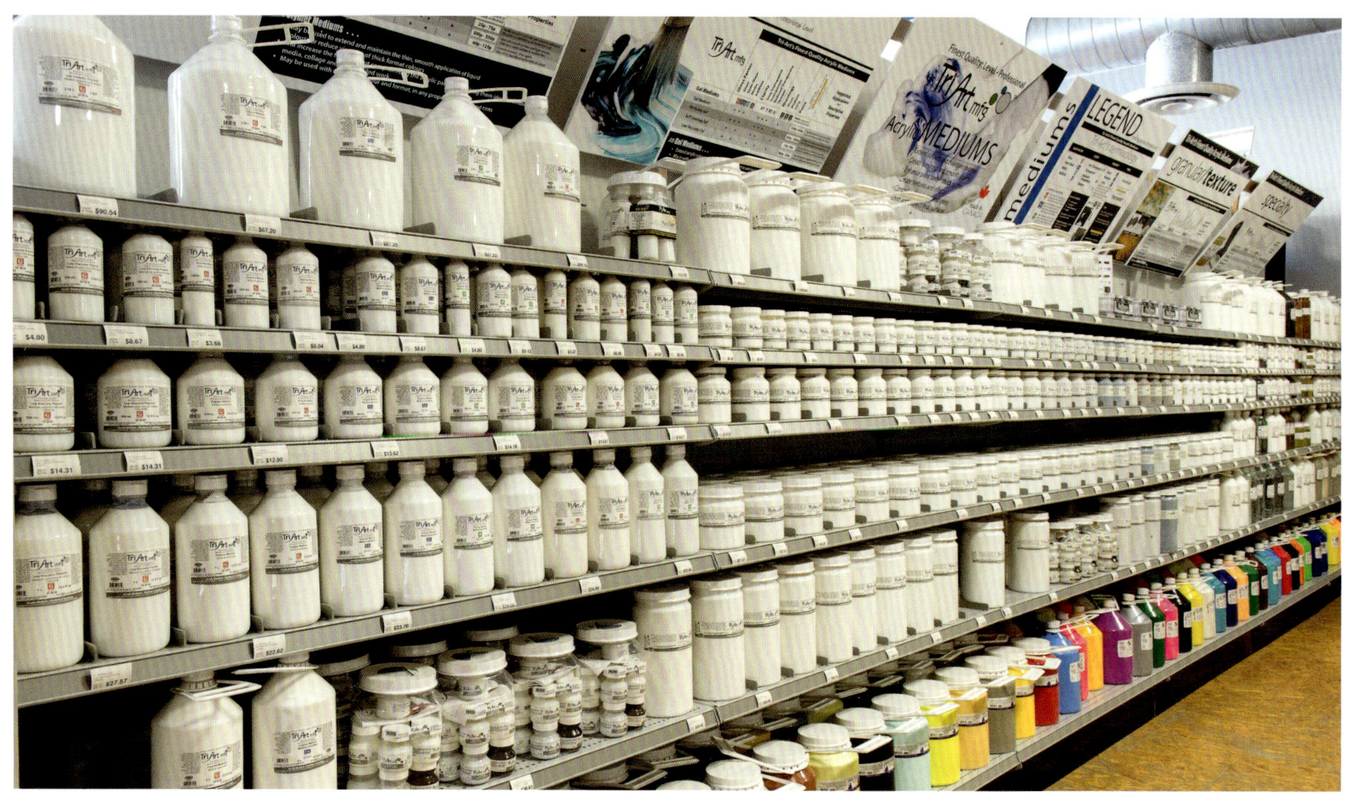

WHAT ARE ALL THESE JARS OF WHITISH GOO?

Mediums are the real manipulators of acrylic paint. Many of you may be familiar with some of the more basic mediums but may still be confused by the quantity and diversity of the acrylic mediums available today. How do you determine which ones to use when? And how do you combine them with your colors and with each other? Before tackling these questions, we need to classify the mediums. Once you know the basic hierarchy and the characteristics of each acrylic-medium family, you will better understand how to incorporate these amazing tools into your personal art-making process.

So let's start putting together your personal technical-knowledge database. This can be as simple or complex as you want it to be, so long as it serves as reference that makes sense to you. As you amass more knowledge, you can build on it.

ABOVE Professional-grade acrylic mediums in every size and format take up more than twenty feet of shelf real estate in this art supplies store.

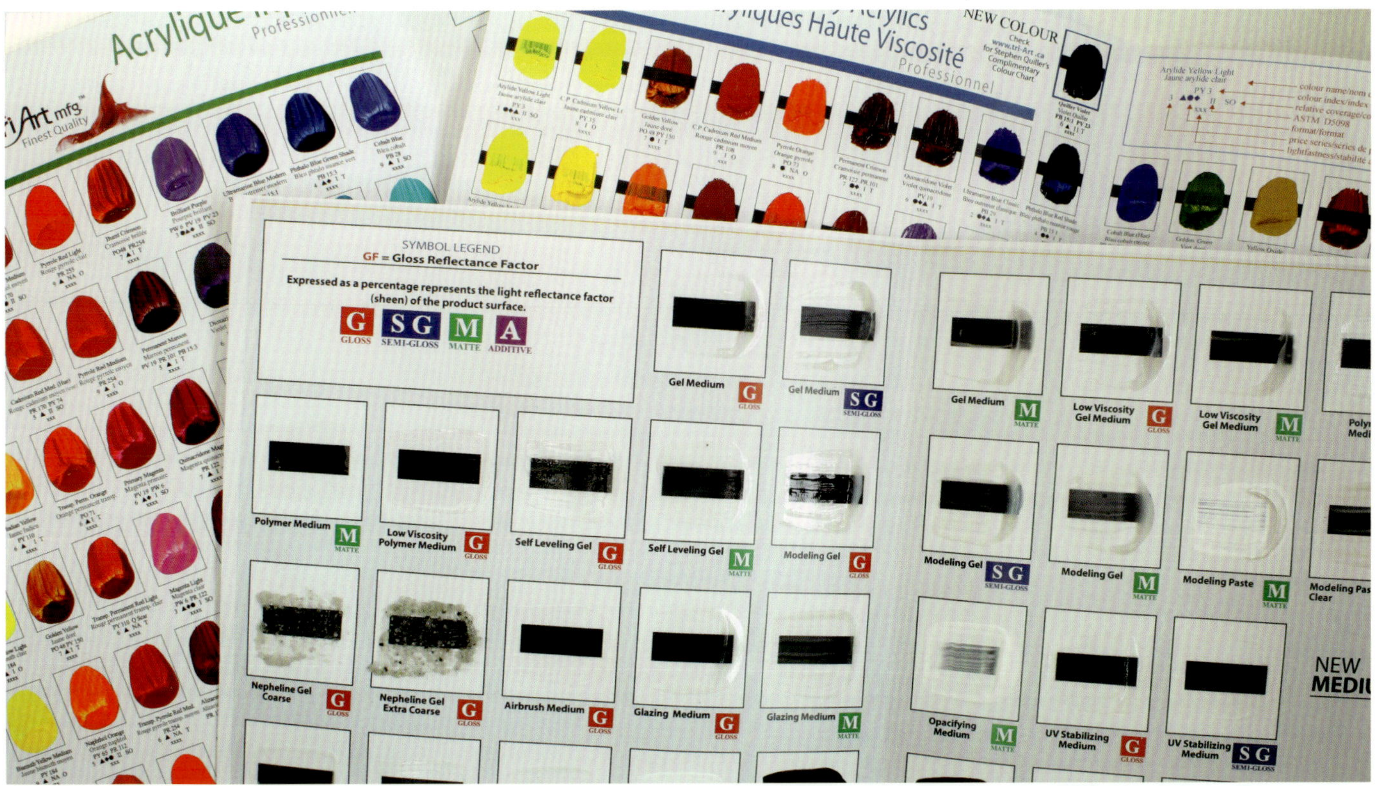

FIRST THINGS FIRST

The prerequisite to learning how to incorporate acrylic mediums into your creative process is to get a firm grip on what they are. To increase your understanding:

- Keep hand-painted color swatch charts
- Read and understand product labels
- Take notes on your experiments for future reference
- Contact the manufacturer if you have specific questions or concerns

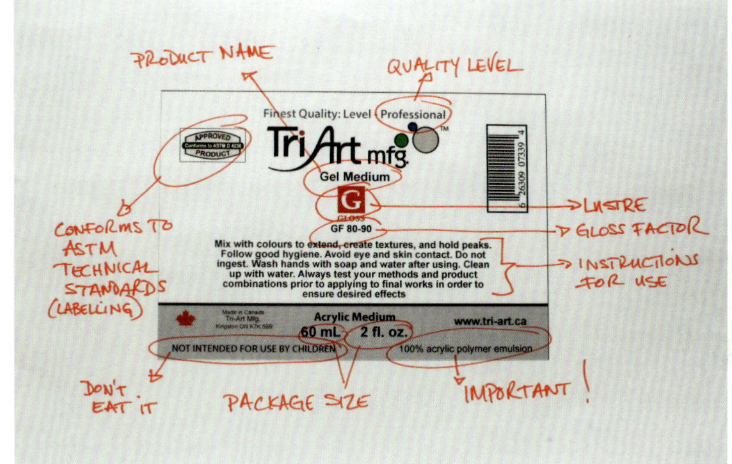

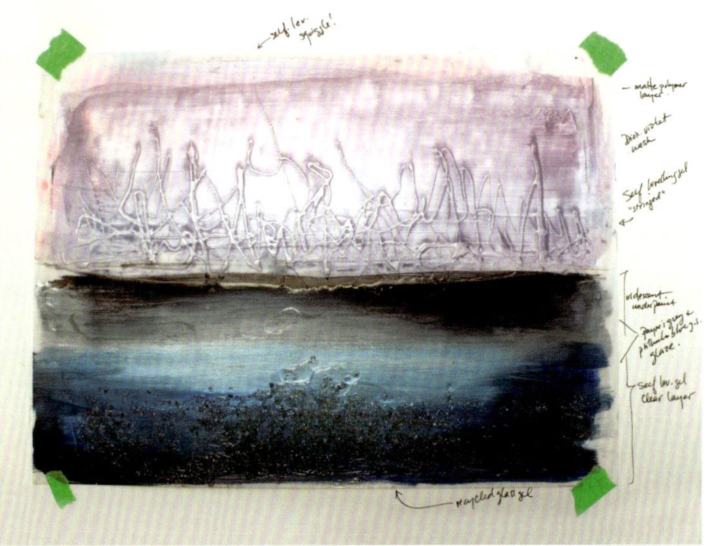

TOP Hand-painted charts of colors and mediums, available from manufacturers, provide you with an at-a-glance description of your painting tools.

CENTER Understand the labels of the products you purchase.

BOTTOM Taking notes on your experiments helps you identify specific mediums' effects.

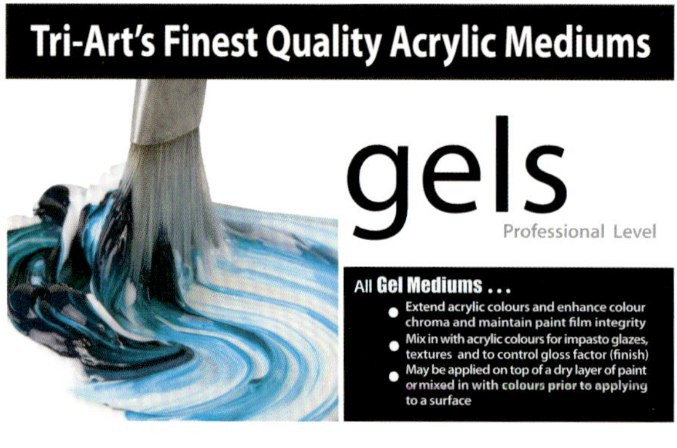

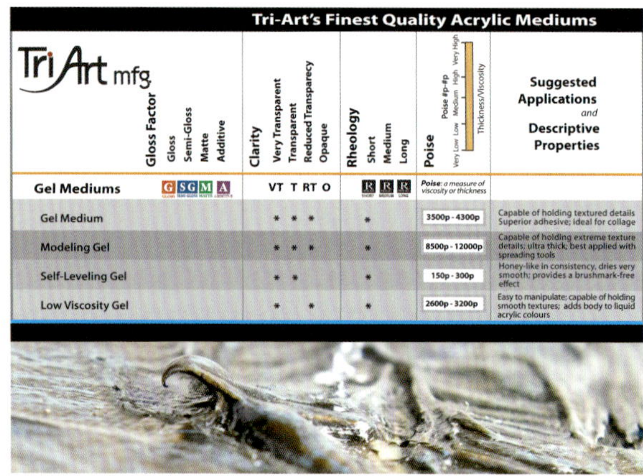

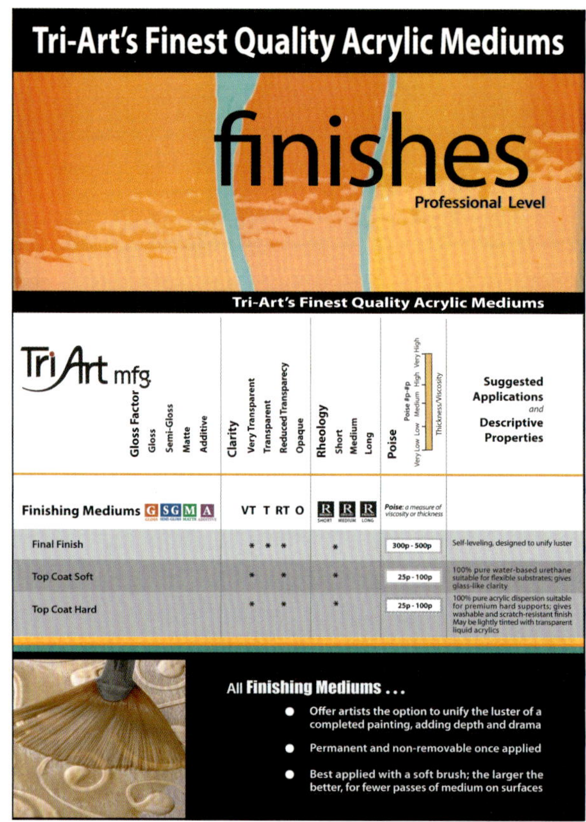

ABOVE Placards like these, prominently placed over displays of acrylic mediums in art supplies stores, help customers choose the right mediums for their projects.

Acrylic mediums are acrylic paint, but without color—well, most of them are without color. Mediums are the cookie dough before the chocolate chips, the vanilla, and the orange zest are added. They are simply acrylic polymer emulsion resin with varying characteristics—thin or thick, gooey or gritty, glossy or matte, transparent or opaque—and their job is to interact with acrylic colors to change their appearance and surface texture.

Don't be intimidated by the science. The terminology used to identify the characteristics of mediums is rooted in chemistry and physics, but you don't need a science degree to grasp the basic concepts. Trust me: I have an arts degree and I totally get it. Delving further into the science will give you a deeper, more specific understanding, but getting the broad strokes is sufficient to begin your education in acrylic mediums. So let's start by taking a look at acrylics—paints *and* mediums—from a technical standpoint.

THE ANATOMY OF ACRYLIC PAINTS AND MEDIUMS

Acrylic paints, whether in a high-viscosity, liquid, or ink format, are made up of these basic ingredients, the specifics of which vary on quality level, brand, and format:

- Acrylic polymer emulsion resin
- Pigment (sometimes)
- Thickening agent
- Biocides and fungicides (preservatives)
- Freeze/thaw stabilizer
- pH stabilizer
- Defoamer
- Surfactant (sometimes)
- Additives

Acrylic mediums have a makeup almost identical to that of acrylic paints, with one primary difference: the absence of pigment. (The exceptions are some colored gessoes and modeling pastes.) They do, however, contain other solids, called additives (see the box below). These additives play a significant role in determining the appearance, texture, and other properties of the individual mediums.

ADDITIVES

Additives can include any of the following:

- Calcium carbonate
- Matting agents
- Opacifiers
- Pigment (in gessoes and modeling pastes)
- UV stabilizers
- UV reactive agents
- Granular particulates (such as nepheline syenite, pumice, glass beads, or other inert aggregates)

RIGHT Granular particulate additives like these give specialty gels their different kinds of body and texture.

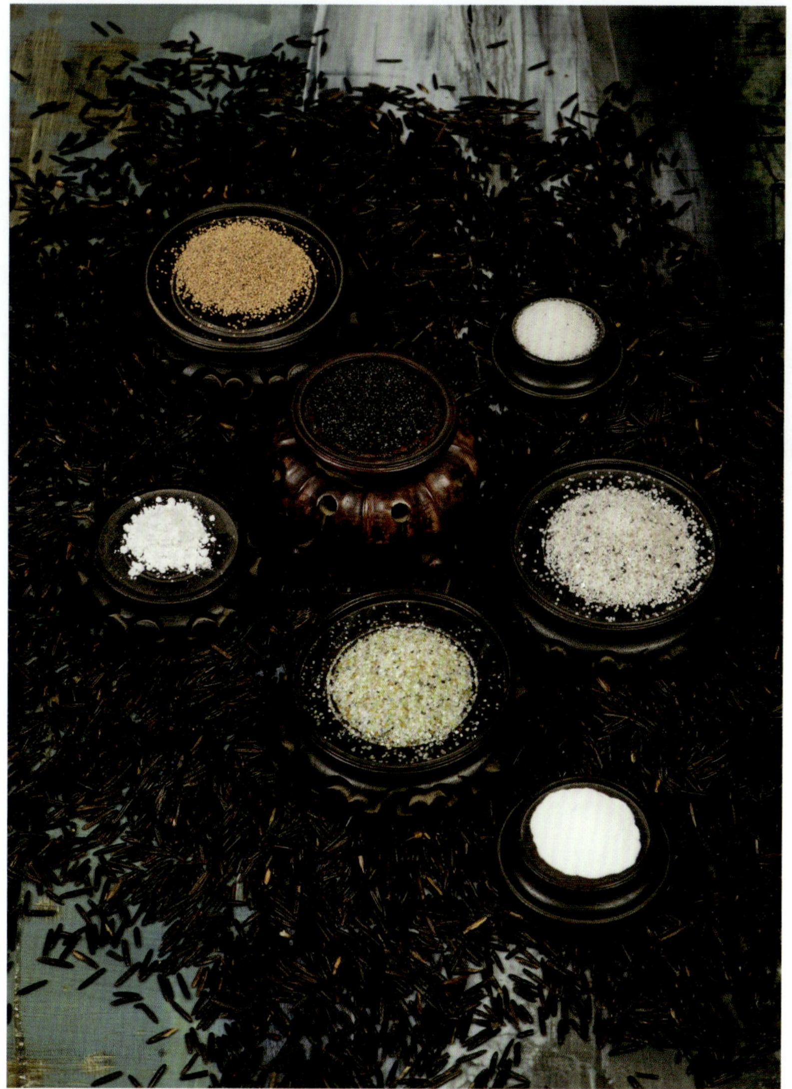

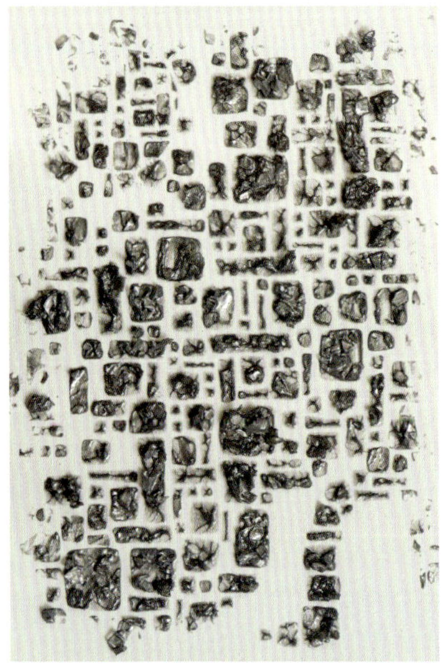
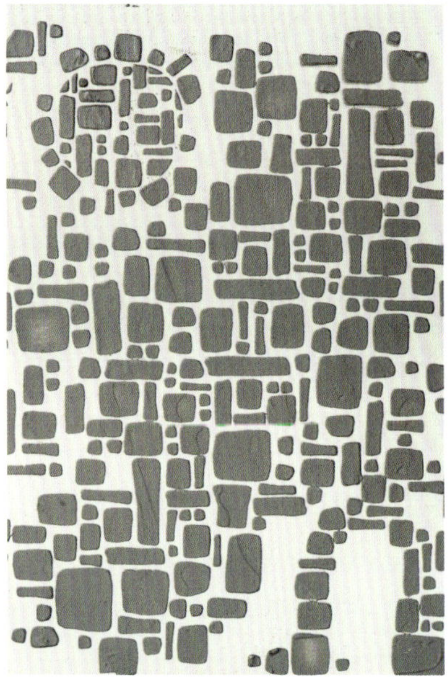
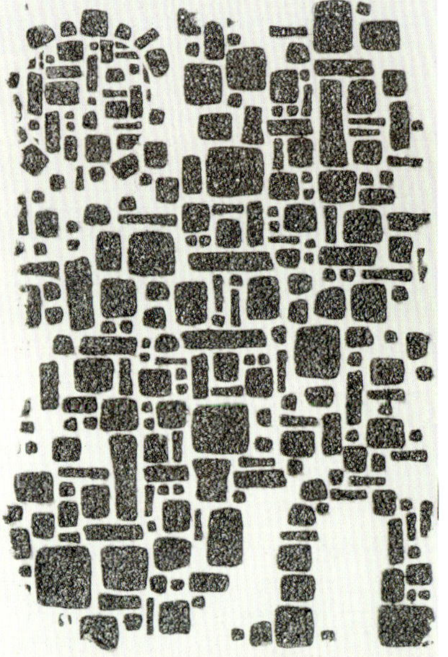
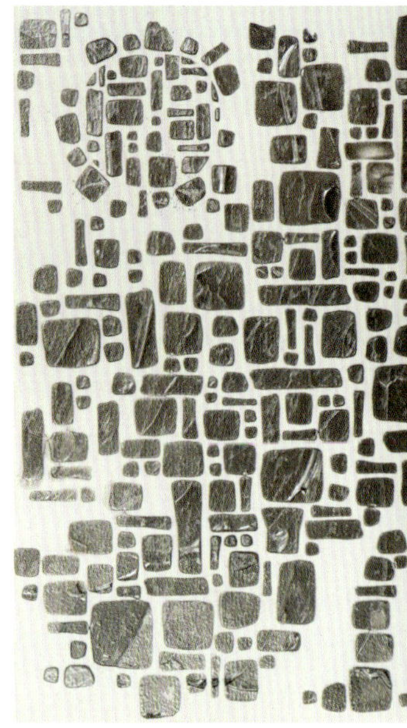
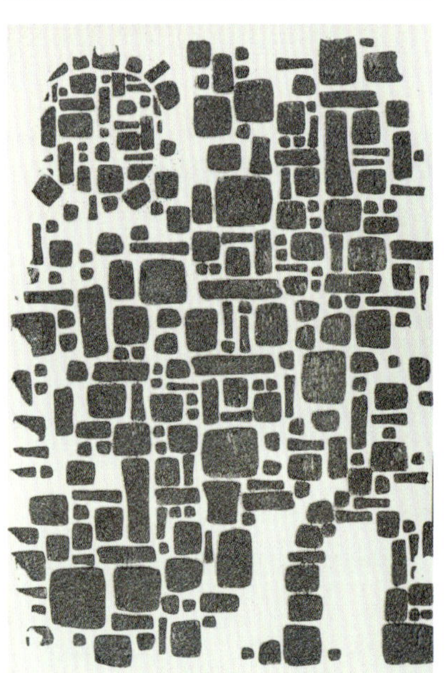
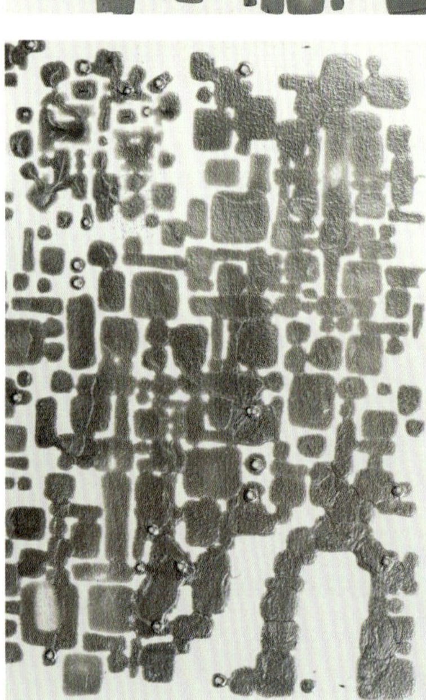
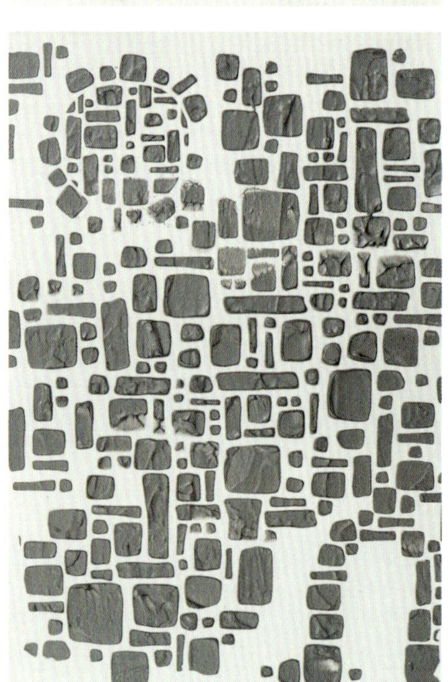

The photos above show how a variety of mediums behave when used with the same color mix and applied with the same method. The color's appearance—its surface texture and luster—changes dramatically when mixed with different acrylic mediums. Liquid mirror is a highly reflective iridescent acrylic color; it is used here in combination with carbon black because its reflective quality augments and thus clearly illustrates the varying sheens and textures of the paint surfaces. The stencil provides consistency in shape and size.

TOP ROW, LEFT TO RIGHT PETE plastic gel + liquid mirror and carbon black. • Clear modeling paste + liquid mirror and carbon black. • Nepheline extra coarse gel + liquid mirror and carbon black. • Gloss gel + liquid mirror and carbon black.

BOTTOM ROW, LEFT TO RIGHT Nepheline fine gel + liquid mirror and carbon black. • Self-leveling gel + liquid mirror and carbon black. • Matte gel + liquid mirror and carbon black.

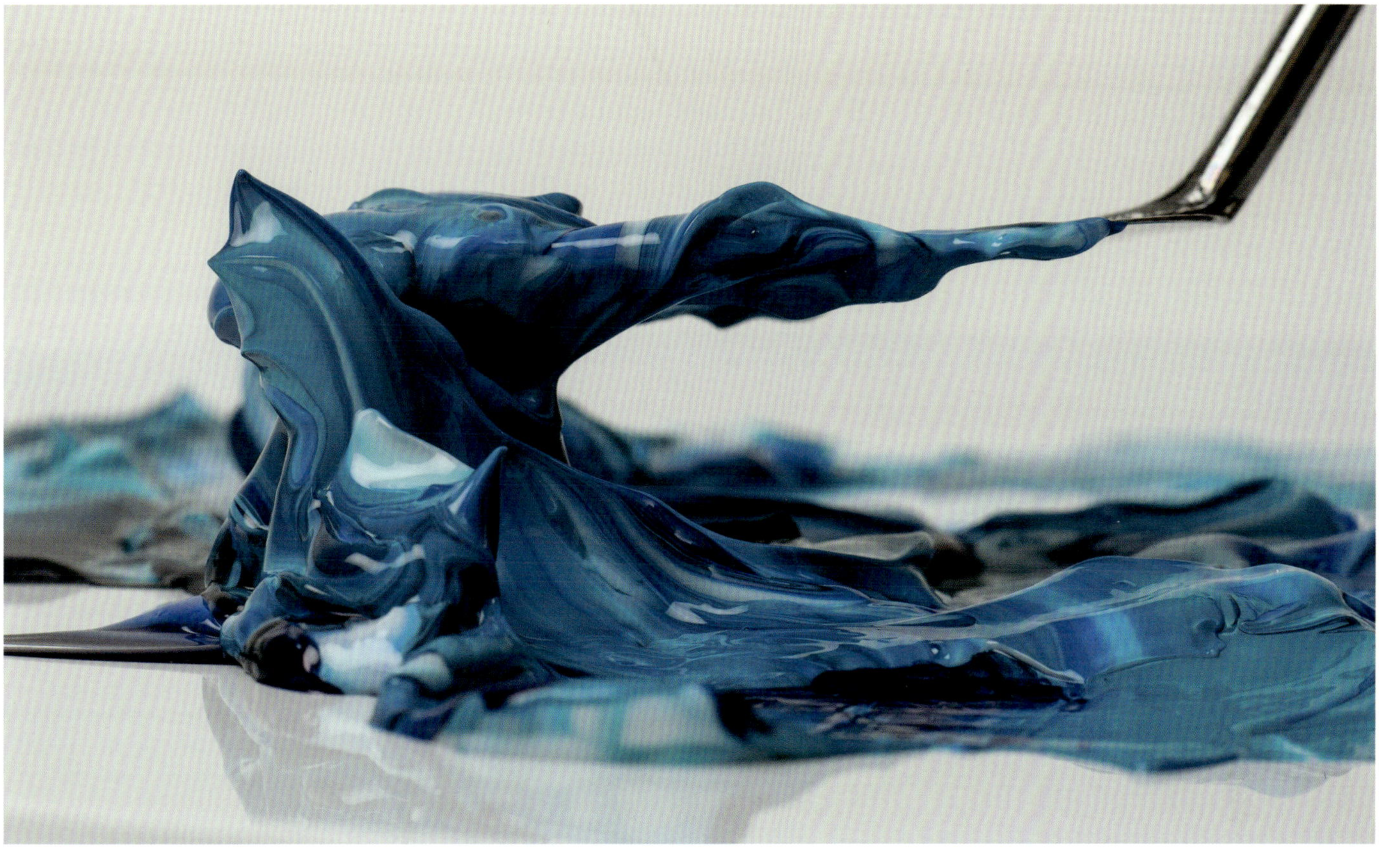

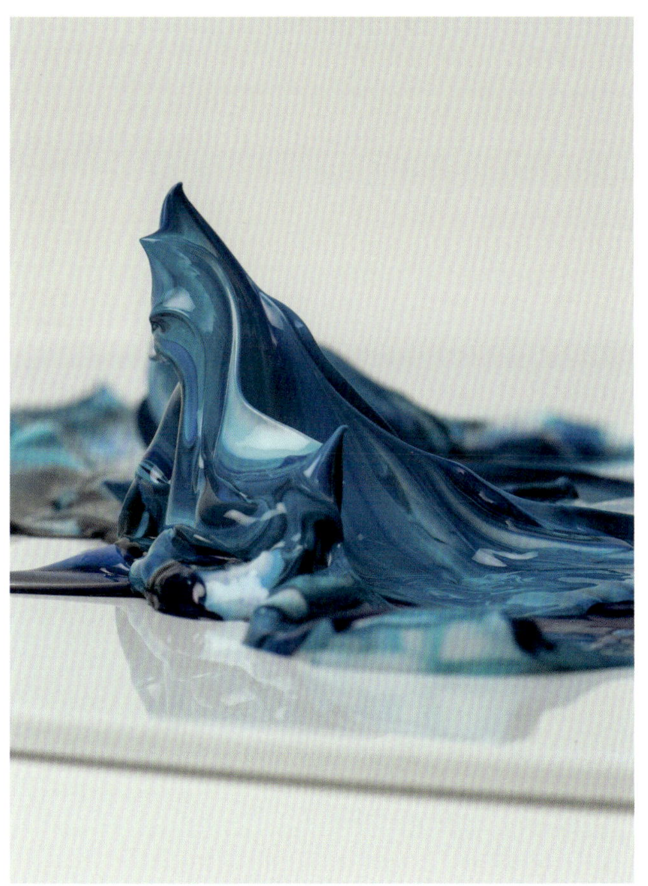

KEY DESCRIPTIVE TERMS

To grasp the nuances of mediums, you must become familiar with their fundamental character, which can be expressed in five key terms: *viscosity*, *rheology*, *luster*, *relative coverage*, and *texture*. Each term describes a particular aspect of the material, and having a good grasp of these terms' meanings will help you make decisions about which mediums to choose and how and when to use them.

ABOVE Mixing high-viscosity gel medium with color.

LEFT Turquoise peak.

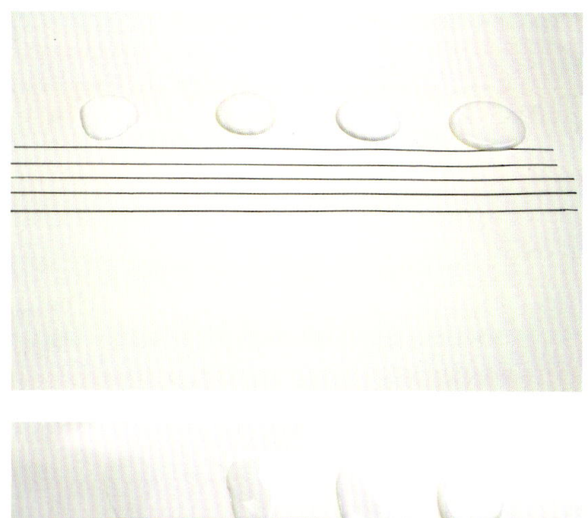

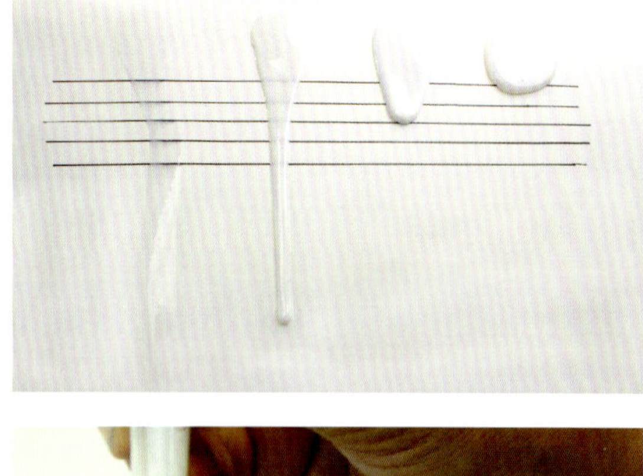

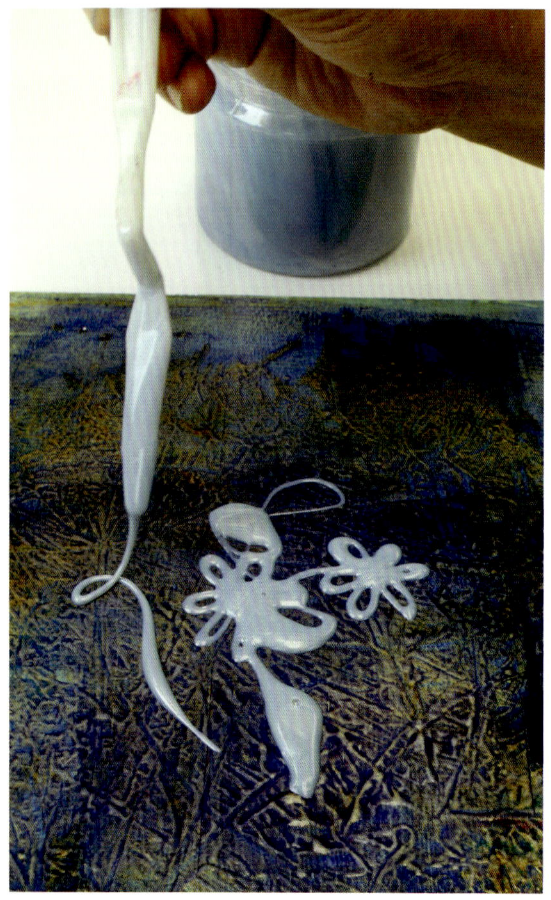

VISCOSITY

Viscosity is the measure of a material's resistance to flow under an applied force. We refer to thick, paste-like acrylics (heavy body acrylics) as high-viscosity paints. A high-viscosity paint or medium is generally thick, but to say simply that it is thick is an oversimplification. A paint can be thick, but then so can a wall or a pile of whipped cream. More precisely, the term *viscosity* encompasses a material's movement as well as its density. The wall may be thick, but it is not viscous, as it is incapable of movement. The whipped cream, on the other hand, is thick but has the ability to flow and thus has a high viscosity.

Describing the viscosity of a material gives us an accurate picture of its malleability and how it will behave in a given technique or application. Viscosity is measured in units of poise (P). The term *poise* is named after nineteenth-century French physicist Jean Léonard Marie Poiseuille. Although the common English word *poise* has nothing to do with the scientific term's origin, it can be helpful to think of a high-viscosity medium as having "poise"—being upright and having a sense of weight.

The higher the poise number, the greater the medium's resistance to flow will be. Polymer (or liquid) mediums are thinner and have a lower poise number than gel mediums. Low-viscosity mediums are suitable for fine detail work, glazing, and creating thin, smooth layers. Higher-viscosity mediums are useful for extending color, glazing, and heavier impasto or textured applications.

FROM TOP TO BOTTOM Globs of four different liquid mediums on a flat surface; from left to right: final finish, glazing medium, polymer medium, and self-leveling gel. • When the surface is tilted at a 45-degree angle, it becomes evident which of these fluid mediums have lower and which have higher viscosities. • Loose strings of viscous tinted self-leveling gel have been drizzled using the flat face of a plastic palette knife as the reservoir.

RHEOLOGY

Rheology is the study of the deformation and flow of matter under applied stress. When it comes to mediums, rheology describes how a medium will behave when you push it around. How a medium behaves tells us which tools to use, how it will mix with colors or other mediums, and, ultimately, how it will look when it dries.

A medium's rheology is described as *short* or *long*. Some highly viscous mediums have a short rheology and will hold sharp, full-bodied textures. Others with a similar viscosity have a long rheology; heavy and honey-like, they tend to ooze into thick pools leaving no visible raised texture. So while both are highly viscous, their differences in rheology dictate their physical appearance. When you dab a palette knife into a medium with a short rheology, you pull it up and get a short, stiff peak that should retain its shape as it dries. The same action performed on a medium with a long rheology will give you a long string of material that will ooze back down onto the surface and eventually level out, leaving no discernable surface texture.

If it helps, you can think of mediums' rheologies in less technical terms: mediums with a short rheology are pert and perky, showing off with brash and dramatic textures. Longer-rheology mediums flow and ooze, retreating into themselves; they are selfish, secretive, mysterious, and seductive.

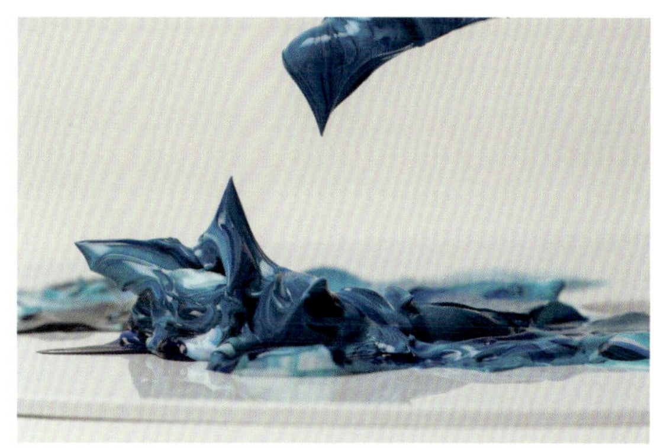

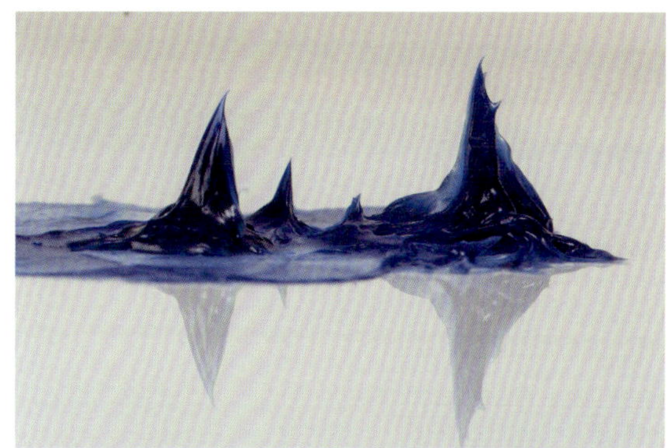

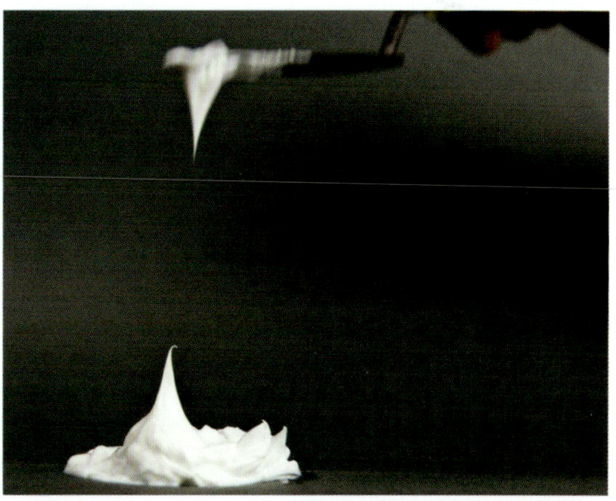

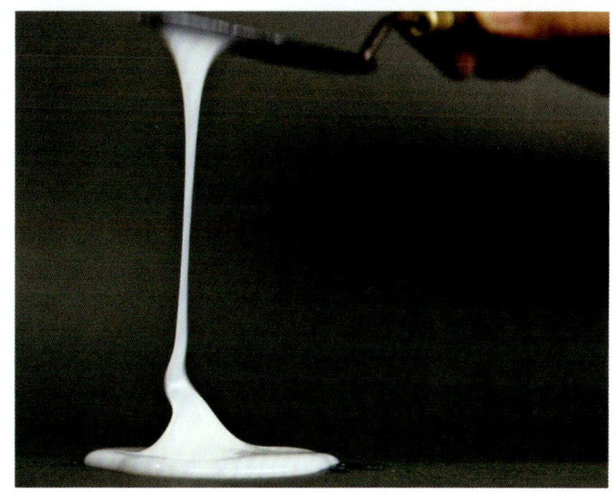

FROM TOP TO BOTTOM Mixing color into a high-viscosity, short-rheology gel. • Blue gel paint peaks: high viscosity, short rheology. • Gel mediums exhibit a short rheology. • Self-leveling gel exhibits a long rheology.

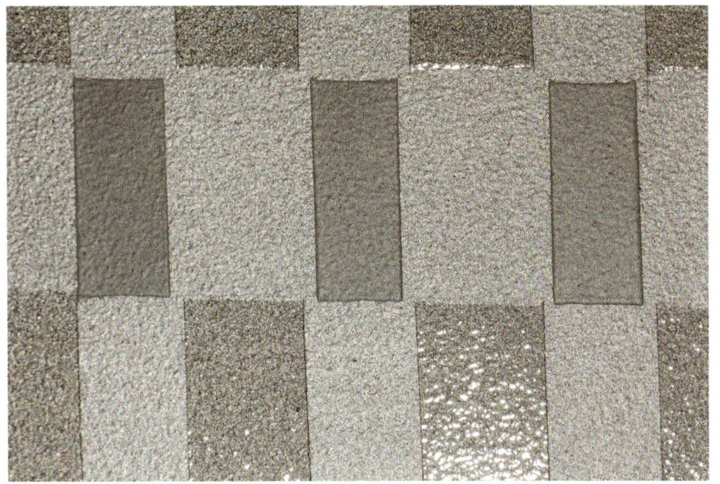

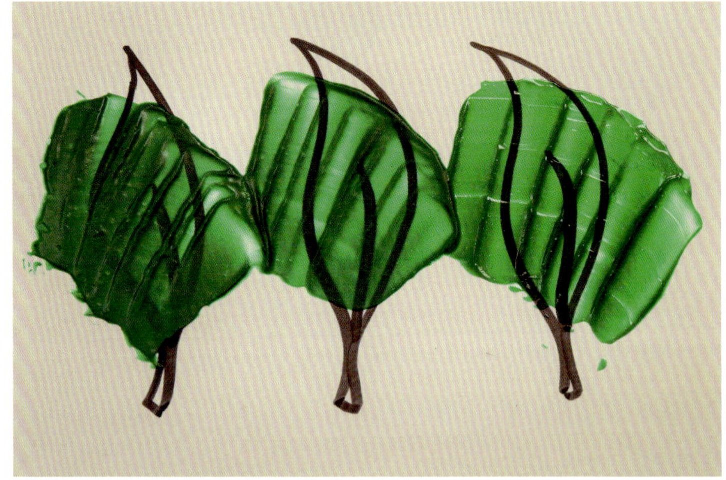

ABOVE, LEFT AND RIGHT Masked areas of matte and gloss gel painted over a semigloss surface clearly show the subtle changes between lusters. • From left to right: matte, semigloss, and gloss gels, lightly tinted with phthalo green yellow shade, show varying levels of transparency. The glossier the medium, the more transparent it will be, so a matte medium makes the color more translucent or semi-opaque.

LUSTER

The surface of a painting has a significant effect on our response to it. A glossy veneer is more than just reflective—it also deepens color tones and imbues them with depth and light. A more satiny finish (what we call semigloss in the art materials world) has less flash than a glossy surface, offering a silky, luxurious sheen. A flat matte surface, which absorbs light, gives the colors and textures of a painting a softer, subtler, more nuanced appearance.

Manipulating the gloss factor of a surface is one of the primary functions of many acrylic mediums. Most families of mediums are offered in gloss, semigloss, and matte formats. By using one or another of these formats—directly mixing the medium with the color or painting the medium over the color as a clear coat or layer—you can change the surface luster.

Sheen, or reflectivity, affects our perception of depth and distance on the painted surface. A glossy surface will show every detail, divot, and irregularity in a surface, including the direction of a brushstroke or the depth of a palette knife's gouge. Besides dulling the surface, a matte medium also gives the surface some "tooth"—a slight graininess that helps scatter the light and also allows dry materials such as graphite, charcoal, or other drawing to grip the surface. The more matte a surface is, the more it appears to recede, whereas a glossy surface gives the illusion of moving forward. Using varying lusters can add subtle contrast to a painting.

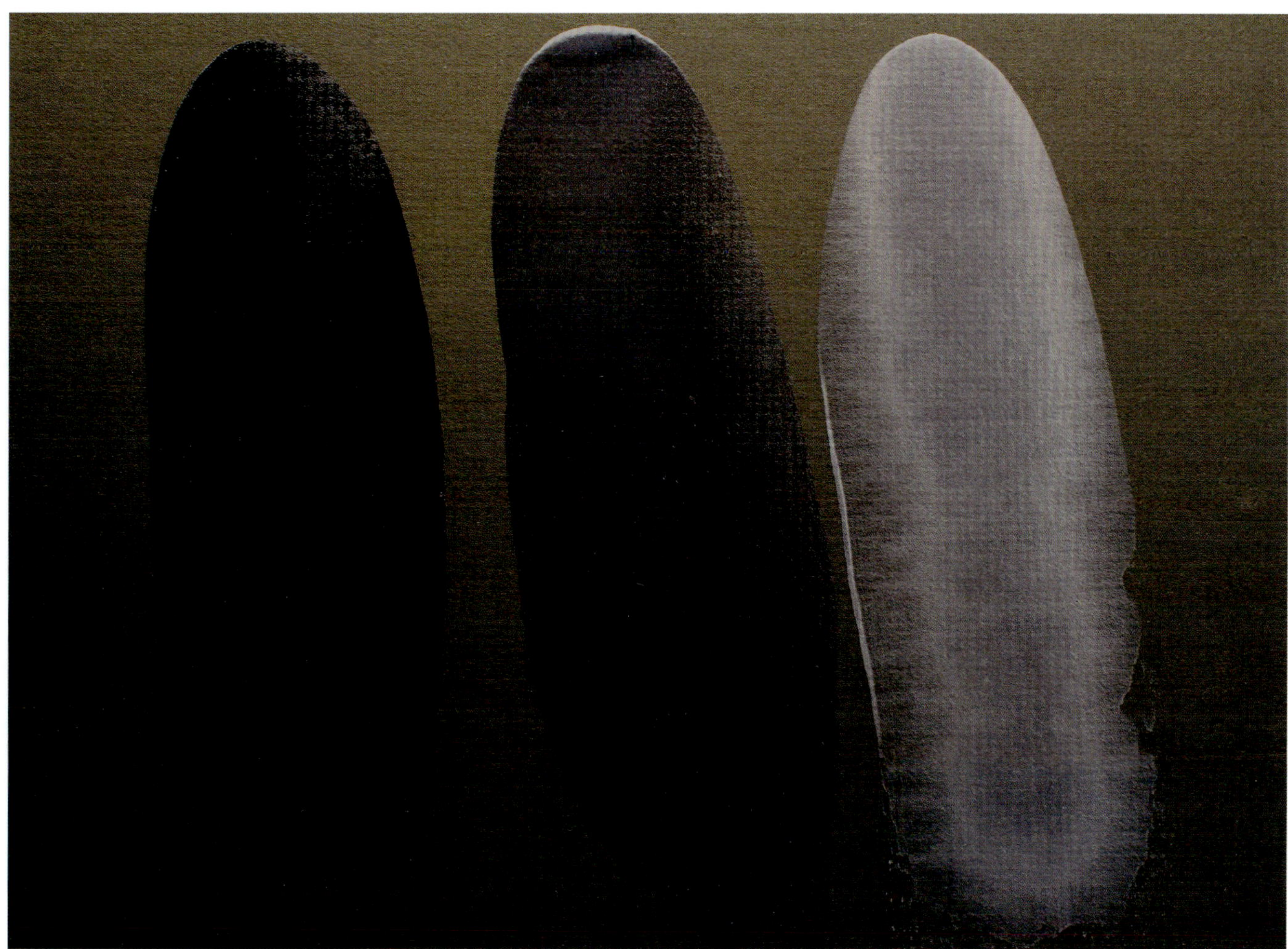

ABOVE From left to right: gloss, semigloss, and matte polymer medium painted on black paper. The effects are transparent, translucent, and semi-opaque. The luster of the medium, when used on a darker surface, has a direct impact on its relative coverage.

Acrylics are naturally glossy; it's what we add to them that modifies the surface from super shiny, to soft satin, to velvety matte. Matting agent, present in semigloss and matte mediums as well as in modeling pastes and other specialty mediums, is one important component affecting the gloss factor of the paint.

Using a medium, you can modify the luster of any color, but it's helpful to know that color pigments themselves can affect a painted surface's luster. This is dictated by the size of the pigment particles—how fine or coarse they are. The coarser the pigment particles, the more matte the color will appear. Professional-grade acrylics, because they contain little or no filler, reveal more of the characteristic luster of the pigment. The cheaper the paint, the more uniform the colors will appear, since matting agent and other fillers have been added to the colors.

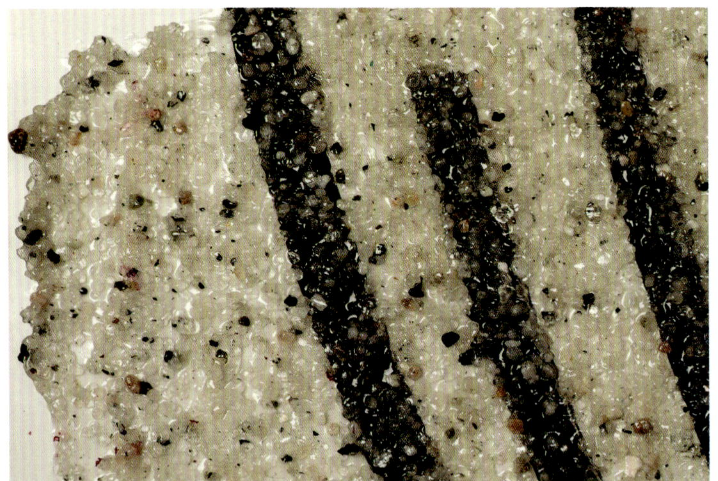

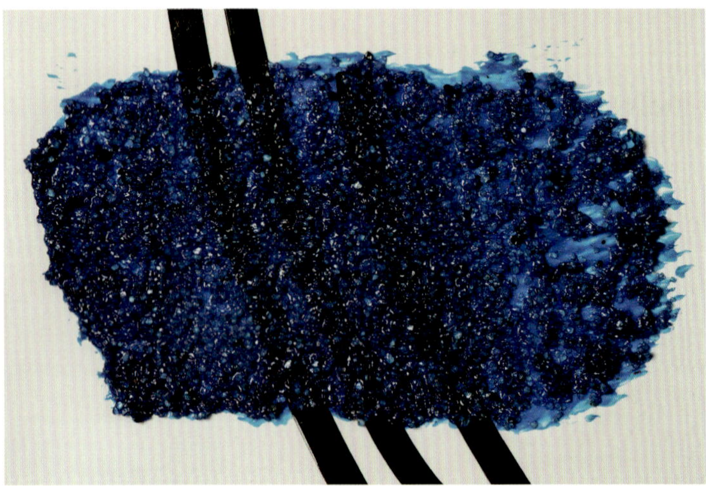

TOP Nepheline gel extra coarse is quite transparent.

BOTTOM When nepheline gel extra coarse is tinted with phthalo blue it remains transparent despite the pigment's deep chroma.

Tip: If you have not used a particular medium, always do a test strip over both a white and a colored background (black is best) to see how it dries.

RELATIVE COVERAGE

Relative coverage pertains to the degree to which a particular material allows light to pass through it. In the acrylic painting world, materials are described as being transparent, semi-opaque (translucent), or opaque. Relative coverage must be heeded to make sure that the medium you choose is the right one for the job.

Because most mediums, in their wet state, are varying shades of white, you must read their labels to discern if they will dry clear, opaque, or somewhere in between. While the majority do become transparent or translucent, some mediums contain pigment and will be opaque when dry. For example, gesso and modeling paste usually contain a quantity of titanium white pigment, but they are also available in a clear format, with no pigment added. (Some brands also offer gessoes and modeling pastes in other colors.)

The relative coverage of a medium will have an impact on the colors you use it with. Transparent gloss mediums are the clearest vessels with which to extend color because they contain no matting agents or other additives. Semigloss and matte mediums all contain a matting agent. The precise material differs from brand to brand, though is commonly a translucent particulate that has no effect on the tint of the product. The addition of this particulate causes the medium to

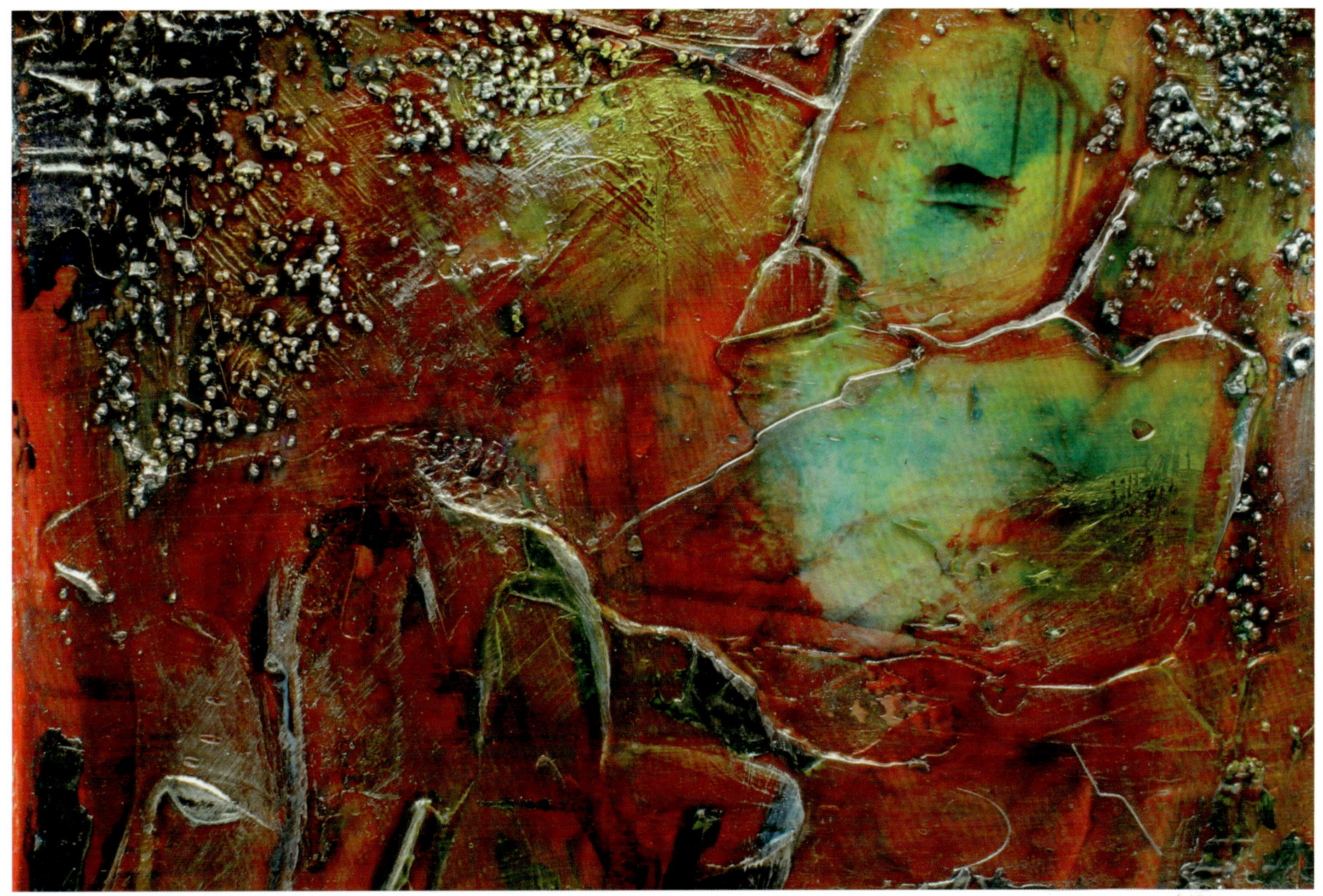

ABOVE A textured surface with glazes.

appear more translucent when dry. This can slightly opacify colors, although the change is subtle—like viewing a photograph through frosted glass. Clear modeling paste and clear gesso contain even more solids, in the form of calcium carbonate, which further add to the haze. Mediums containing an opaque pigment, such as titanium white or carbon black, will cover completely.

TEXTURE

Impasto is probably the first word that springs to mind when mediums are mentioned, and with good reason.

Mediums come is a vast array of texture-producing formats. From sharp relief, to smooth pools, to erratically cratered surfaces, mediums are masters at shaping color and modifying paint surfaces.

Some mediums can be manipulated to produce texture, but others contain components that make them inherently textured. How the surface of a medium looks and feels when dry is due in part to its viscosity and rheology, but also in part to any particulate matter that's been added to it. Mediums can contain granules, fibers, and other fillers that produce specific kinds of textured surfaces. There's no unit of measure that defines these textures; instead, mediums are described, texturally, according to how they feel to the touch when dry: smooth, fibrous, granular. Granular mediums are also usually described as having fine, coarse, or extra coarse particles.

OPPOSITE Gloss gel plus colors on a brush resting on a highly textured gel medium surface that has been amplified with dry-brushed reflective metallic color and dark washes.

SHRINKAGE

Acrylics shrink as they dry. The shrinkage is due to the evaporation of water as well as of volatile compounds such as propylene glycol, fungicides, and antibacterial agents, which are needed only when the acrylic is in its wet state.

The viscosity of a medium is not usually a good indicator of how much a medium will shrink. A thin medium and thick medium may contain the same quantity of water; if so, they will both shrink by about the same extent. The extent of the shrinkage has primarily to do with the quality of the medium. The lower the quality, the more water it contains, and the more water it contains, the more it will shrink.

As every medium from every brand has a different formulation, it is impossible to give an exact percentage for the shrinkage or even a reasonably accurate estimate. From my observations of the brand I primarily use, I'd guess that the shrinkage of a professional quality gel medium is somewhere around 10 to 15 percent. In other words, it's visible but slight. I have not experienced any circumstance (in my own art-making) where this makes any difference at all to the overall painting. Perhaps there are situations where it does matter, but I just have not discovered them.

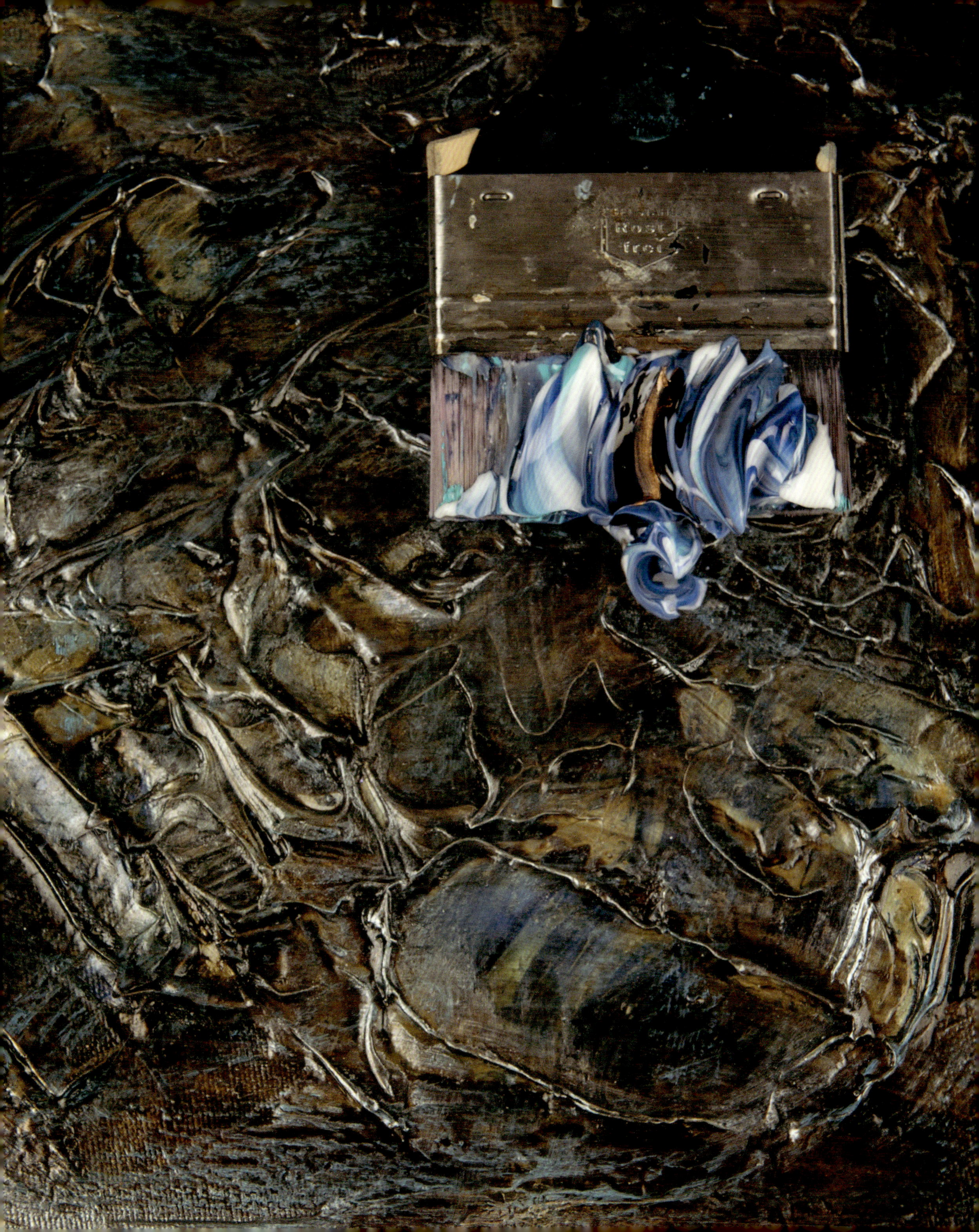
Rost
frei

CHAPTER 2

CATEGORIZING MEDIUMS

NOW THAT THE KEY TERMS HAVE BEEN DEFINED, we have a good foundation to begin exploring various kinds of mediums. Categorizing mediums is not as straightforward as organizing colors, and every retailer has its own way of displaying them. For the purposes of this book, mediums have been loosely arranged according to how they function in the construction of a painting: grounds come first, followed by extenders and texture manipulators and, finally, finishing mediums. Within this framework, mediums are further organized first by viscosity (thin to thick) and then by how generally familiar they are: mediums that are more commonly used are discussed before specific-use and specialty mediums.

As the variety of acrylic mediums continues to expand, the ways they are organized in retail environments—and, ultimately, in studios and classrooms—is likely to change.

Categorizing mediums isn't as easy as organizing colors.

OPPOSITE The soft teeth of a silicone-tipped Catalyst tool pull color through an acrylic gel.

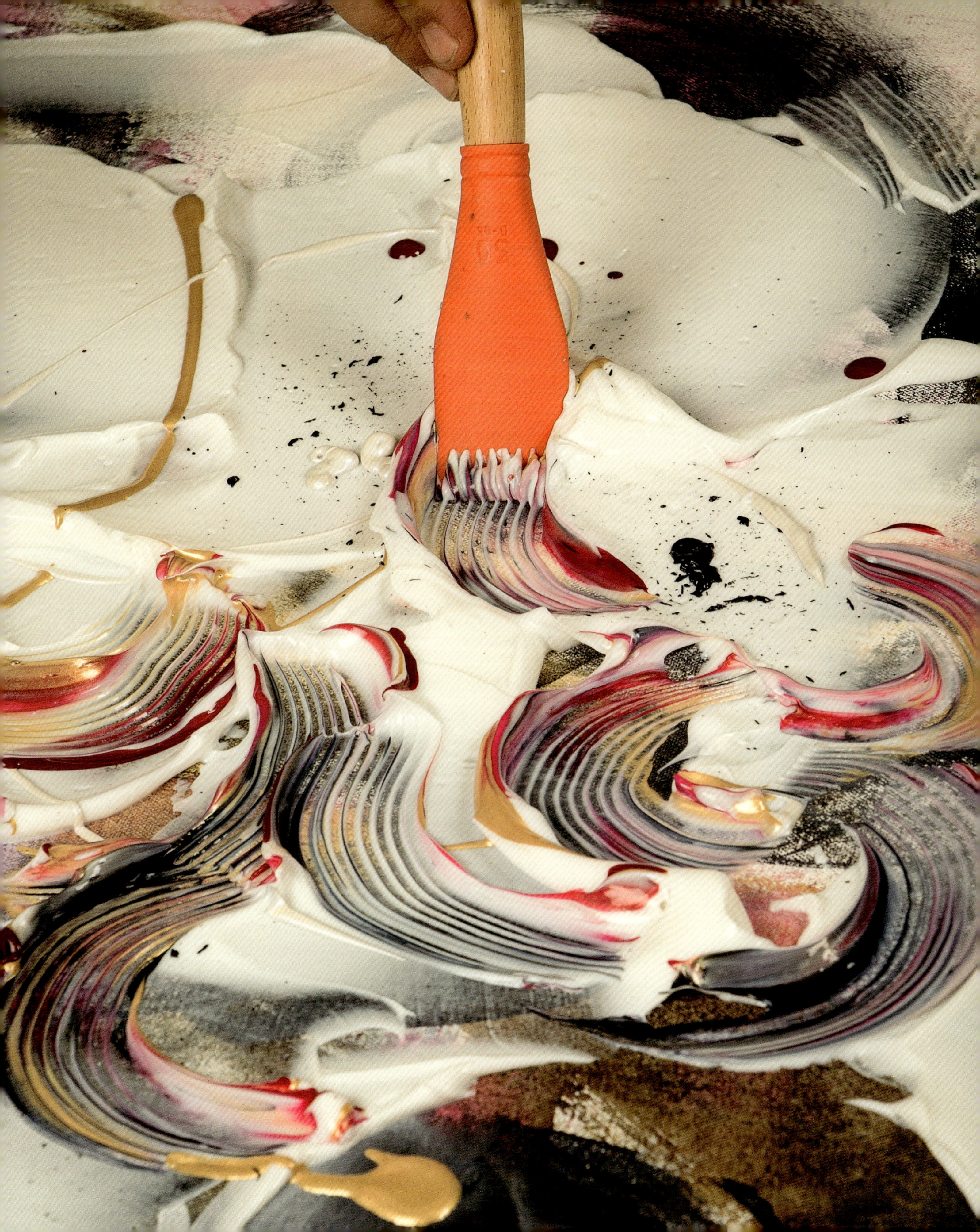

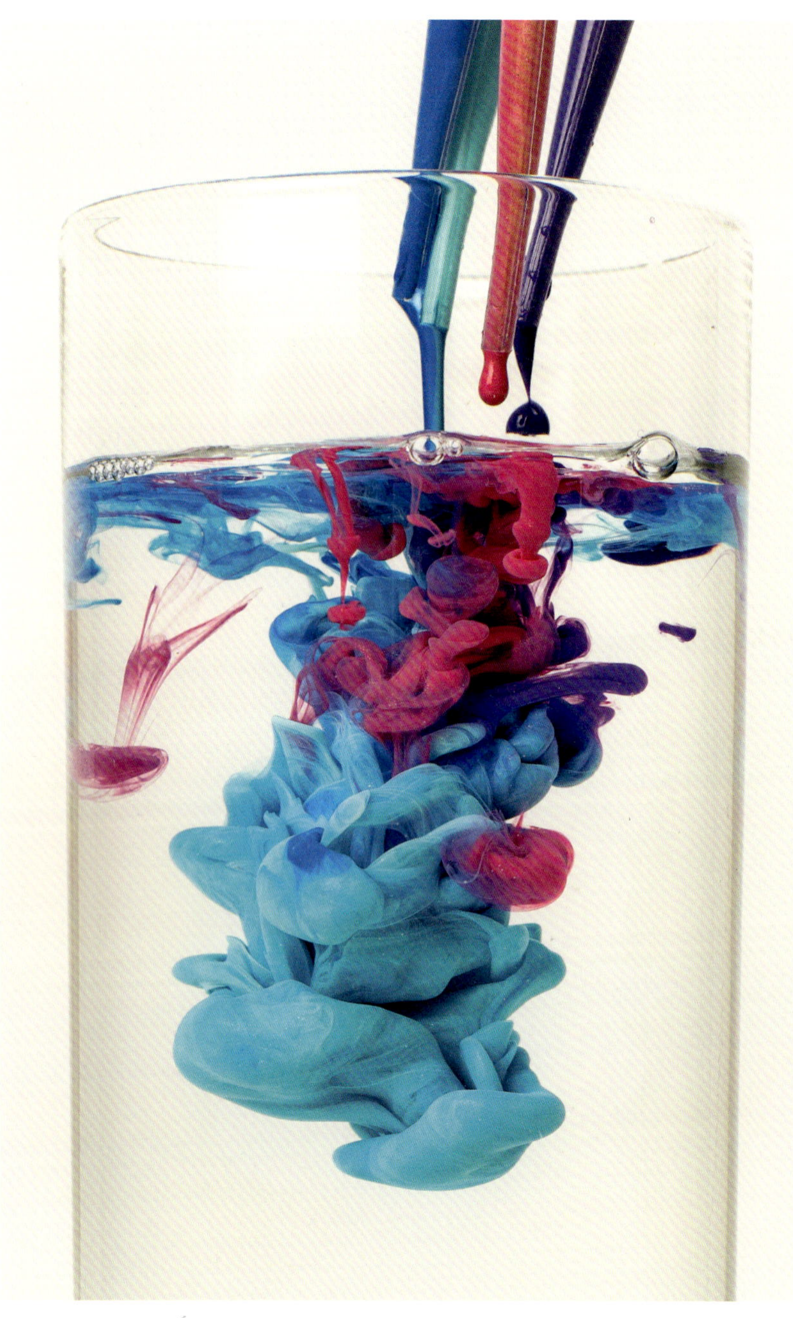

ABOVE Diluted liquid acrylic colors in water—a mesmerizing capture.

WATER: THE FIRST MEDIUM

Water can be thought of as the first medium. It is an integral part of every acrylic paint and medium. There is no acrylic paint without water in it, and water plays an essential role in the acrylic painting process. Acrylic paints are water miscible, meaning that paint and water can be mixed together in any proportion.

When mixed with acrylic paint, water can do the following things. (Keep in mind that these can be either pros or cons!)

- *Increase the porosity of the paint film.* When a lot of water has been added and the paint film has cured, the pathways caused by evaporation will allow subsequent layers of paint to sink in and bind with the diluted paint film. Too much water will dilute the binder and result in poor adhesion between layers.
- *Reduce the viscosity of a medium.* The most common way to thin a color or medium is to simply add water. Elsewhere, you'll find all sorts of recommendations regarding just how much or how little water should or should not be added to any acrylic. But the actual quantity of

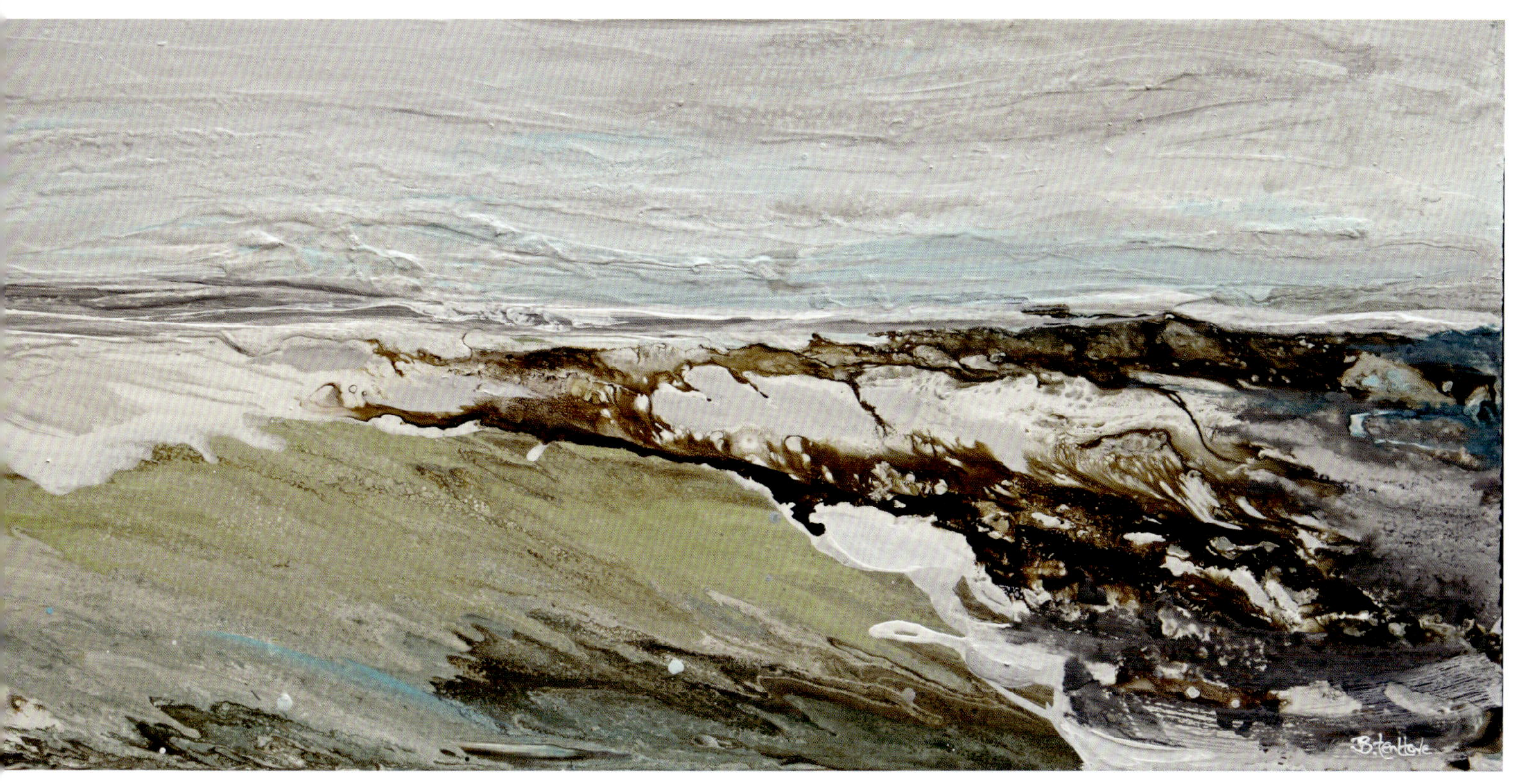

ABOVE **Beth ten Hove, *Tide's Out,* 2017, acrylic on canvas, 48 x 24 inches (122 x 61 cm). Collection of the artist.**

Tide's Out was painted using a technique in which all the colors were premixed with water and poured directly onto the surface of the canvas in a single application. The paint was then mixed and spread by lifting and rotating the canvas manually. To finish the painting, artist Beth ten Hove sprayed alcohol onto the surface to enhance the cellular effect and used a palette knife to apply liquid acrylics in the sky and along the horizon line.

About her practice, Beth ten Hove says, "I live and work on a lake that calls my name every day, inviting me into her deep knowing of light and shadow, movement and stillness. My response to her call is in itself 'water play,' as I mix my paint generously with water directly on the surface of the canvas. My canvas lies flat on the table in my studio, like the lake on a calm day, quietly waiting for the dance to begin. I am fascinated by how water alters pigment and responds to the movement of the canvas."

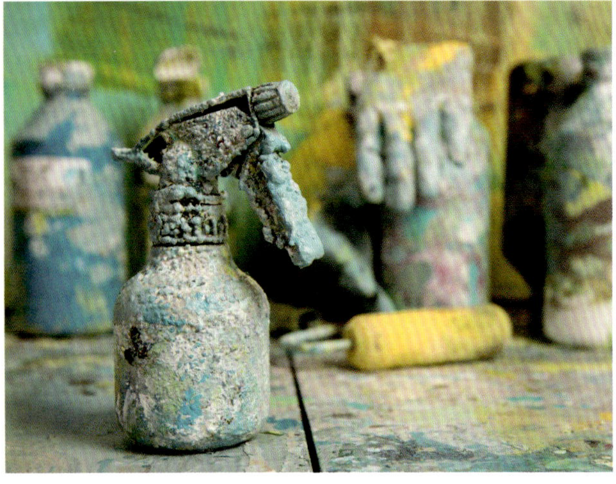

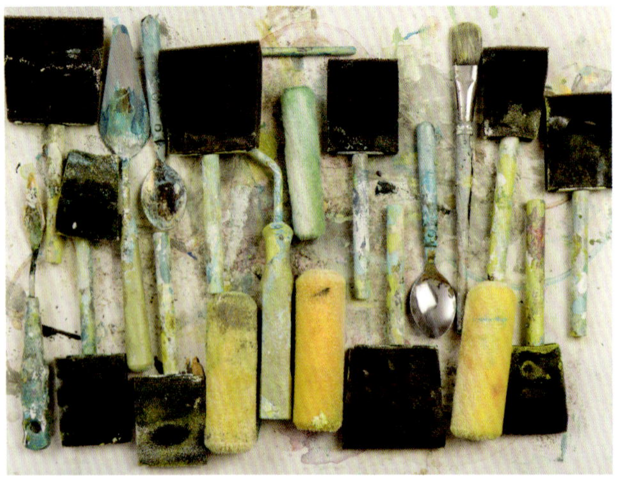

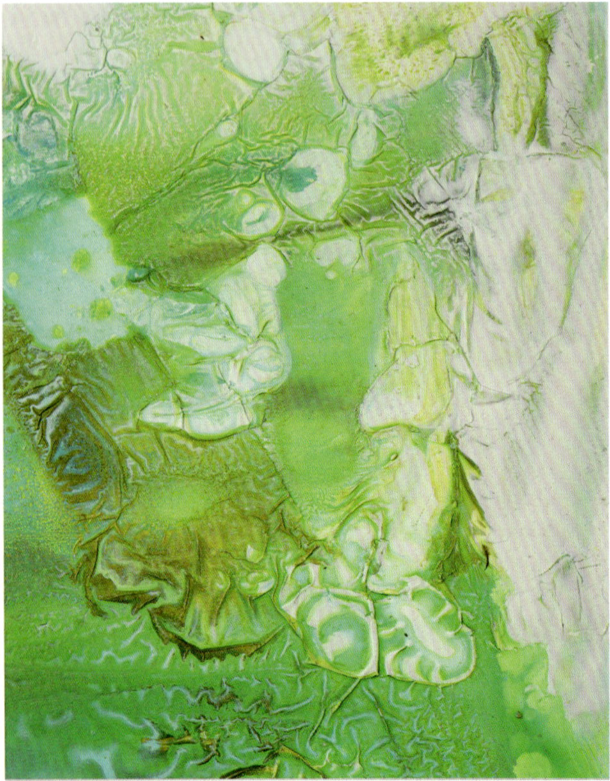

water that *can* be added is "as much as you want." The quantity that ***should*** be added is determined by how you want the paint to behave, how strong a paint film you need to maintain, and other factors.

- *Make the surface very matte.* Water is the most aggressive of the volatiles; through capillary action, it forms microscopic pathways through which it can exit the paint film. The more water in the paint, the more pathways there are, resulting in a surface with a high saturation of "holes." The more holes, the more matte the surface will be, as the holes give the light a place to hide, decreasing the overall reflectivity and gloss.
- *Cause the surface to craze and buckle.* Paint dries through evaporation, and water evaporates first and most quickly. If too much water is added to the mix in a thick paint application, it can cause uneven drying, resulting in a cratered, buckled, or crazed surface.
- *Delaminate partly dried paint.* When barely dry to the touch, acrylic paints remain very susceptible to water. Even if the paint looks and feels mostly dry, a drop of water left on the surface for a few seconds or minutes can rehydrate the paint, making it easy to lift off with a painting tool. This will reveal the underlying paint layer and can give the surface a distressed appearance.
- *Decrease the integrity of the paint film.* The dried (cured) acrylic polymer is what holds the paint film

FROM TOP TO BOTTOM Beth ten Hove's water bottle is encrusted with the history of her palette. • Beth's tools—an artist's painting tools laid out to dry. • Adding a very watery wash over partly dried paint can cause buckling and crinkling as water seeps under the driest top layer and gets trapped there.

together and makes it strong and flexible. When extra water is added to the acrylic polymer emulsion, the acrylic molecules are pushed away from each other, effectively weakening their eventual bond. This will produce a paint film that is more fragile, more prone to tearing, less scratch-resistant, more absorbent, and weaker in all ways.

- *Opacify partially dried medium.* Even a fully cured acrylic paint film is porous. When water is pushed into those pores, it does little more than bring microscopic amounts of dust with it, which can make the paint film very slightly duller. When the paint film is dry to the touch but not fully cured, the effect is more dramatic. Water can flow more readily into the pores and can slightly rehydrate the paint film from the inside. Joining with the uncured acrylic, the water gives the paint film a whitish, milky, semi-opaque to fully opaque appearance. In most cases, given good ventilation in a relatively dry environment, the water will eventually evaporate, and the paint film will clarify again.

While thin mediums can be used to reduce the viscosity of thicker ones, water is actually most effective at this. A generous amount of water will break down any medium's ability to hold texture. Depending on the density of the solids in a given medium, the addition of water will reduce the binding action of the acrylic polymer and dilute the bulking action of the thickener, causing the medium to become less viscous. To avoid clumping, water should be added in small, incremental amounts for a smooth blend.

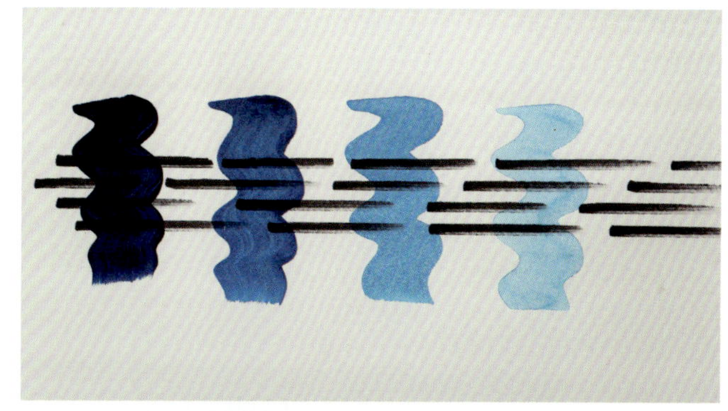

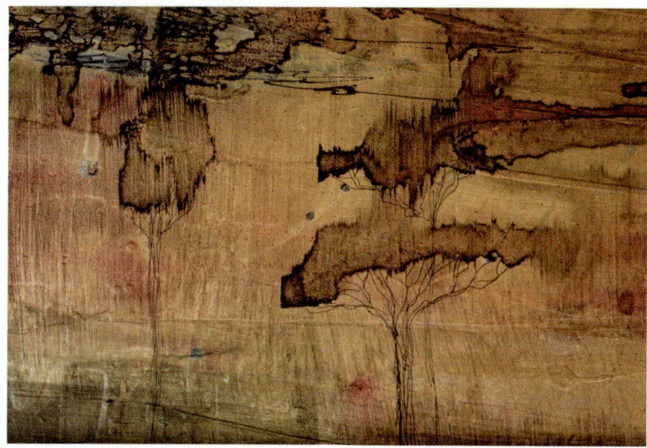

TOP AND BOTTOM Phthalo blue in incremental dilution with water. From left to right: the full-strength color followed by dilutions in which the water content is gradually increased. • Incidental water marks. A watery wash of black over a rosy copper-hued paintscape was left to partly dry, then lightly removed with a soft straight-edged tool. The remaining outlines give the impression of trees.

KEY CHARACTERISTICS OF WATER

KEY TERM	WATER
VISCOSITY	very low
RHEOLOGY	short
LUSTER	none
RELATIVE COVERAGE	transparent
TEXTURE	none

DELAMINATING PARTLY DRIED PAINT

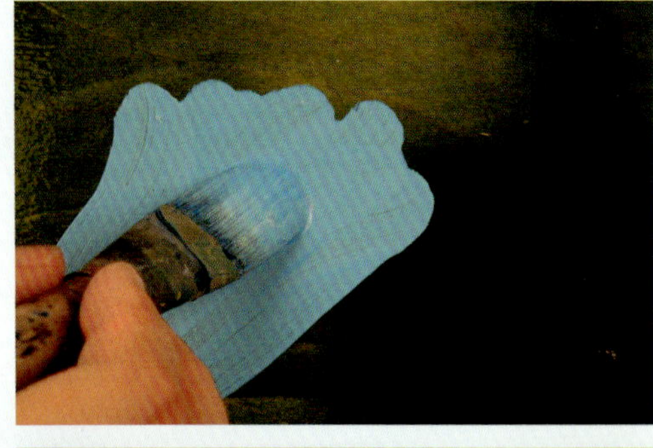
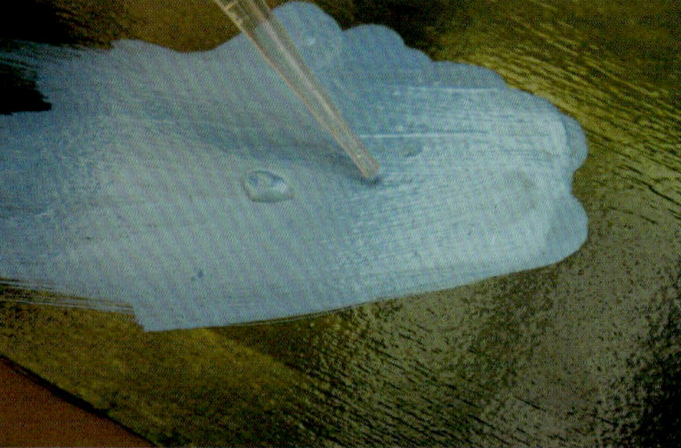
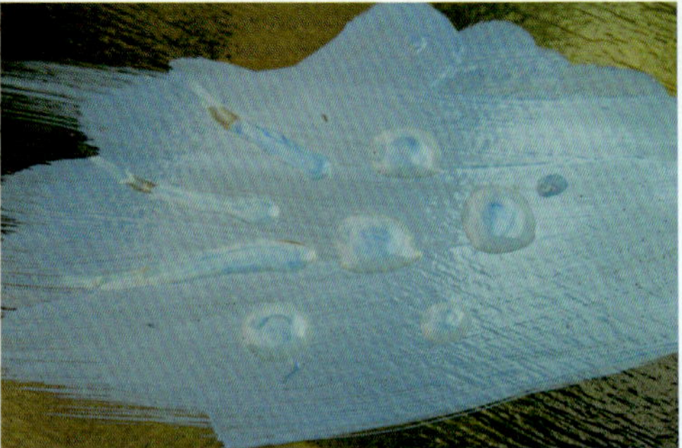
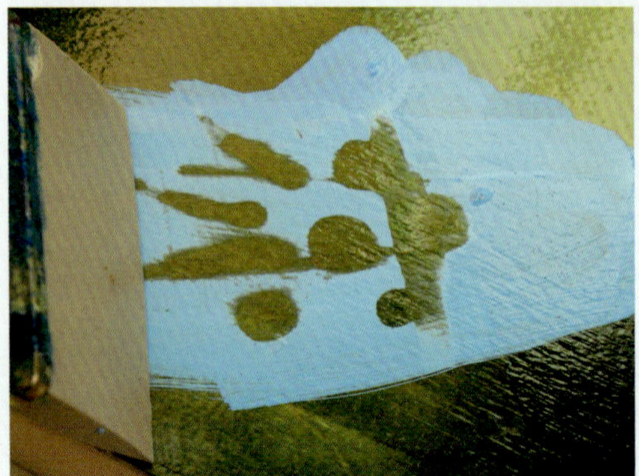
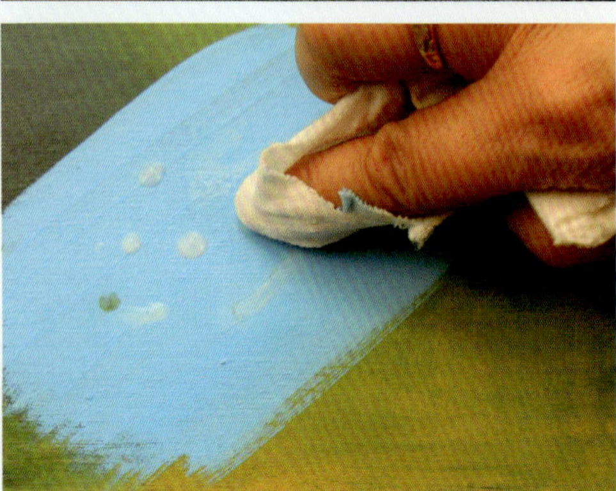

TOP, LEFT TO RIGHT The paint is applied to an already-painted surface. • When it is partly dry, drops of water are dropped onto it. • The water drops reactivate the paint and pull it away from the surface.

BOTTOM, LEFT TO RIGHT The reactivated paint wipes cleanly away from the surface with a flat, soft tool. • Wet paint can be rubbed off with a cotton rag.

WATER—OTHER USES

Other uses for water include these:

- Reducing the adhesive quality of the medium
- Increasing the flow of viscous paint
- Breaking down the texture-holding ability of thicker paints and mediums
- Causing some separation in mixed pigments
- Increasing the open time
- Cleaning tools

RIGHT A small color shaper creates a pattern with watered-down liquid color.

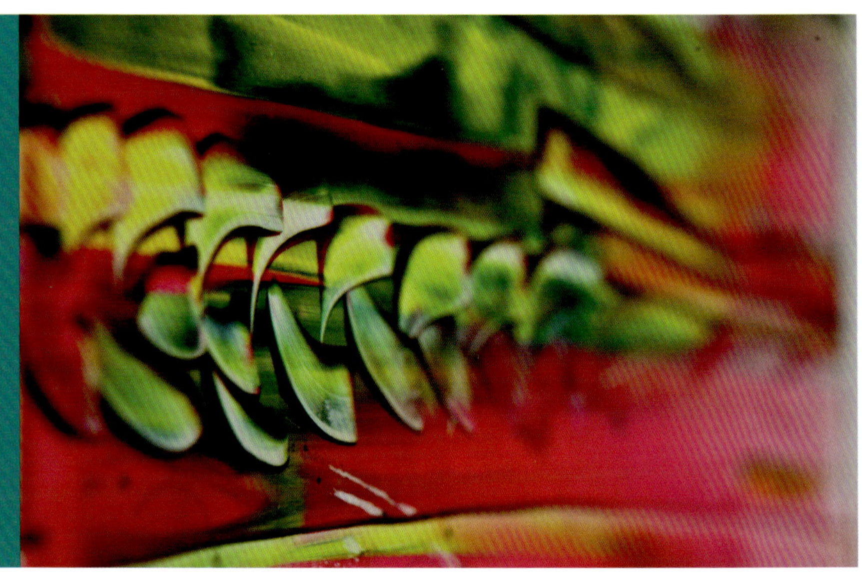

PROCESS

FLUID PAINTING

Fluid painting is a process in which the acrylic colors float and flow within a watery base. Although this style of painting can be achieved using various low-viscosity mediums or pouring mediums, water is the medium used here.

Water is liberally added, by either pouring or spraying, throughout the painting process. The water breaks up the color, which produces swirling, undulating, and usually unpredictable patterns.

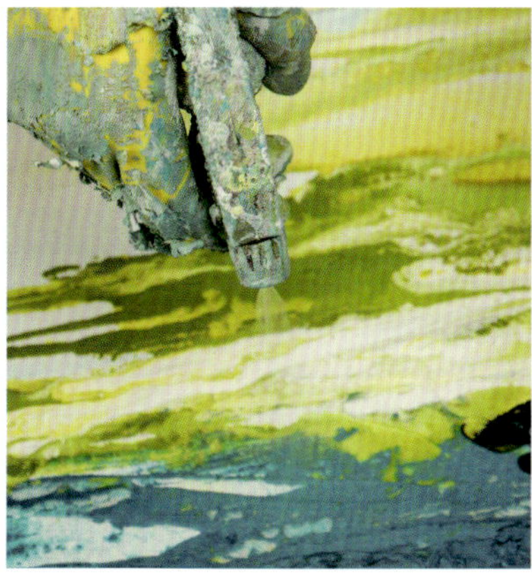

1
Keeping the paint hydrated is the key to fluid painting.

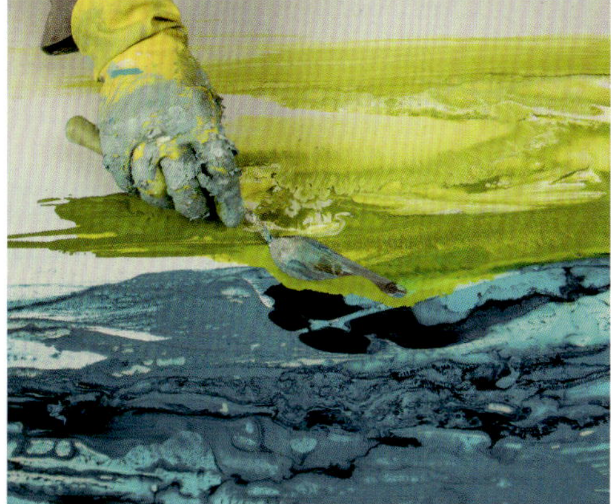

2
When liquid color is manipulated with a palette knife (helped by water on the surface), the vibrant yellow undertone of the green gold begins to show—and flow.

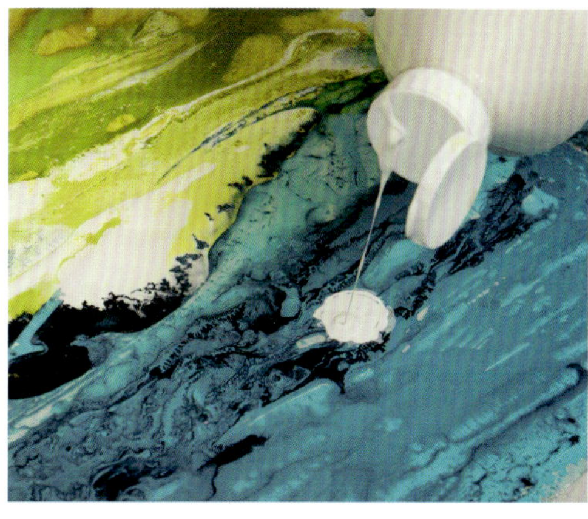

3
Liquid acrylics appear thick and heavy when poured into this watery base.

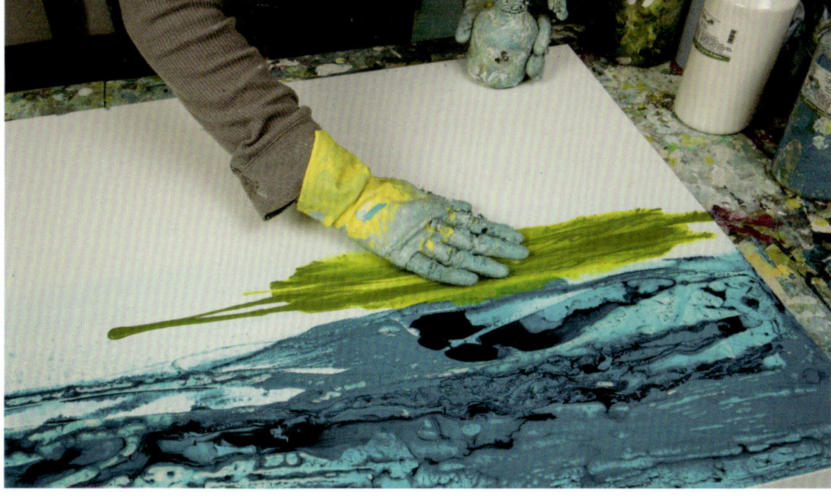

4
The liquid color is spread out.

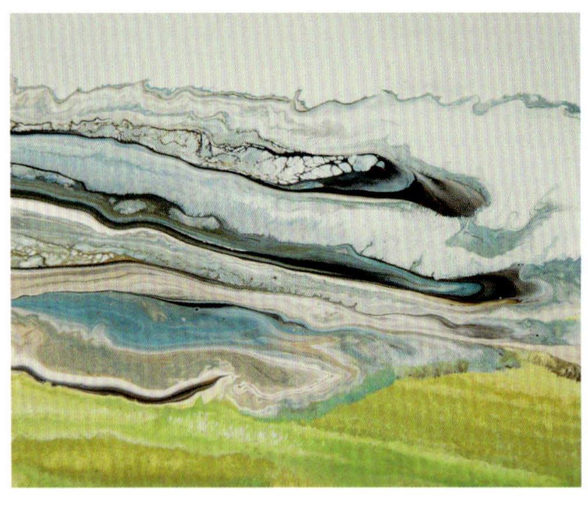

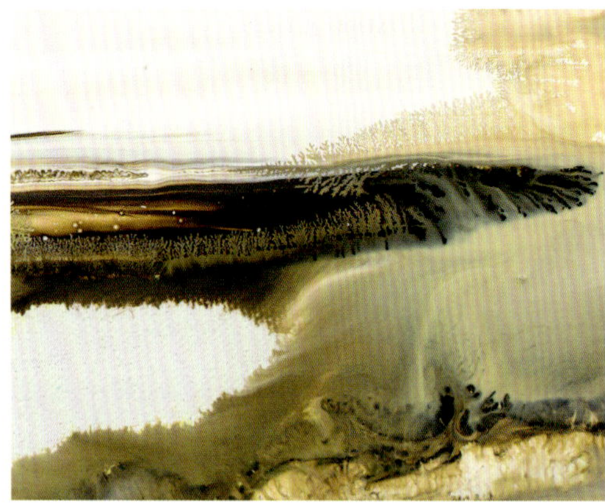

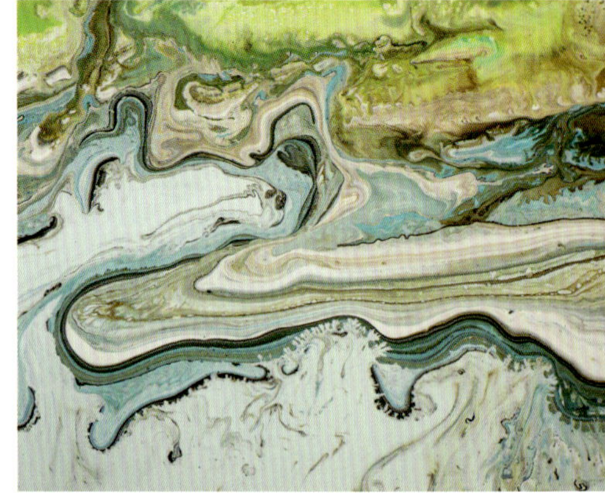

5
The mixed colors begin to separate and blend; because they are not overmixed, striations of color remain clearly delineated.

6
The conflicting viscosities and densities of colors mixing into each other mimic the appearance of dendritic agate.

7
Now, the surface begins to look like an aerial photograph, with rivers of watery color creating sinuous shapes.

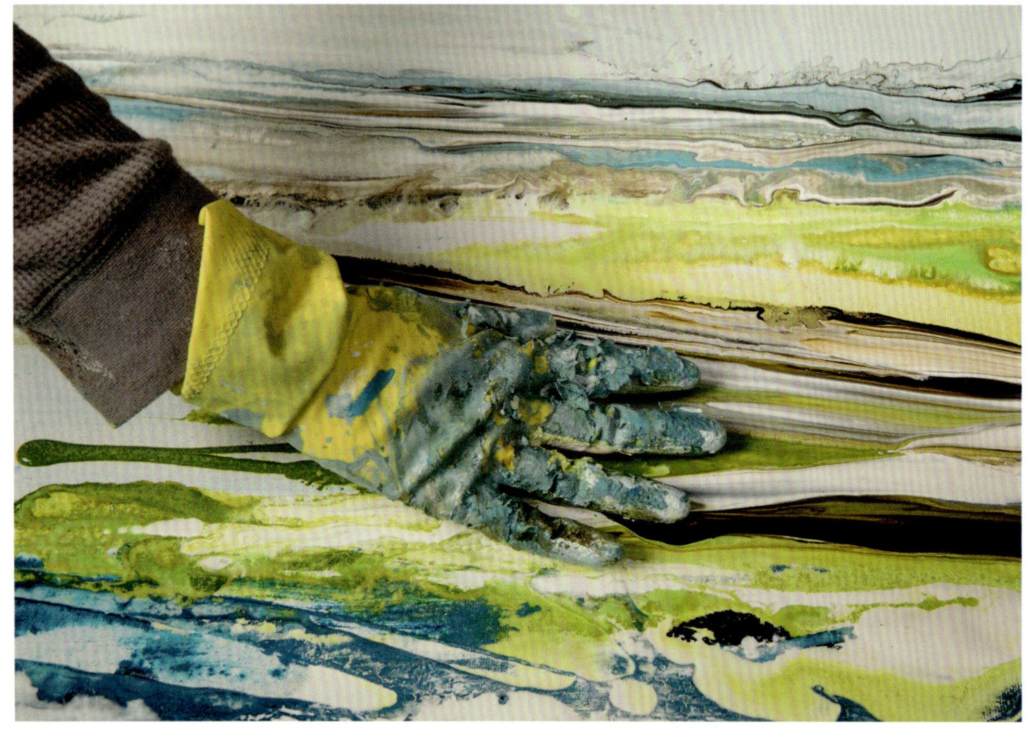

8
Sweeping a hand through the paint encourages the flow.

GROUNDS

A ground is the foundation of an acrylic painting. Its purpose is to provide a semi-absorbent, flexible, and (in some cases) slightly toothy surface for acrylic media. Grounds also protect the support—canvas, wood panel, or other—from the migration of the paint material. A ground is the starting point for building a strong and stable acrylic composition.

Acrylics are a friendly paint, easy to get along with and usually not too fussy about where they end up. They will make fast bonds with many types of supports without a ground to facilitate the union. Some supports, however, require a go-between, something to ease relations and keep the interaction harmonious. It's important to know the difference!

Once you have established what kind of support you'll be painting on, deciding what to use to prepare that support depends on several factors. Grounds can be used to seal the support, act as a barrier against chemical or water migration, create an acid-free buffer zone, add tooth, or provide a reflective, textured, or colored surface.

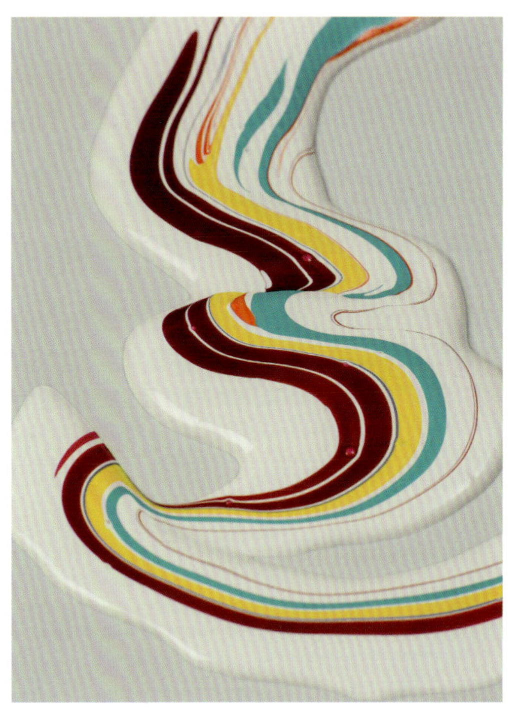

ABOVE The thick, opaque white boldness of gesso holds strains of liquid color.

GESSO

The most common, best-known grounds are the primers, and the quintessential priming ground is gesso. The term *gesso* refers to a paint mixture consisting of a binder (in this case, acrylic), chalk (calcium carbonate), a matting agent, and pigment—usually titanium white, although gesso has now evolved beyond the basic white. Clear gesso contains no pigment; buff (or canvas) gesso is made with unbleached titanium white pigment; black gesso with carbon black; brown gesso with burnt umber; and gray gessoes with some combination of titanium white and carbon black. You can create a colored gesso of your own using clear gesso (sometimes called transparent gesso) tinted with any pigment. Beyond their smooth, matte appearance, gessoes are tough, absorbent, and sandable.

When choosing a gesso, be sure to purchase one with 100 percent acrylic resin. Some gessoes are not 100 percent acrylic, so read the label carefully. If it says "for acrylic" but does not explicitly state that it contains acrylic, it should be avoided, as it will not have the necessary properties to properly prime your support. Inferior gesso can crack or crumble or may dehydrate the paint layer (causing the paint film to become brittle) and generally offers a poor surface for painting. It is much better to pay a little more for a quality gesso than to have to deal with the problems a bargain brand can cause. This is the base—the founding structure—for your painting, so get things right at the beginning.

BELOW A professional-grade gesso should be flexible, smooth, and supple. These strips of black and white gesso were peeled off of a nonstick palette and rolled to show their flexibility.

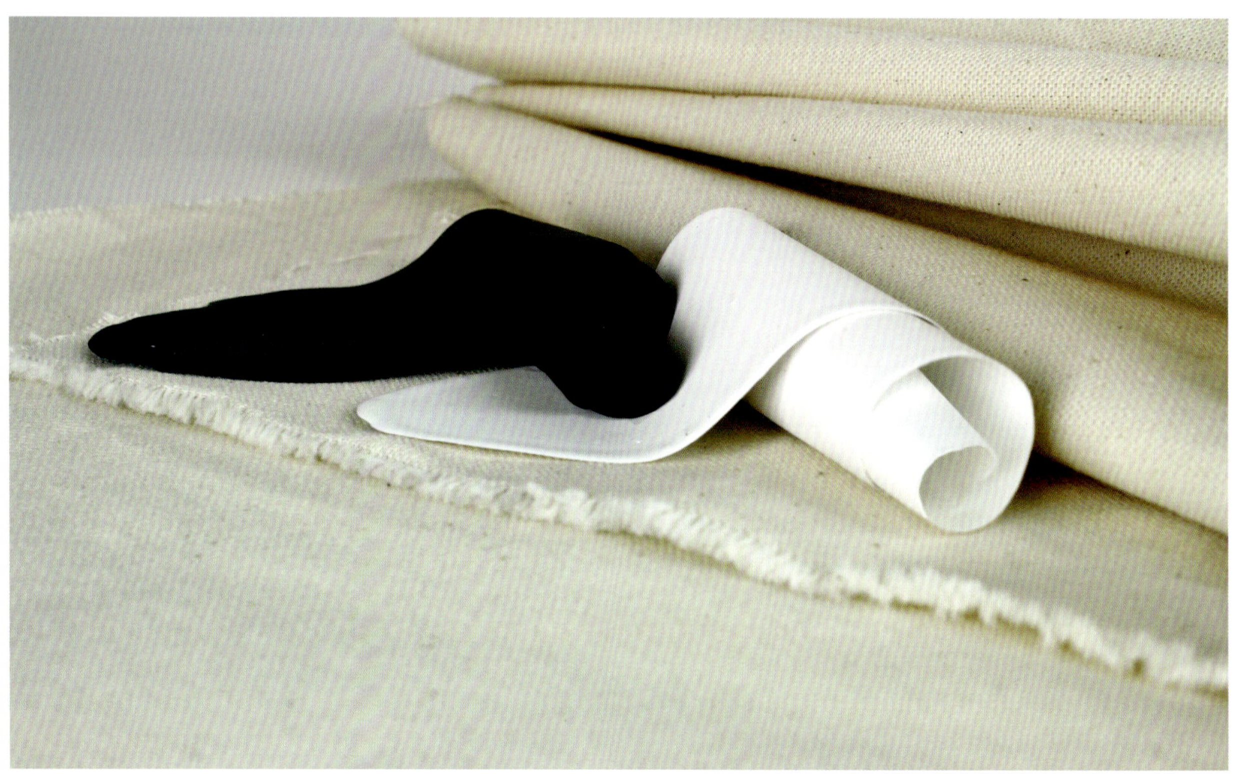

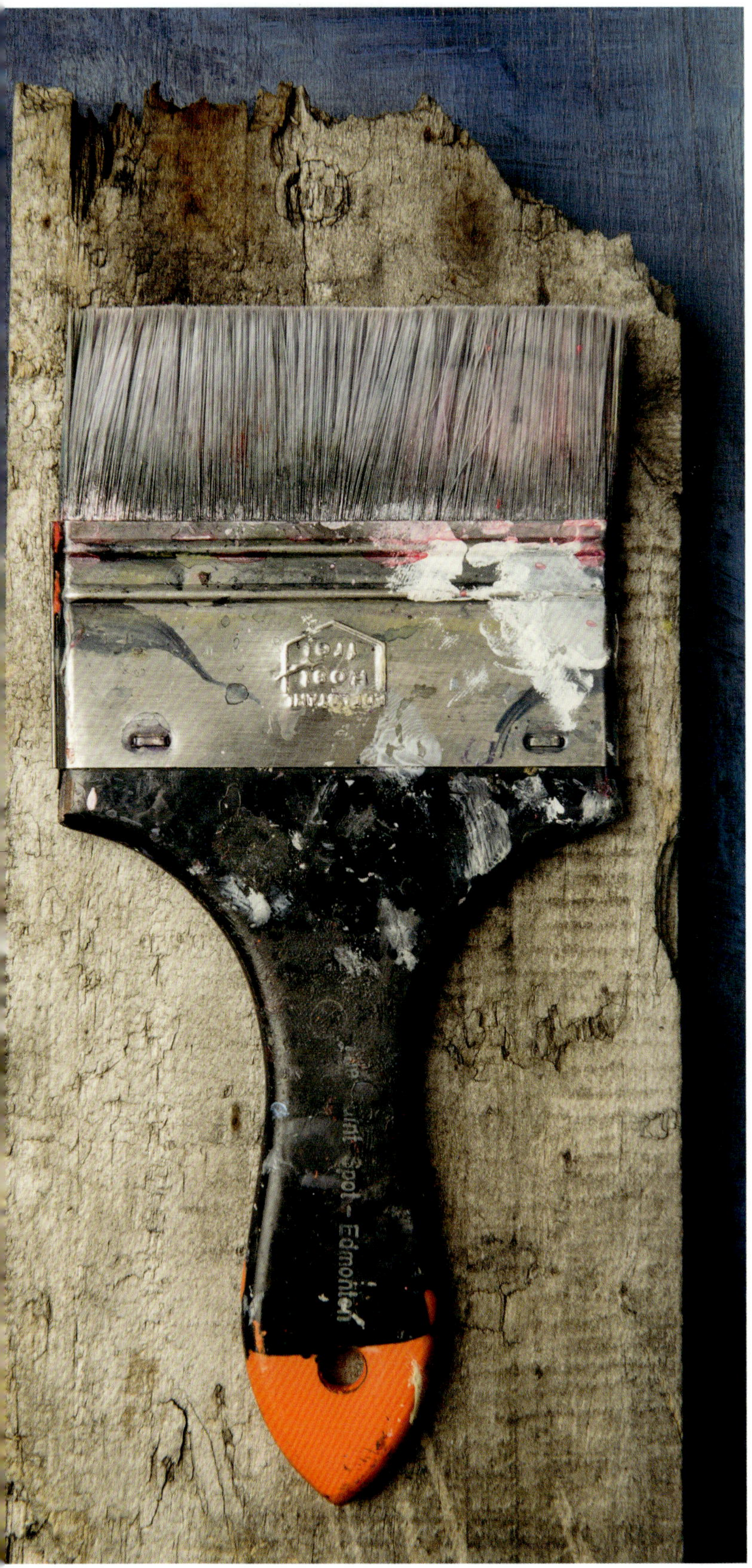

The purposes of gesso are manifold, but its primary function is to create a bond between the substrate and the paint. It adheres to the substrate, creating an effective seal. The calcium carbonate in the gesso acts as a neutral pH buffer zone, preventing any acids from the painting materials from boring into the substrate and causing it to degrade. The solids in the gesso also absorb the paint, and this absorption fuses the two layers together. The smoother the application of the gesso, the stronger this bond will be. Laying down several layers of gesso and sanding each when dry provides the strongest bond.

KEY CHARACTERISTICS OF GESSOES

KEY TERM	CLEAR GESSO	COLORED GESSO (WHITE, GRAY, BROWN, BLACK, ETC.)
VISCOSITY	medium	medium
RHEOLOGY	medium-long	medium-long
LUSTER	matte	matte
RELATIVE COVERAGE	translucent	opaque
TEXTURE	smooth	smooth

LEFT Short, tightly packed, synthetic Taklon bristles give this wide, robust brush a sturdy durability that's great for pushing paint around. When used for applying gesso, this sort of brush produces a very smooth surface with minimal visible brushstrokes.

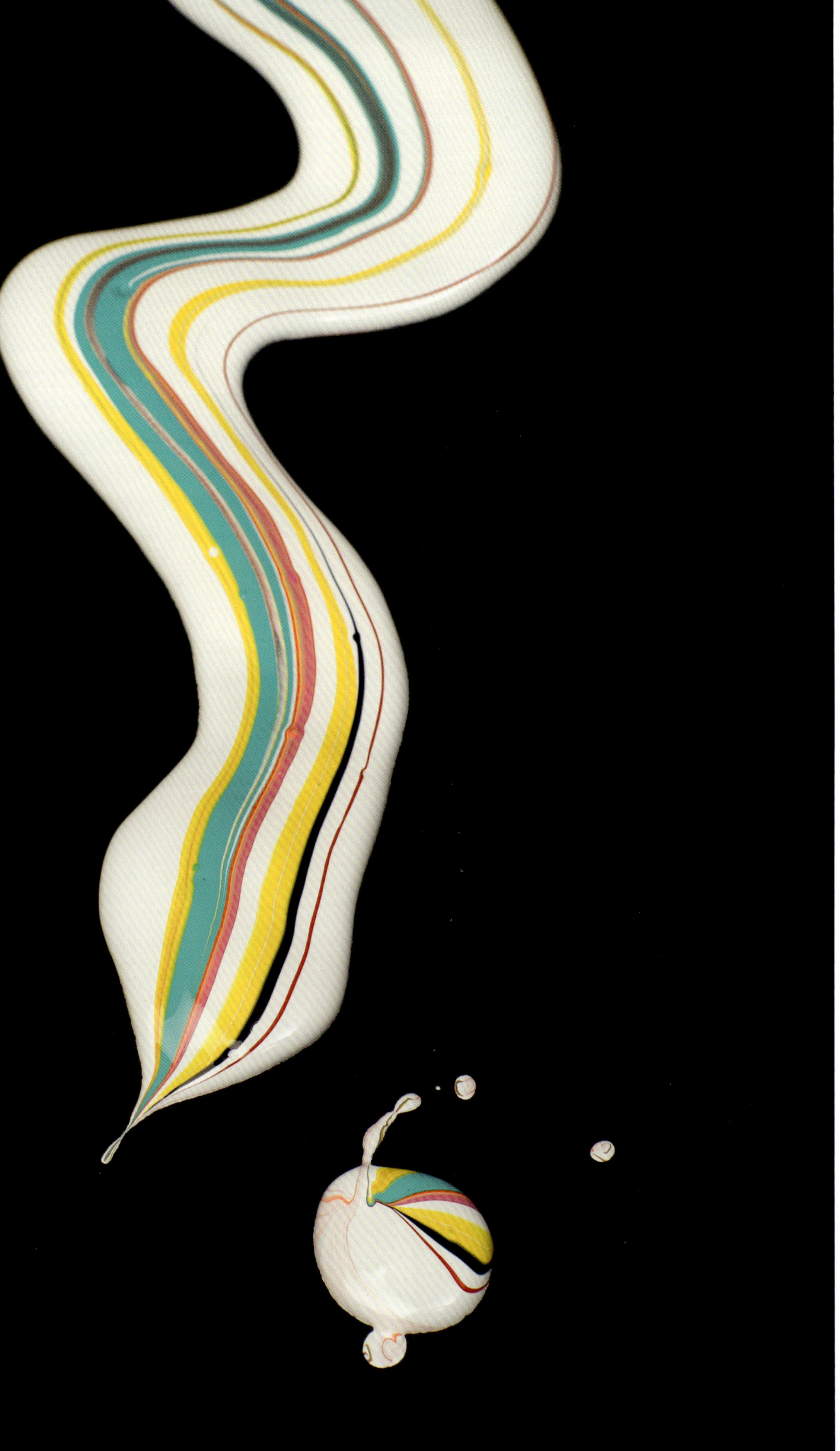

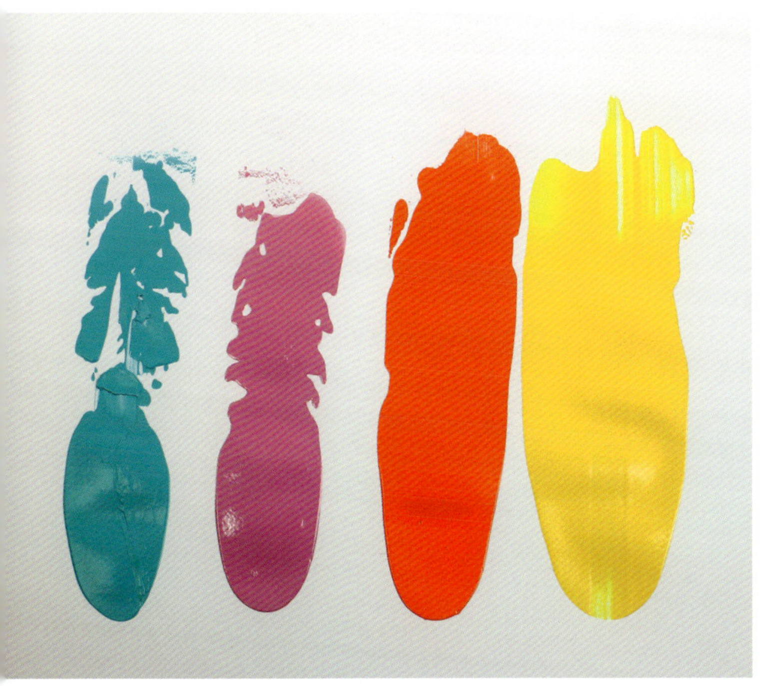

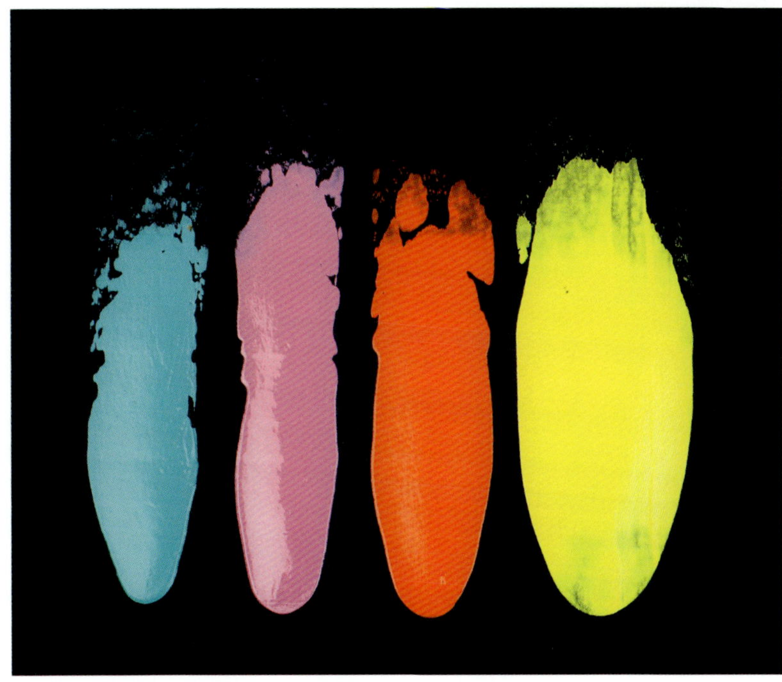

OPPOSITE The opaque boldness of black gesso really makes colors pop.

ABOVE Four swatches of the same colors on white and black grounds seem quite different. On a white ground, the colors appear duller, darker, and more lifeless. The black ground brings out their bright, true nature.

Those solids in gesso also perform another function: they provide tooth. This facilitates a uniform absorption of paint and also creates a finely grained surface for preliminary drawing. White gesso (the traditional form) reflects light, which is particularly effective for glazes and transparent colors. The alternative gesso colors provide different types of light effects: black absorbs light but makes the colors painted on it "pop"; a gray gesso shows the true value of the colors that are laid on top.

All of photo-realist painter Barry Oretsky's paintings (see page 61) are done on a black ground. He explains: "My reasons for this start with how the human eye takes in color. Historically, Realist painters would give their canvases a wash of mid-range scumble in order to subdue the white and still see their draft. I determined that I had to start from an even darker base in order to control the light values of my mixtures adequately. As my practice progressed, my studies in color theory gave me a better understanding of how the human eye perceives color. The human eye takes in the fastest wavelength of color first, which is white light traveling at 186,000 miles per second. Any color coming in after is always darker, so if one places a color value on a white surface, one cannot actually see the true color. Painted on a dark surface, one sees exactly the color mixed. This way the painter is in control of the paint and not vice versa."

MODELING PASTES

Modeling pastes contain ingredients similar to those in gesso but are very viscous and significantly denser. The various brands have slight differences, but all are designed to be exceptionally tough and to create dramatic textures. All are sandable. The massive quantities of solids in a modeling paste reduce the ground's flexibility, so modeling paste should be applied only to rigid surfaces to prevent structural damage. Modeling paste, either clear or opaque (containing titanium white pigment), is harder and more dense than gel mediums (see page 69) and can be carved into once fully cured. These are the grounds to turn to when producing heavy base textures and sculptures.

KEY CHARACTERISTICS OF MODELING PASTES

KEY TERM	CLEAR MODELING PASTE	MODELING PASTE
VISCOSITY	high	high
RHEOLOGY	short	short
LUSTER	matte	matte
RELATIVE COVERAGE	translucent	opaque
TEXTURE	smooth	smooth

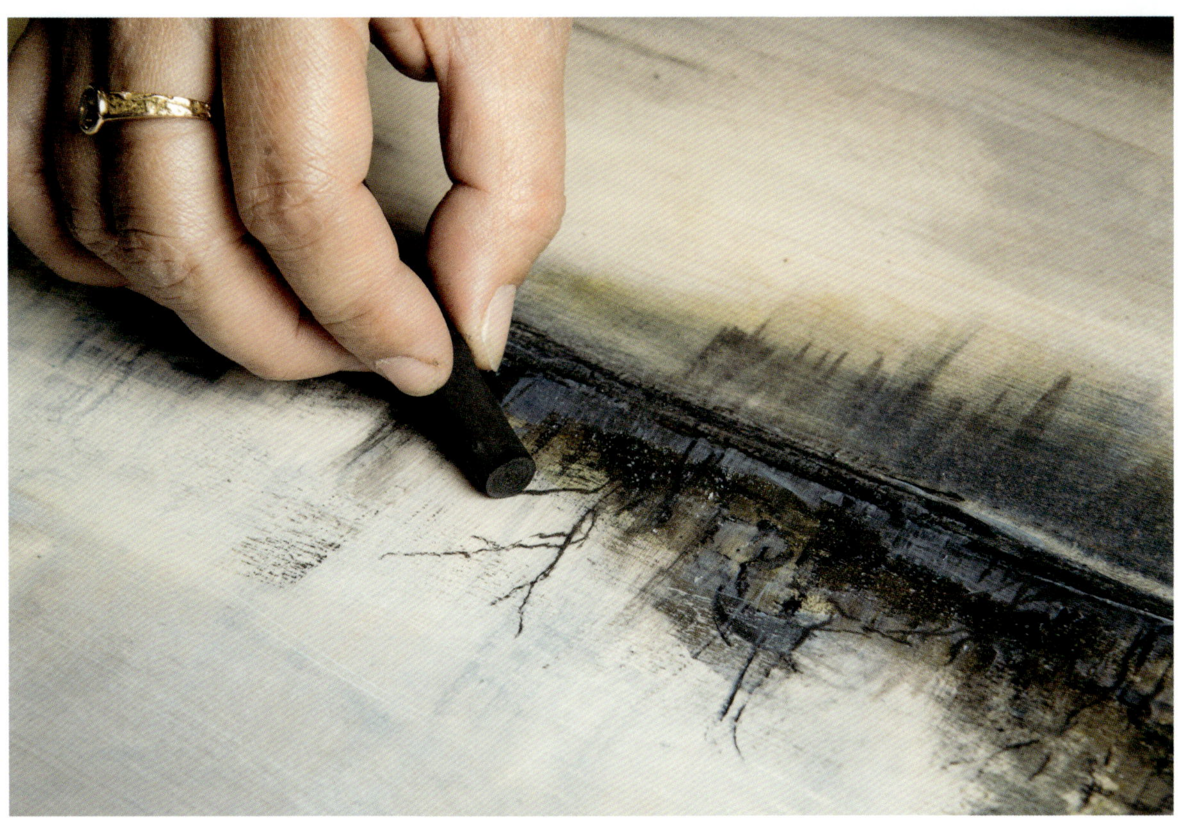

DRY MEDIA GROUND

ABOVE Dry media ground's light, uniform tooth gives purchase to vine charcoal, allowing delicate drawing like this.

Considered a bridge between wet and dry media, dry media ground (also known as pastel ground) is absorbent, matte, and very toothy. This viscous, semi-opaque ground dries to a sandpaper-like surface. As the name implies, it offers an ideal base for all types of dry art media, including graphite, charcoal, pastel, and colored pencil. Because it is translucent, dry media ground will not fully obscure the substrate. If you desire a colored ground, it can be tinted with acrylic colors in any proportion.

KEY CHARACTERISTICS OF DRY MEDIA GROUND

KEY TERM	DRY MEDIA GROUND
VISCOSITY	high
RHEOLOGY	short
LUSTER	matte
RELATIVE COVERAGE	translucent
TEXTURE	granular, fine

LIQUID MEDIUMS

Liquid mediums come in many forms and serve a huge variety of purposes. Whitish and opaque when wet, they all dry clear. Beyond basic liquid medium, this category of mediums includes glazing mediums, low-viscosity polymers, UV-reactive polymers, polymers with added UV protection, polymers designed to work in conjunction with digital imaging, and polymers that make paints more flexible, make paint films harder, increase adhesion during laundering, and so on. The list goes on and on, with many acrylic brands adding more liquid mediums as the demand for new materials grows.

Some manufacturers call these polymer mediums or fluid mediums. Others call them gloss or matte mediums, which gives you no clue as to a medium's format, only its finish. All these different names can be confusing, so I refer to this group simply as liquid mediums. Ranging in viscosity from very thin and watery to a thicker consistency similar to that of shampoo, liquid mediums all flow with ease.

KEY CHARACTERISTICS OF LIQUID MEDIUMS

KEY TERM	POLYMER MEDIUM (LIQUID MEDIUM)	GLAZING MEDIUM
VISCOSITY	low	low
RHEOLOGY	short	short
LUSTER	gloss to matte	gloss to matte
RELATIVE COVERAGE	transparent	transparent
TEXTURE	smooth	smooth

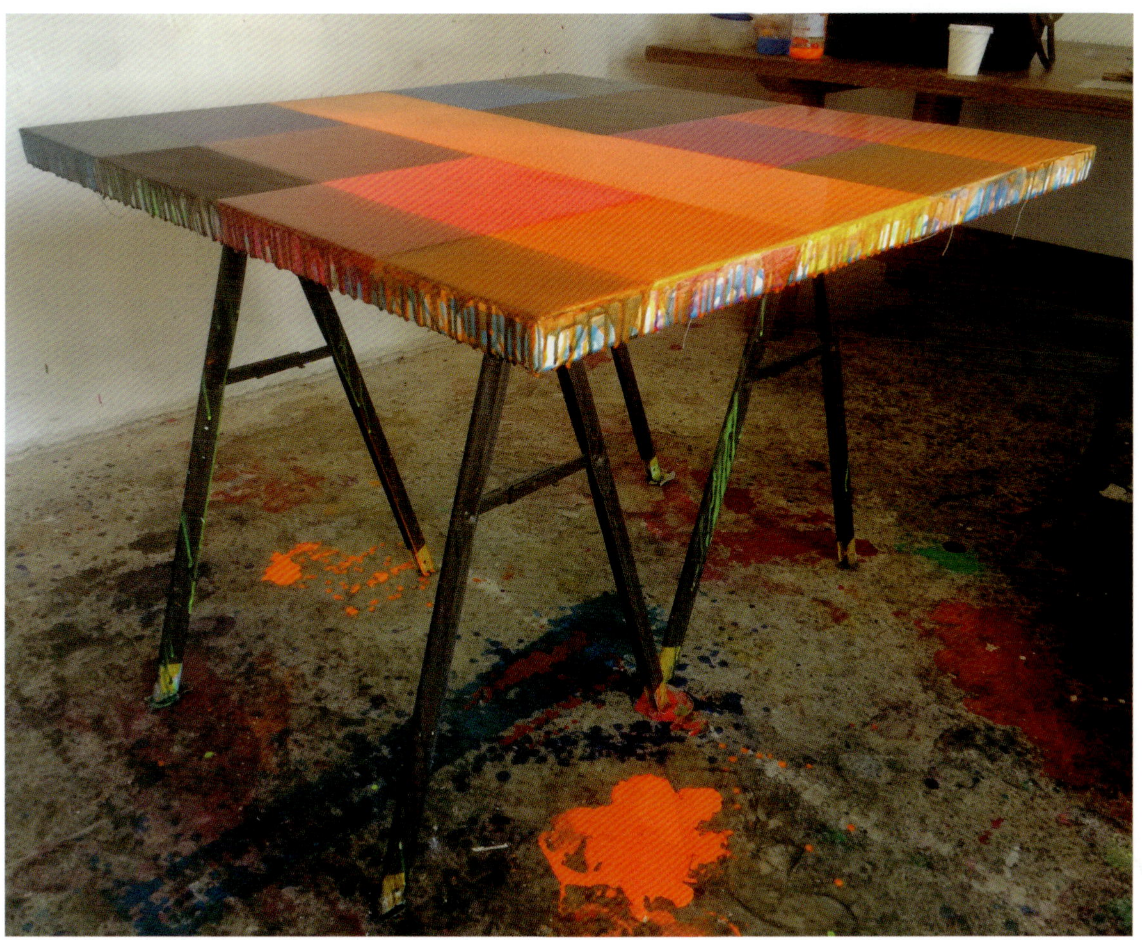

ABOVE This studio shot shows the state of Hester Simpson's painting *Pink Orange Blue Brown* (page 61) after the fluorescent orange band had been poured.

Liquid mediums are available in gloss, semigloss, and matte formats, providing luster control as well as color extension and manipulation. Whether they are added directly to a color or layered on top, these mediums give the painter control over varying degrees of gloss and transparency. They can also be tinted with color in any proportion to create glazes. The more medium that's added, the more transparent and desaturated the color will be, although the addition of the medium will bring light into the surface and create a feeling of depth. From extending drying time to providing a vehicle for glazing, liquid mediums are probably the most useful mediums to have on hand. Liquid mediums are also ideal vehicles for additives and enhancers that further modify the paint film, giving you more application possibilities. They act as adhesives, clear coats, extenders, and all around acrylic band-aids.

PROCESS

WORKING WITH LIQUID MEDIUMS

This sequence illustrates part of Hester Simpson's process when creating her painting *Pink Orange Blue Brown*, which incorporates multiple overlapping bands of transparent and translucent color.

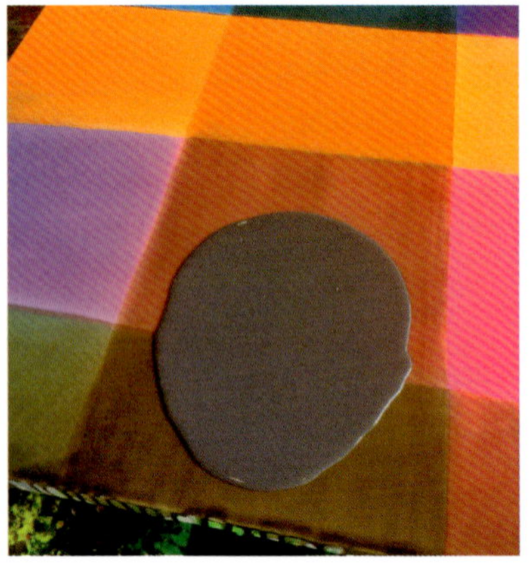

1
After laying down the multicolored, gridded background, Simpson poured the brown band, consisting of raw umber fluid acrylic mixed with water, retarder, and fluid matte medium.

2
She then guided the flow of the brown band.

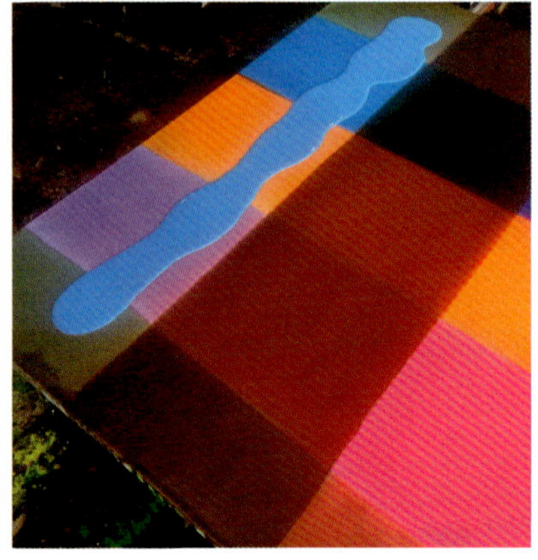

3
Next, she poured the blue band—a mixture of manganese blue hue fluid acrylic, water, retarder, and fluid medium.

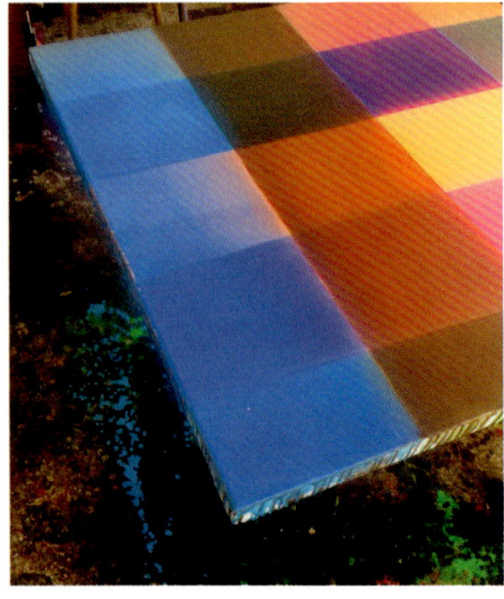

4
As before, she carefully guided the flow of the paint on the canvas.

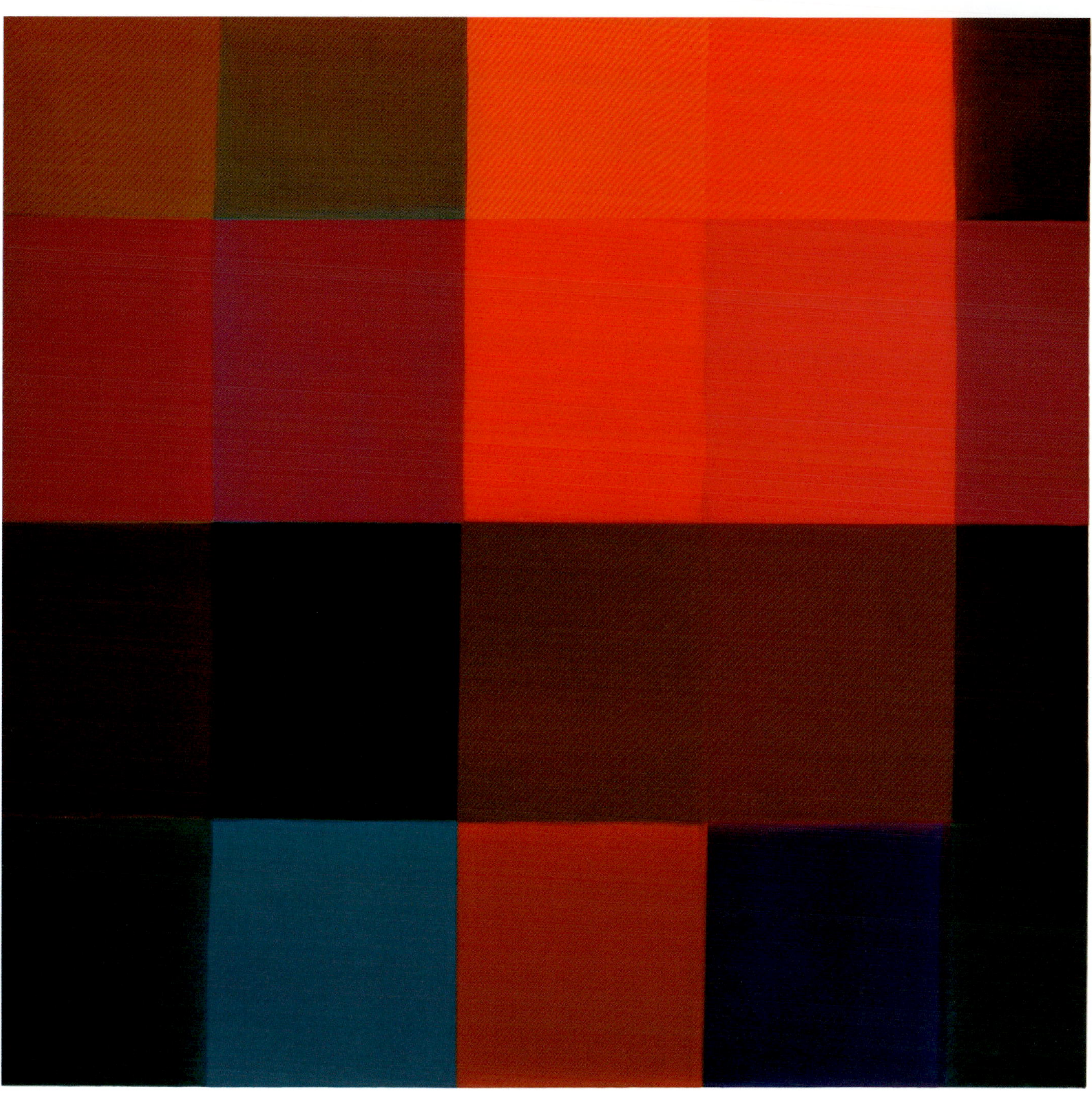

About her work, Hester Simpson says, "The use of acrylic mediums in my paintings over the last eighteen years remains simple. I use fluid acrylic colors mixed with liquid matte mediums and/or gloss medium, thinned with water to a creamy consistency. For larger paintings I add retarder for longer working time. I then run the mixture through cheesecloth into a clean container. Sometimes this mixture is ready to use immediately, but often it must sit for a day or so before air bubbles have settled. Bubbles are one of my greatest enemies, as I want my surface to be perfectly smooth."

ABOVE Hester Simpson, *Pink Orange Blue Brown*, 2017, acrylic on canvas mounted on wood panel, 36 x 36 inches (91.4 x 91.4 cm). Photo by Malcolm Varon, New York City; courtesy of Ricco/Maresca Gallery, New York.

CHOOSING YOUR FIRST MEDIUMS

For your first foray into the world of acrylic mediums, I suggest starting modestly, just to get the feel of things. It is not always an easy transition to go from straight color to working with the awesome power of acrylic mediums. You might want to begin with an assorted set of mediums in small, "trial size" jars. While you may not necessarily want every medium in the set, you may be surprised to find that a medium you never thought you would use has the potential to become a favorite. If you would rather go with a larger jar or two instead, choosing mediums from these basic categories will give you a good foundation:

1. **Polymer medium (liquid medium).** From extending drying time to providing a vehicle for glazing, polymer mediums are probably the most useful to have on hand. They can be used as extenders, adhesives, final clear coats, and much more.
2. **Gel medium.** The consistency of gel mediums ranges from very soft to stiff (with the holding capability of modeling paste). Gels are always fun, but they can take some experimentation to really understand.
3. **Self-leveling gel medium.** It's one thing to work with a viscous gel that will hold a texture but quite another to manipulate one that will not. Because self-leveling gels exhibit a consistency unlike that of acrylic colors, they encourage the use of alternative tools like palette knives and spreading tools. They're great for stretching your experimental muscles.
4. **Granular, textured, or another custom medium.** You may never use it again, or it may become a staple on your art table. Either way, you should try at least one of the more unusual mediums. As when cooking with an unfamiliar spice, you can't really know whether you'll like it without first giving it a taste.

Of course, this list will not suit everyone, but it does provide a broad enough starting point. Most mediums currently on the shelves are variations on these four formats, ranging from gels filled with inert materials to provide texture to polymer mediums infused with UV-reactive components that glow under black light. First, acquaint yourself with the basic elements, then find the specific mediums that have the ideal properties for your process.

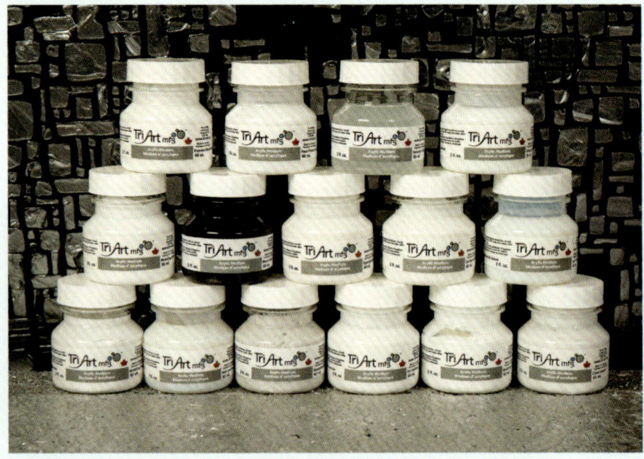

A sample kit of assorted mediums will get you started on your experimental path. These 2 ounce (60 ml) jars give you just a taste of their potential.

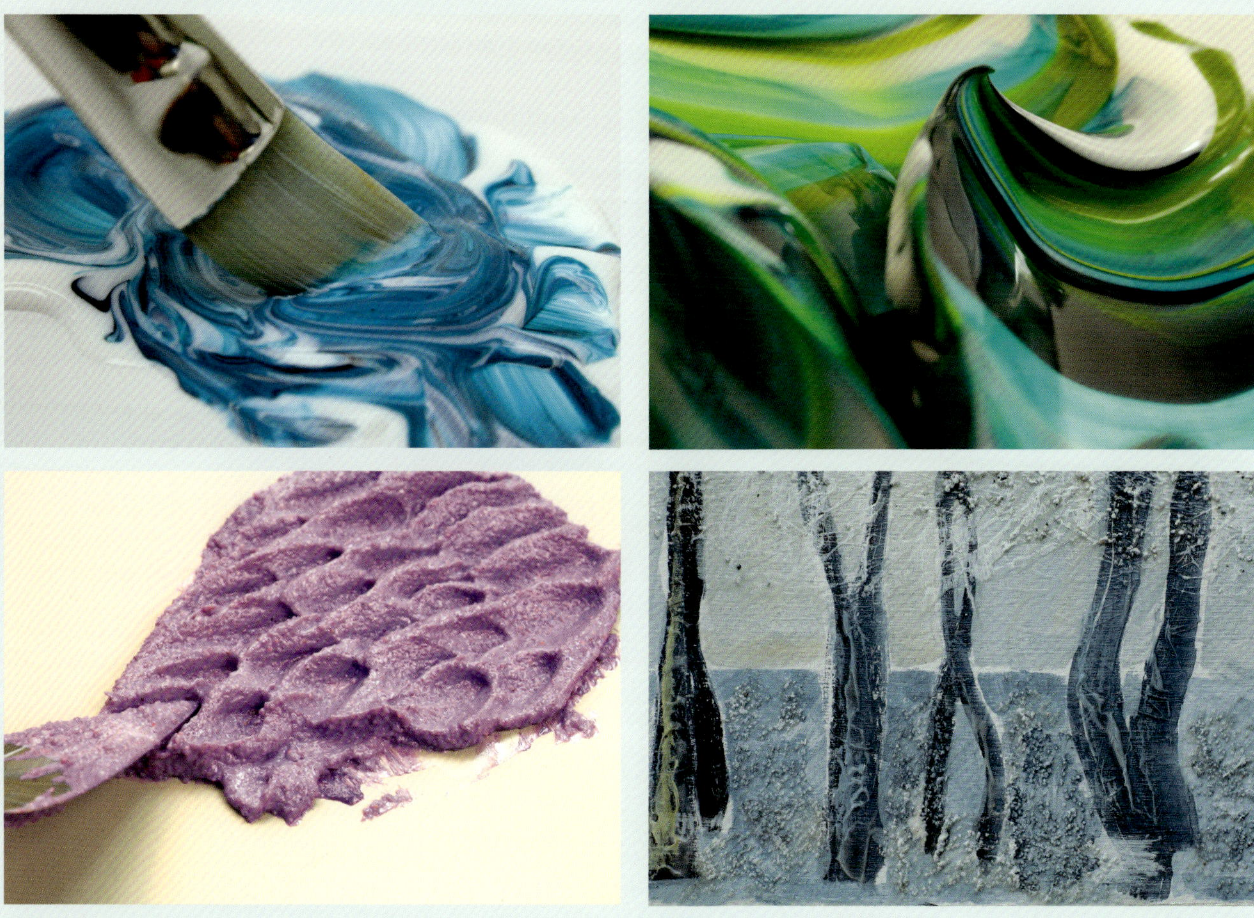

TOP, LEFT Phthalo turquoise (liquid acrylic) loosely mixed with matte polymer.

TOP, RIGHT Phthalo turquoise and green gold (liquid acrylic) lightly mixed into gloss gel.

BOTTOM, LEFT A tiny drop of dioxazine violet is enough to tint this quantity of nepheline gel (coarse).

BOTTOM, RIGHT Even though its palette is minimal, this composition gets some pizzazz from a combination of disparate mediums. It doesn't take much to suggest dimension and texture. Granular mediums, gel medium, and strings of self-leveling gel are swept with a bit of white. A light wash of Payne's grey bumps up the texture of the granular medium.

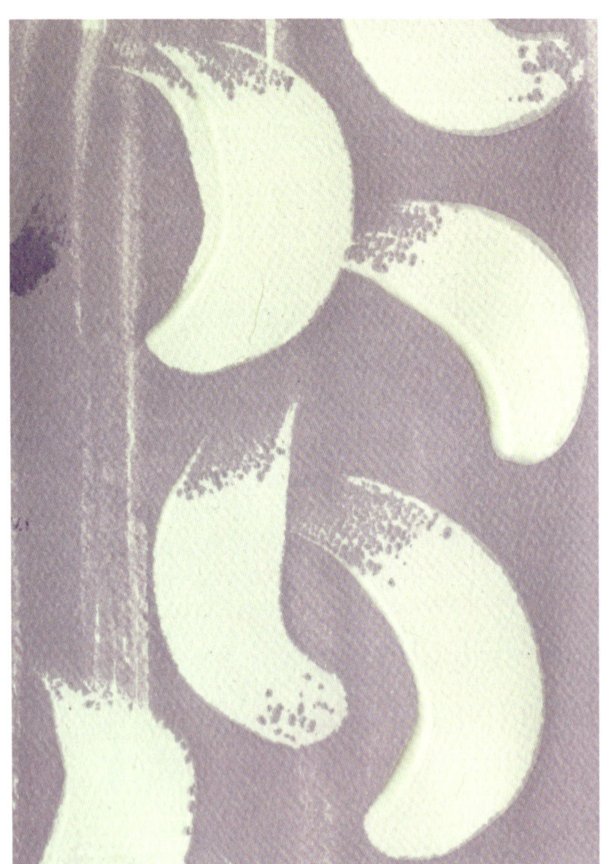

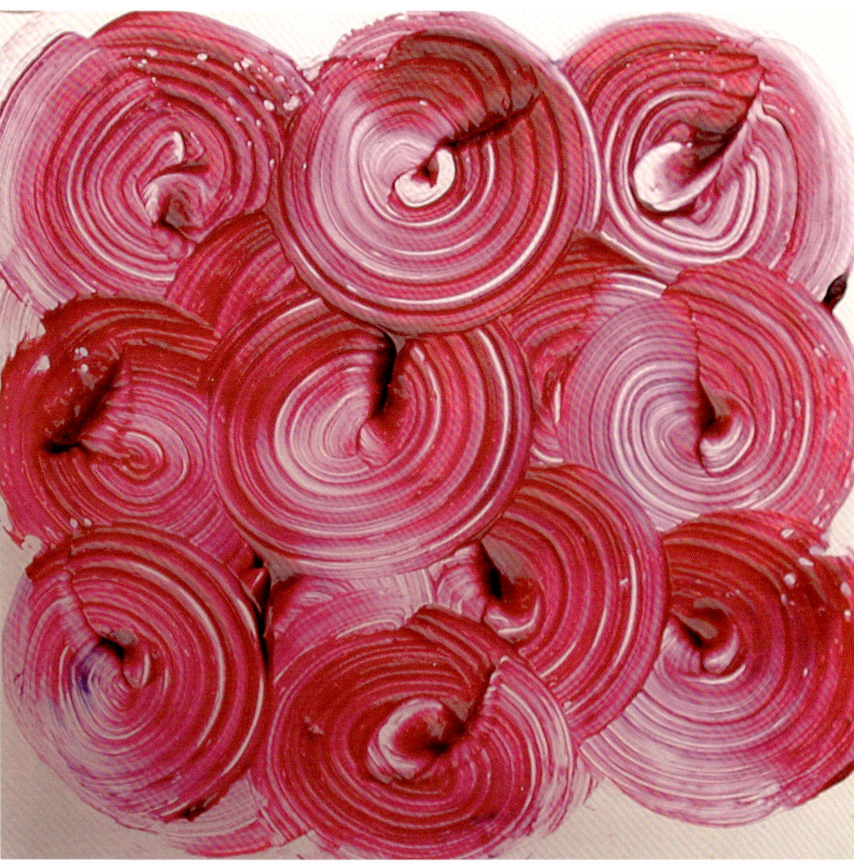

In essence, liquid mediums are liquid acrylic paint without pigment. They are the ideal vehicle for extending colors for smooth applications. They will thin and increase the flow of thick-format acrylic colors and are perfect for producing fine detail work. Liquid mediums are also strong adhesives and can be used in collage and craft applications such as decoupage. Most liquid mediums will maintain brushstrokes and light textures (some to a greater extent than others) and will dry to a flexible, water-resistant film. They can also be added to resoluble water-based paints such as watercolor and gouache to increase water resistance, flexibility, and permanence.

ABOVE, LEFT A dry liquid gloss medium (white strokes) will completely repel a thin wash of paint on a porous surface, acting as a resist. (The surface here is watercolor paper.)

ABOVE, RIGHT Very lightly tinted liquid medium applied with fan brush.

GLAZING MEDIUM

Glazing medium is the ideal liquid medium for creating smooth and luminous glazes with a gloss or matte finish. It is very thin and free-flowing with good adhesion and flexibility, and it retains very subtle brush marks. Its consistency allows for veils of layered colors to be applied while maintaining a smooth and even surface. For thinner glazes, I recommend using glazing medium with liquid acrylic colors or acrylic inks and applying the mixture with a wide, soft-haired brush. Some glazing mediums contain a small quantity of retarding agent, which makes them stay wet and workable longer—an advantage when glazing large areas of a painting.

LEFT Here, whisper-thin glazes of phthalo blue have been built up to create a zone of saturated color. The single brushstroke below shows the dilution of the glaze, and the dots below that track the number of layers applied.

All these liquid mediums have different formulations so that they can perform different tasks. They are generally available in gloss, semigloss, and matte formats, and in varying viscosities. Some also contain additives that make them suitable for other purposes. For example, there are UV-stabilizing and UV-reactive mediums, mediums that facilitate digital printing, mediums that opacify colors, and so on. Some of these will be covered in the sections on mediums for specific uses (page 196) and special-effects mediums (page 84).

Some primary uses for liquid mediums are as follows:

- *Lubricating the paint.* Adding wet medium to a color keeps the paint from drying out and increases the open time.
- *Extending color.* Liquid medium can be used to stretch acrylic colors. Extending higher-viscosity acrylics will decrease their viscosity, while extending liquid colors will increase the volume without changing the consistency of the paint.

- *Increasing flow.* Because adding liquid medium reduces the viscosity of higher-viscosity colors, it increases the flow, allowing for a single paint application to go further and to be applied more smoothly.
- *Increasing the transparency of colors.* Adding more acrylic medium further separates the individual pigment particles from one another, creating a more transparent, desaturated color that pulls in the light and increases luminosity and depth.
- *Adding depth.* A layer of untinted gloss medium—applied over a finished painting or layered within the painting—pulls light into the surface. This creates both depth and luminosity.
- *Glazing.* Use glazing to mix and intensify colors in a controlled, systematic manner. Thinned, transparent, or translucent layers of color, laid one on top of the other, create optical mixing and the gradual intensification of tones.
- *Adhesion.* All acrylics are sticky, and liquid mediums are very efficient glues. They can easily be spread over a porous or semi-porous surface, either smooth or rough, to bond other materials to it.
- *Creating fine textures.* Liquid mediums have varying viscosities, but even the thinnest of them can leave subtle, residual texture.
- *Unifying luster.* Because they are thin, liquid mediums are the obvious choice for changing the reflectivity of a surface. Their low profile provides optimum transparency, allowing any underlying color, detail, or texture to show through. The medium's luster—gloss, semigloss, or matte—will then dictate the appearance of the surface without affecting any other aspect of the work.
- *Sealing.* Collage elements, water-sensitive media, or dry media can be effectively sealed on the surface of a painting or painted object with any liquid medium. Although liquid mediums are porous and therefore not entirely waterproof, they will make the surface water resistant.

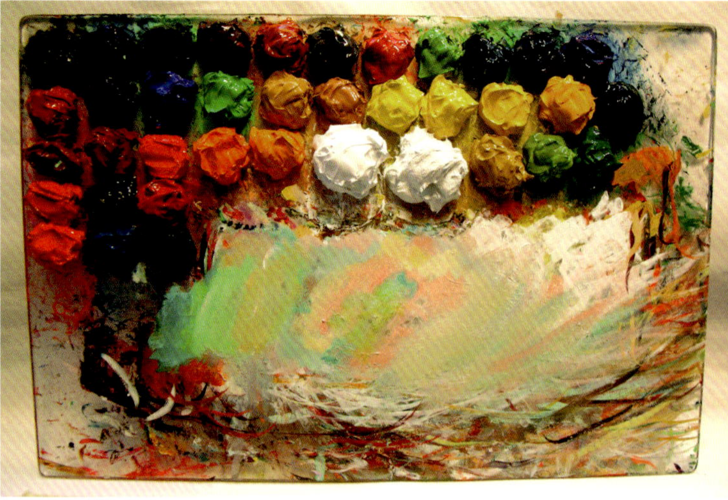

ABOVE Barry Oretsky's palette.

"My palette consists of forty-one dollops of pure colors. I lay them out in a specific order so that I may access them blindly, as one would a keyboard. Over the course of these many years, I have experimented with most available brands of fine art acrylic paint. Each brand produces a different color-mixture language."

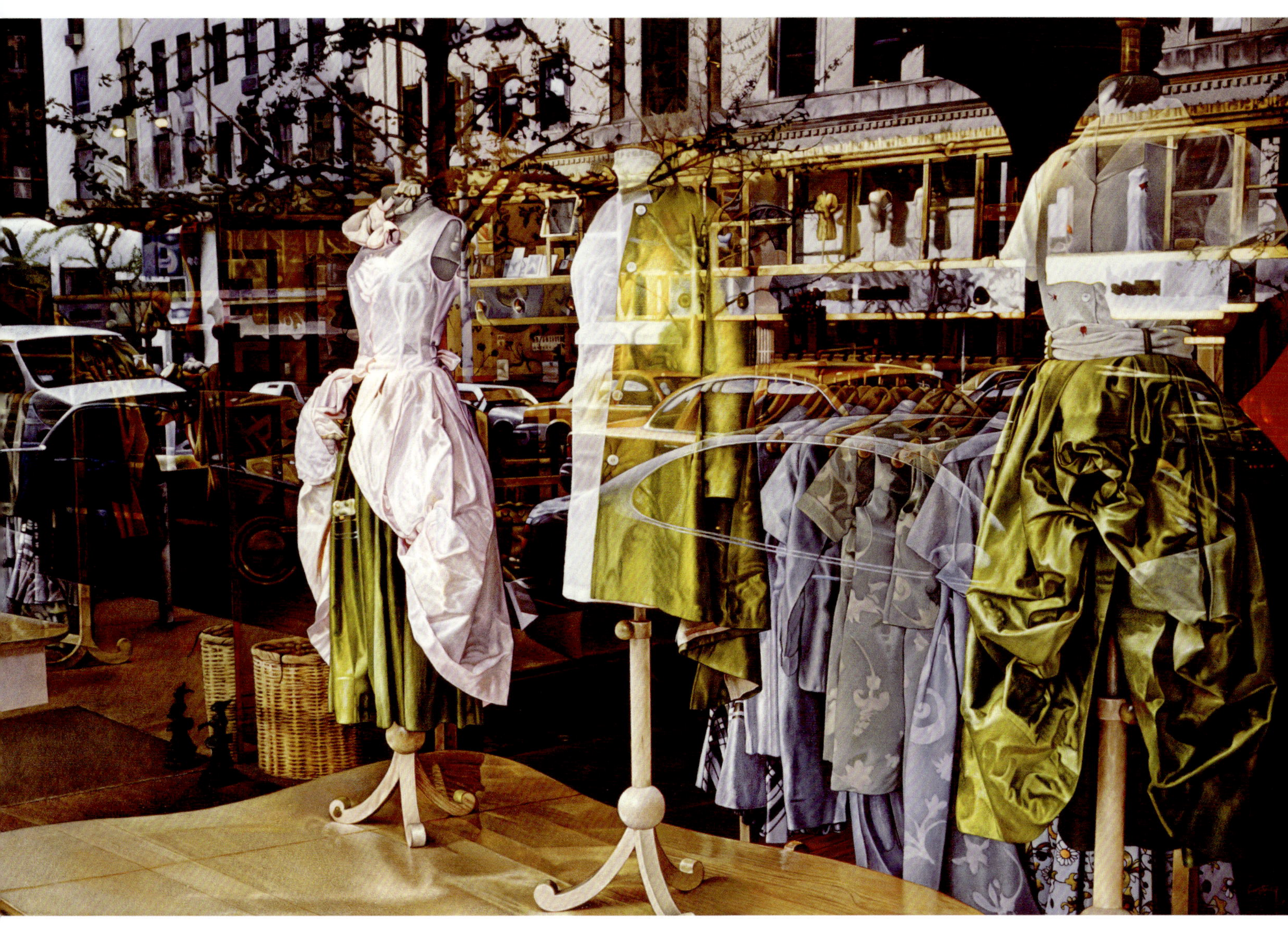

ABOVE **Barry Oretsky, *The Anatomy of Folds,* 2005, acrylic on canvas, 36 x 42 inches (91 x 107 cm). Private Collection, Toronto.**

Barry Oretsky adds a thin layer of liquid medium over every application of color; thus, there are literally thousands of these veils of medium over the full surface of the painting. After he completes the color work, he applies three layers of gloss medium over the painting to even out the surface, creating even more luminous depth.

About his artistic process, Barry Oretsky says, "Since 1972, I have been using acrylic paint exclusively. I use it because it is nontoxic and dries quickly. My work is detailed and multilayered, and large-scale paintings have taken me up to two years to complete. In other, slower-drying media, the production time would be extended exponentially. Each layer of paint is glazed with acrylic gloss medium, and there are many layers of paint building every element in the painting. There are thousands of these separation glaze layers in this painting. Acrylic paint has allowed me to complete a greater number of works.

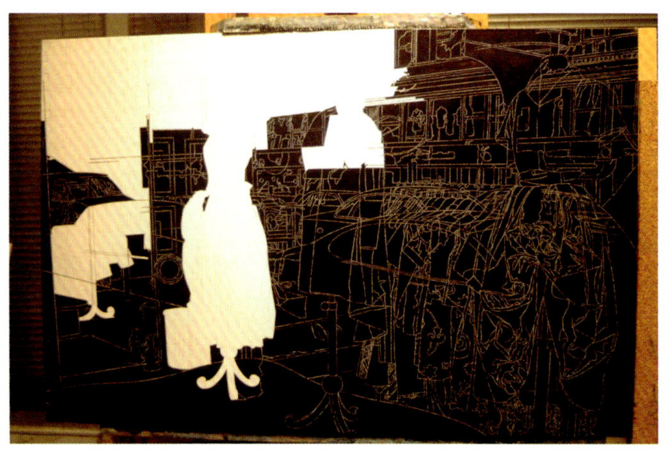

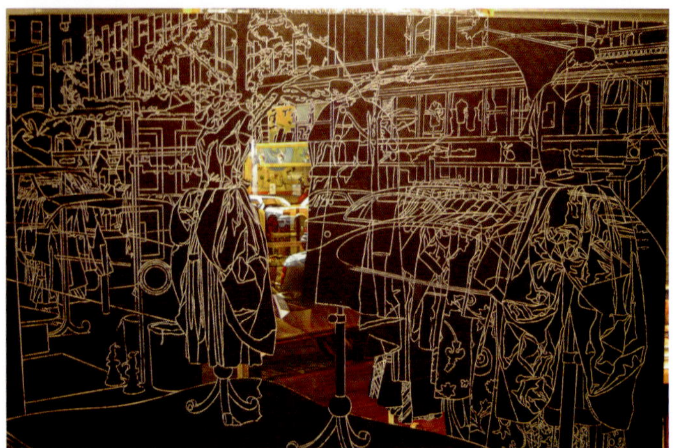

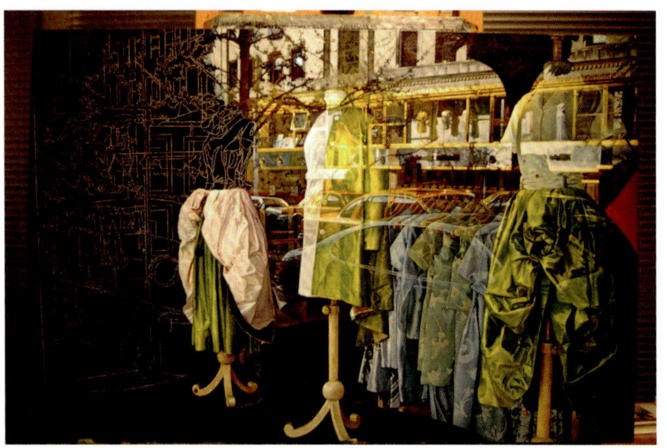

PROCESS

BUILDING LUMINOSITY

Barry Oretsky talks us through his process in creating *The Anatomy of Folds*:

1

"I began with an initial draft of the image from my photo resources, creating a pencil rendering of it on the canvas."

2

"The canvas started with a black line on a white surface, which I inverted by hand to create a white line on a black surface. I painted with a mixture of Mars black, permanent crimson, and phthalo blue green shade acrylic paint on a primed canvas just up to the pencil line on each side, leaving a drawing of thin white lines."

3

"After deliberation, I decided on the best point of entry into the painting, and the journey began. Usually, I first create the environment that the main elements exist in, then work from the outside inward, saving the primary focuses until the end of the painting, because these are impacted by all the peripheral light they exist in."

4

"At this stage, the main concern was to control the adrenaline rush and stay centered until the end. Most mistakes are made at the end, by my anticipating the conclusion of the painting. At the end of the painting, I used three even coats of gloss medium over the entire painting to even it out."

THIN MEDIUMS VERSUS WATER

Thin mediums and water can be used for some of the same purposes, but there are times when mediums and water will produce quite different results. Being aware of these differences will let you know whether to use a medium or water for a technique, depending on the result you are hoping to achieve. The table below summarizes the similarities and the differences.

WATER OR LIQUID MEDIUM?

WATER	LIQUID MEDIUM
Can be mixed in any proportion with color	Can be mixed in any proportion with color
Desaturates intensity of color	Desaturates intensity of color
Decreases the strength and flexibility of paint film	Increases the strength and flexibility of paint film
Increases porosity of paint film	Decreases porosity of paint film
Reduces the gloss of the paint film	Affects the gloss of the paint film according to luster format (gloss, semigloss, or matte)
An aggressively water-thinned paint film should not be used as a final layer.	A liquid medium can be used as a final layer.

GEL MEDIUMS

What is gel medium? Gel is by definition a densely viscous, semisolid material. Because gel mediums possess the same consistency as high-viscosity acrylic colors, they are the ideal material with which to extend those colors without losing body and/or texture-holding capabilities. They are also the medium of choice for producing texture.

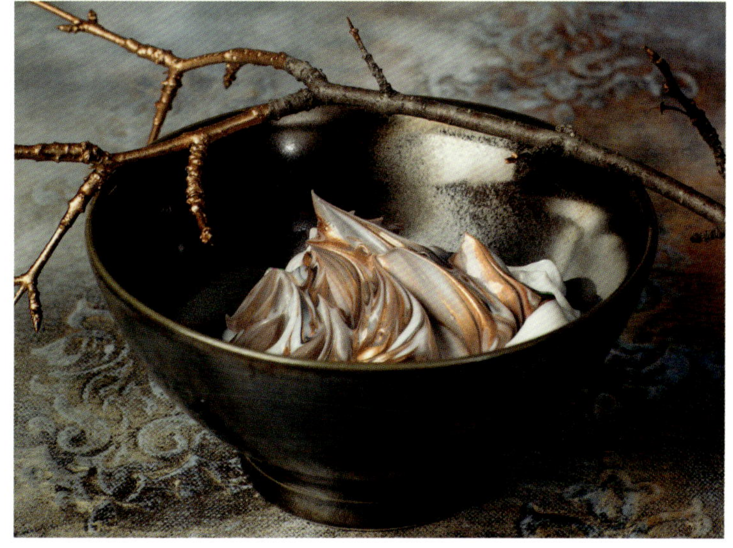

RIGHT Modeling (or extra heavy) gel holds dramatic peaks. This mixture of gloss gel medium with iridescent copper and graphite gray liquid acrylics almost looks good enough to eat.

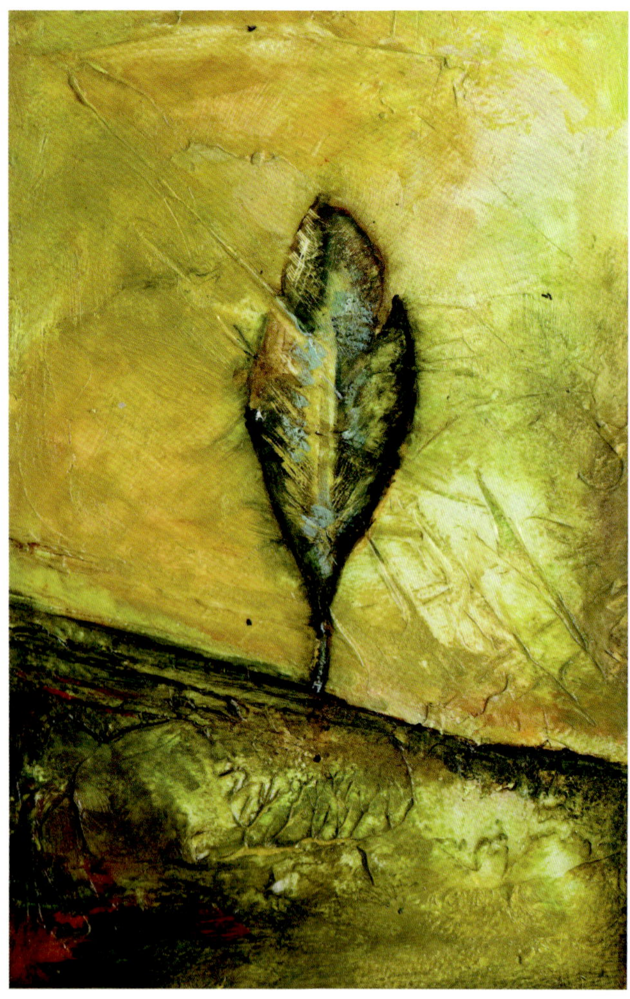

ABOVE Even student-grade colors can be given a boost by combining them with gel medium and applying them in layers, as in this sketch by Connie Morris.

Gels have a broader format range than liquid mediums. Their rheologies range from short to long; they vary in viscosity; and they come packed with a staggering amount of additives. (The working properties of a gel are dictated largely by its rheology.) When wet, gels are whitish, but they will dry clear unless they contain colored additives. Although this viscous stuff will hold a sharp peak, when used in a large quantity the sheer weight of the gel will make it appear somewhat softer and give it a certain flowing quality.

KEY CHARACTERISTICS OF GEL MEDIUM

KEY TERM	GEL MEDIUM
VISCOSITY	high
RHEOLOGY	short
LUSTER	gloss to matte
RELATIVE COVERAGE	transparent
TEXTURE	smooth

Every acrylic brand has its own names for its gel varieties, but the descriptions are usually similar, and it is easy to determine a gel's counterpart from brand to brand. Common names are gel, soft gel, heavy gel, modeling gel, and extra- or super-heavy gel. You get the picture.

All gels are creamy and buttery, and they range from very soft to very dense and gummy. Basic gels can create a range of textures from stiff peaks to smooth finishes in a range of lusters. Although they are thick, they are also very easy to manipulate with brushes and other painting tools. Their short rheology allows for very crisp, delicate to heavy impasto textures. These are some primary uses for basic gels:

- *Lubricating the paint.* Adding wet medium of any viscosity to a color keeps the paint from drying out and increases the open time.
- *Extending color.* Thick mediums can also be used to

stretch acrylic colors. The consistency will be thick, but easy to manipulate. Extending liquid acrylics with gels will bump up the viscosity, while extending higher-viscosity colors will increase the volume without changing the paint's consistency.

- *Increasing viscosity.* Liquid acrylics and acrylic inks are easily plumped up with the addition of a gel medium. The texture-holding capability of the gel will override the liquid state and give the mixture a robust viscosity.
- *Increasing the transparency of colors.* Adding acrylic gel to a color separates the individual pigment particles from one another, creating a more transparent, desaturated color that pulls in the light and increases luminosity and depth.
- *Glazing.* When gels are used for glazing, the transparent or translucent layers of color, laid one on top of another, create optical mixing and gradually intensify tones. Because of gels' thickness, use a gloss gel rather than a matte gel when creating glazes of this volume.
- *Adhesion.* All acrylics are sticky, and gels, like other mediums, are very effective glues. Gels will hold objects and heavier collage elements on a painted surface.
- *Creating dramatic textures.* Even the thinnest gel can create a substantial texture. The more viscous the gel, the sharper the texture.
- *Unifying luster.* Gel mediums can also be applied in a thin film to change the reflectivity of a surface. This low profile allows for optimum transparency, allowing any underlying color, detail, or texture to show through. Care must be taken, however. If your intention is to produce a uniform layer, spread the gel with a straight tool for the smoothest result. The gel's luster (gloss, semigloss, or matte) can then

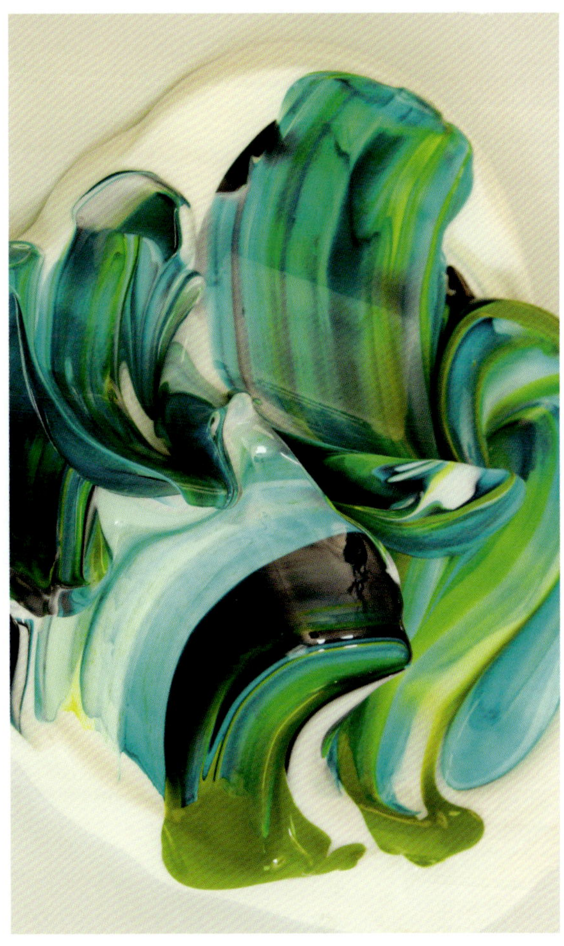

ABOVE Phthalo turquoise and green gold (liquid acrylics) lightly mixed into gloss gel.

Tip: Thicker gel medium applications can trap moisture inside the paint film, causing it to lose transparency and appear milky. To avoid this, work in successive layers, building volume gradually and allowing time for the paint film to cure and clarify between applications.

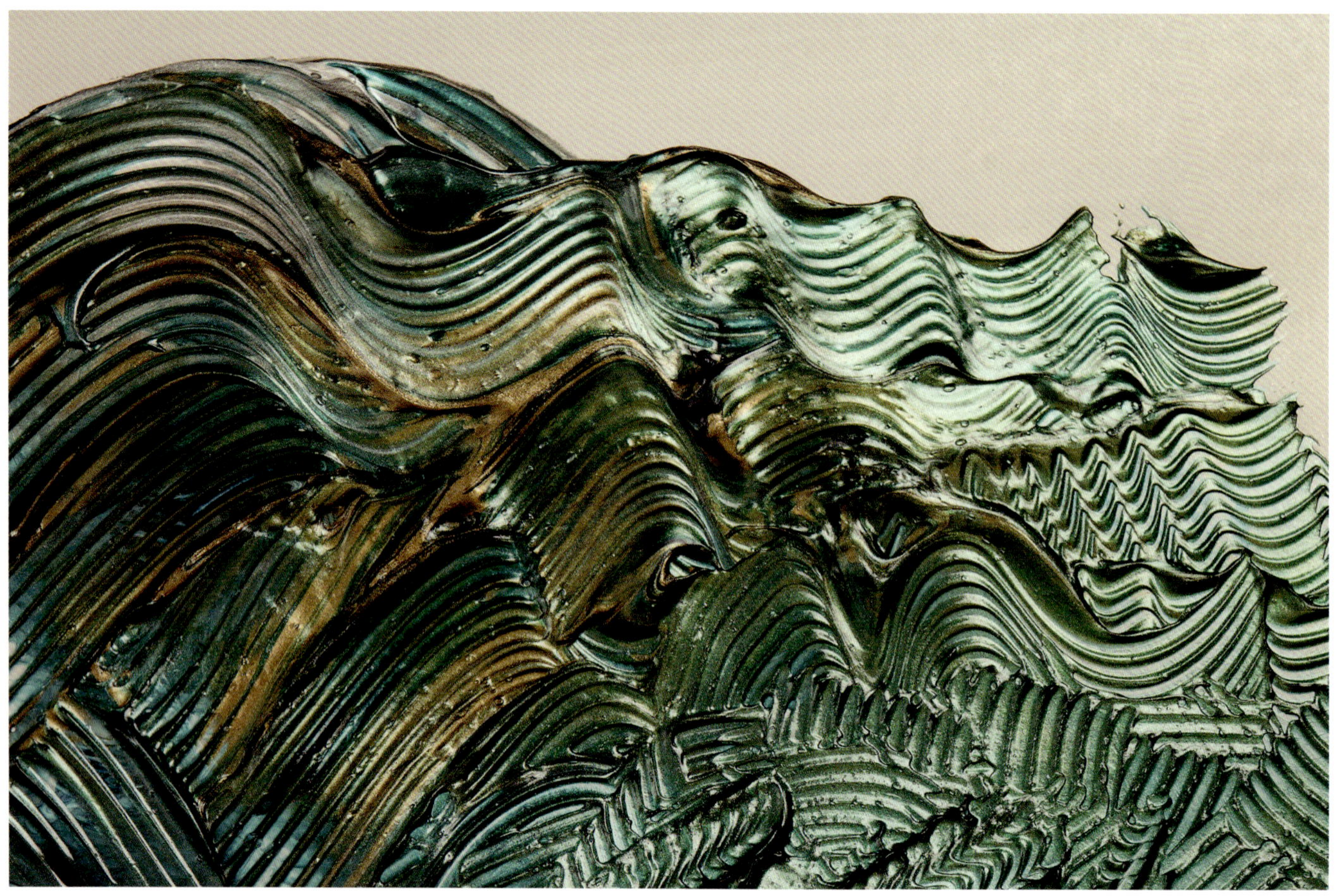

ABOVE Gloss gel plus phthalo turquoise and iridescent gold manipulated into sensuous corrugated ridges with Catalyst painting tools made by Princeton Artist Brush Co.

dictate the appearance of the surface without overly affecting any other aspect.

- *Sealing.* Gel mediums can effectively seal water-sensitive media, dry media, or loose bits of collage elements on the surface of a painting or painted object. Although not entirely waterproof, since all acrylics are porous, the gel will make the surface water resistant.

The ubiquitous gel is the king of functionality, the most actively experimented with and top-selling acrylic medium. Although mediums are vastly diverse, the term *acrylic medium* is often understood as referring to the basic gloss gel medium. It's the undisputed mascot of the acrylic team. This is in part because gel medium's thickness resembles that of high-viscosity acrylic colors, but is primarily due to the sheer scope of application possibilities gel medium offers.

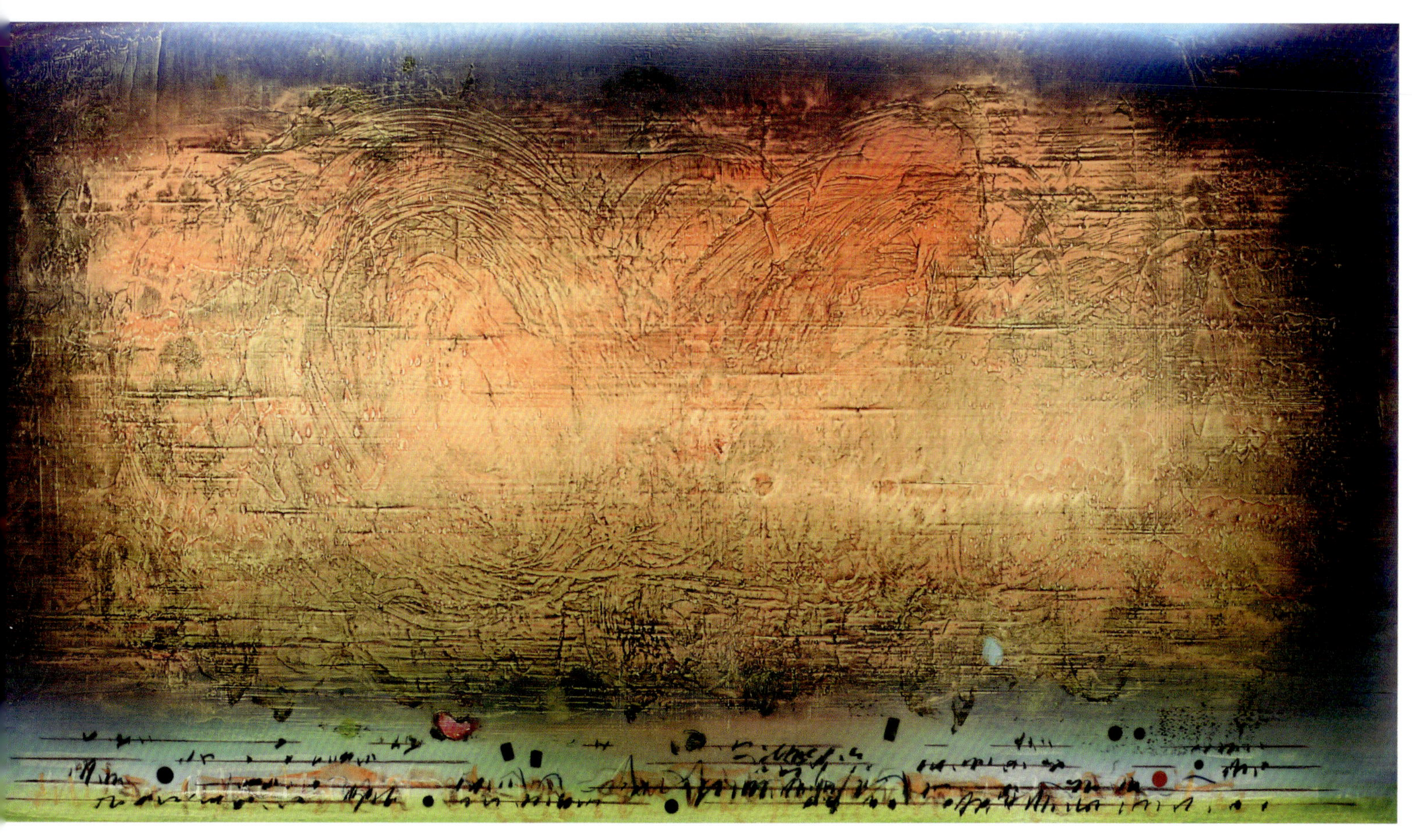

Alice Teichert, *Soundscape,* 2013, acrylic and crayon on canvas, 40 x 72 inches (102 x 183 cm). Courtesy of Oeno Gallery, Bloomfield, Ontario, Canada.

Alice Teichert used gel medium extensively in Soundscape. *Here's what she says about the painting: "I did not choose this title beforehand. The title came to me after I created close to thirty layers of color glazes using pigmented gloss gel medium interspersed with other textural elements that emphasize a fluidity of depth and glow. By mixing gloss gel medium with a tiny amount of pigment, I obtain a mixture that is perfectly spreadable. Using oversized and custom-designed spreading tools, I have created my own layering techniques over the years, applying multiple layers with a sense of precision while being fast enough so as not to leave any streaks behind. Time is always of essence here, because the acrylic medium dries fast. Layer upon layer—each layer is rather immediate but requires drying time before it reveals its desired transparent nature to the point that the canvas support still reveals itself through the glazes. Eventually, the multiple layers of superimposed transparencies together reflect a space of holographic appearance that resonates back to a rhythmic annotation (in graphite) at a baseline of this pictorial environment. Everything has a melody."*

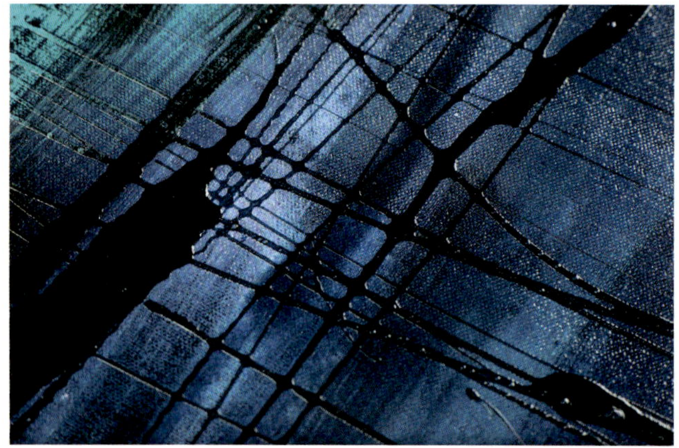

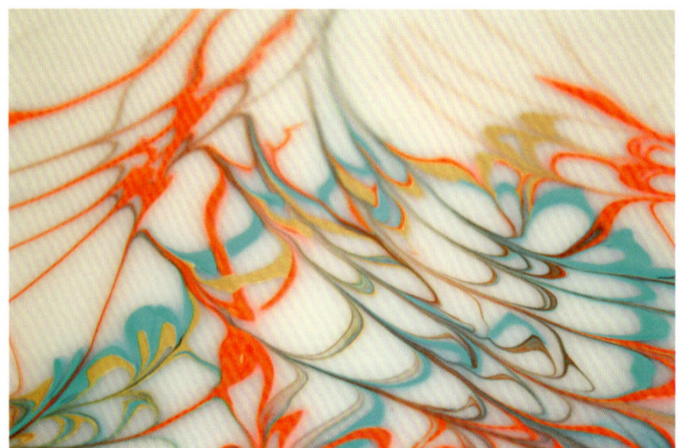

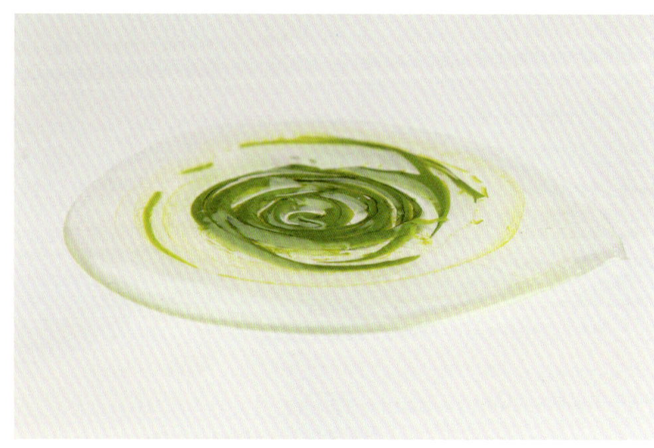

TOP Strings of self-leveling gel on a surface, resting under a layer of glaze.

CENTER When a bamboo skewer is swirled through liquid acrylic colors in a sea of self-leveling gel, the color stays suspended, producing a marbled pattern.

BOTTOM A swirl of liquid acrylic color in a pool of self-leveling gel.

SELF-LEVELING GELS AND POURING MEDIUMS

Self-leveling gels have a high viscosity and a medium to long rheology. Rather than being peaky and buttery, these gels are viscous but smooth. They do not hold texture but instead ooze and flow into thick, level pools. The long rheology of self-leveling gels gives them a tarry consistency. Sticky and gooey as honey, a self-leveling gel can be drizzled in long strings to produce smooth-edged relief lines or be left to merge with itself, forming a surface that looks like syrupy liquid glass.

Some self-leveling mediums are called pouring mediums. Pouring mediums tend to be runny, with a viscosity similar to that of a liquid medium, but they have a longer rheology as well as an inability to hold even the slightest texture.

String gel and tar gel are similar to self-leveling gel. They differ somewhat in viscosity and rheology but behave in essentially the same manner, and both have the same handling properties.

KEY CHARACTERISTICS OF SELF-LEVELING GELS AND POURING MEDIUMS

KEY TERM	SELF-LEVELING GEL	POURING MEDIUM
VISCOSITY	high	medium
RHEOLOGY	long	medium
LUSTER	gloss to matte	gloss
RELATIVE COVERAGE	transparent	transparent
TEXTURE	smooth	smooth

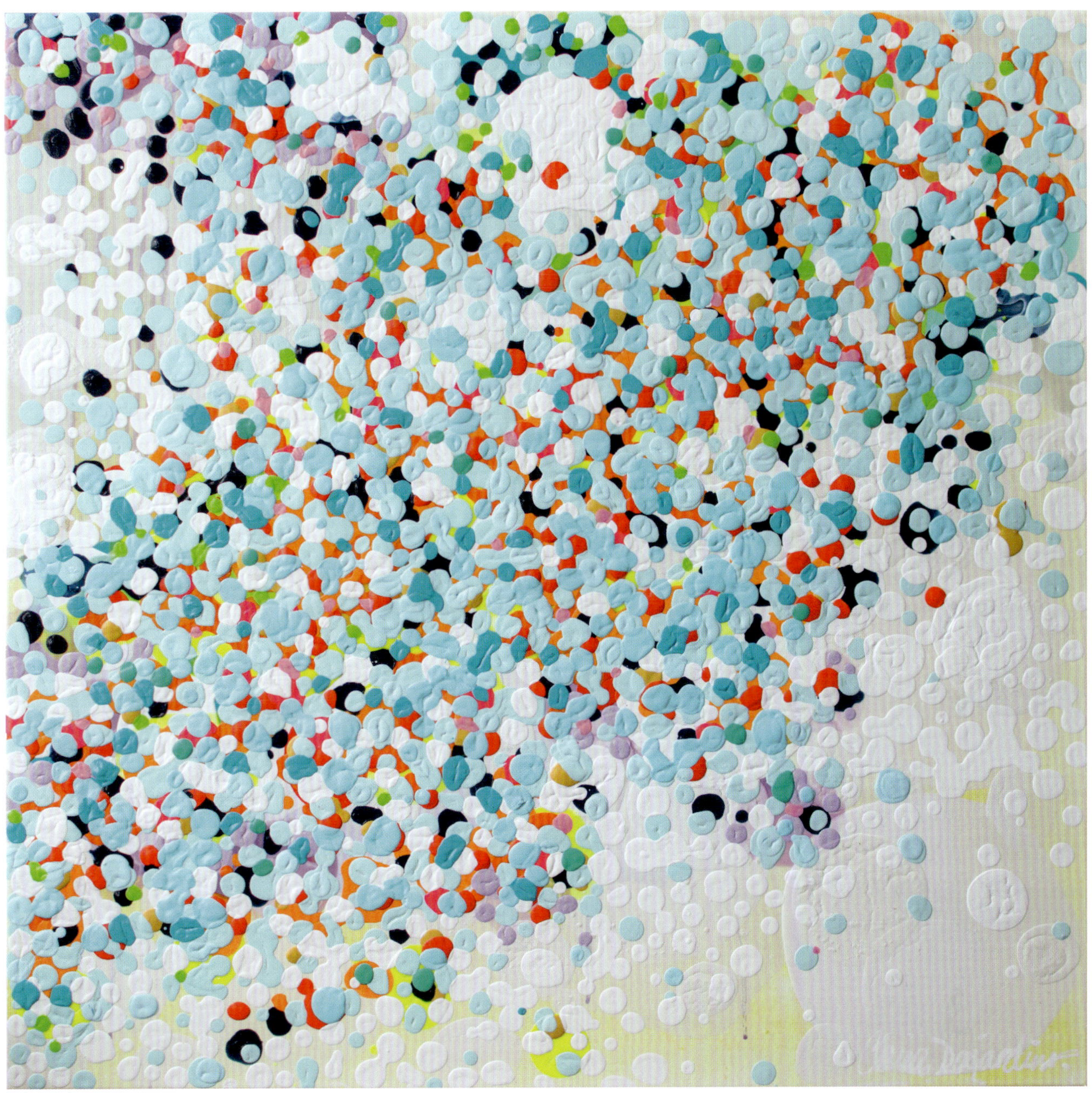

ABOVE Claire Desjardins, *Blue Crush,* 2013, acrylic on canvas, 36 x 36 inches (91 x 91 cm). Private collection, Munich, Germany

Canadian artist Claire Desjardins regularly uses various kinds of polymer mediums. For *Blue Crush*, she mixed heavy- and soft-bodied acrylic paints with a matte polymer medium. Each small drop was poured onto the canvas, one at a time, layer by layer. She created visual interest by using warm and cool colors in groups. The shadows created by the matte medium—as important to the painting as the colors and forms—are softer and more subtle than they would have been if gloss or pouring medium had been used instead.

About her painting, Claire Desjardins says, "All of my work is an attempt to decipher the chatter in my head, to put forward a less awkward side of myself, to navigate through my everyday chaos toward calm. My paintings are personal. The outside world is referenced quite indirectly through the filter of my emotions. Through my work, I find myself expressing sentiments or reactions of which I am consciously unaware."

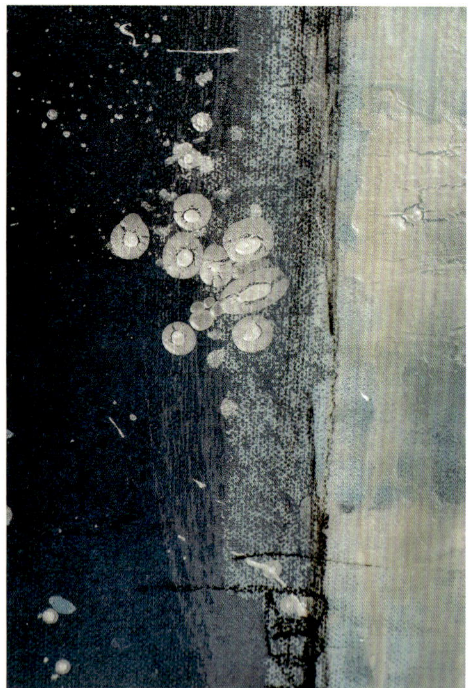

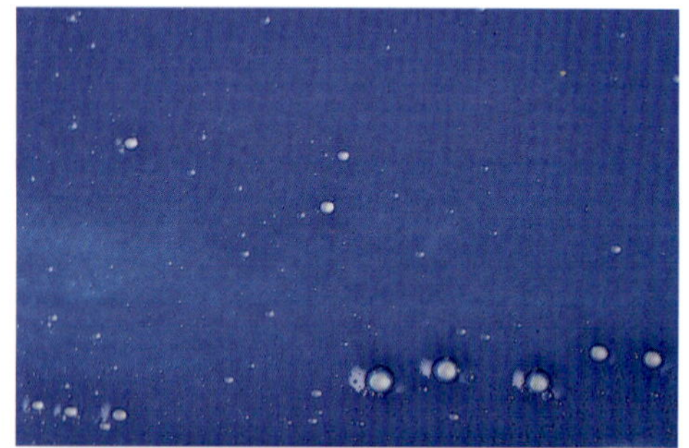

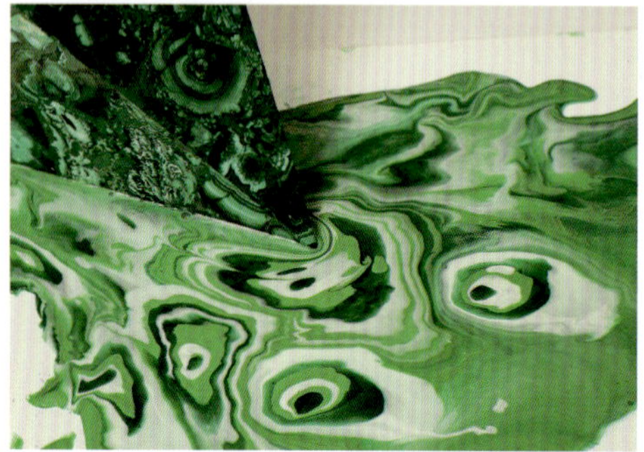

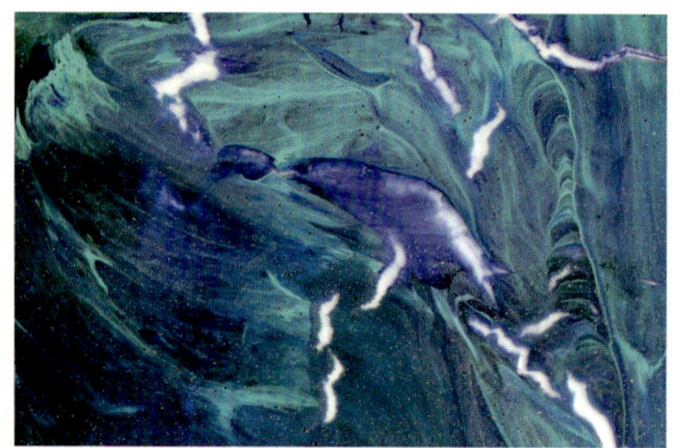

Self-leveling and pouring gels have a tendency to dry more slowly than regular gels. They therefore need to be given time to set and cure before applying additional layers. Twelve to forty-eight hours, depending on the humidity level, should be sufficient. In very humid environments, it is prudent to wait an additional day. Although a thick application of self-leveling gel will dry relatively thin, it will be thick enough to capture a great deal of light and will have the appearance of a lacquered surface.

You can tint self-leveling gels by adding color in any proportion, although you must allow the mixture to rest after vigorous mixing to let air bubbles rise to the surface. If you don't do this, you'll end up with a paint surface full of trapped air bubbles. Be advised that significant quantities of added color (high-viscosity acrylic color in particular) may affect the self-leveling properties of the medium.

Pouring mediums will not generally accept too large an addition of pigment. Refer to the manufacturer's suggestions when tinting this type of medium. The proportion is usually around 1 part color to 20 parts pouring medium.

Use a bubble level to determine whether the surface to which you are applying a self-leveling gel or pouring medium is perfectly level. These materials, which flow readily and are slow-drying, are heavily influenced by gravity. If poured on an uneven surface or one that is not level, the medium can pool or overflow.

Very heavy pools of self-leveling gel can dry unevenly, as trapped air tries to escape and thinner areas dry at faster rates. The surface may craze, forming craters and crevices in an otherwise smooth paintscape.

LEFT, FROM TOP TO BOTTOM Liquid mirror iridescent liquid color dropped into still-wet self-leveling gel produces pale auras, giving the drops an extra glow. • Stirring color into a self-leveling gel takes a lot of force and introduces a lot of air. If you do not allow the medium to rest for a few minutes or even hours after mixing, air bubbles will be trapped in the mixture. • Playing with mimicry: the flow of color and pouring medium mirrors the surface of malachite. • Sometimes a paint film will craze—usually when it dries unevenly. Thinner areas dry more rapidly and pull away from the mass.

TEXTURED GELS

Some thick mediums have their own built-in textures, created by granules of nepheline, pumice, or quartz; synthetic fibers; crushed gemstones; glass balls; or other materials. These mediums can really extend the textural possibilities of your paintings.

Why do such mediums exist? Because their gritty, inherently textural quality changes the way colors travel on, around, and through a surface. They allow for broken color mixing, bas relief, and extra-dimensional optical effects. These mediums require some experimentation, as they can be a little finicky and hard to control. Textured gels that contain larger particulates are easier to handle when some regular gel (of similar luster) is added; this will not have much impact on the textured gel's appearance other than to make it slightly more transparent.

KEY CHARACTERISTICS OF NEPHELINE GEL AND PUMICE GEL*

KEY TERM	NEPHELINE GEL	PUMICE GEL
VISCOSITY	high	high
RHEOLOGY	short	short
LUSTER	semigloss	matte
RELATIVE COVERAGE	translucent	opaque
TEXTURE	granular, fine to coarse	granular, fine to coarse

*Although pumice gel is not covered in this book, it is a popular granular medium and is included in this table for purposes of comparison.

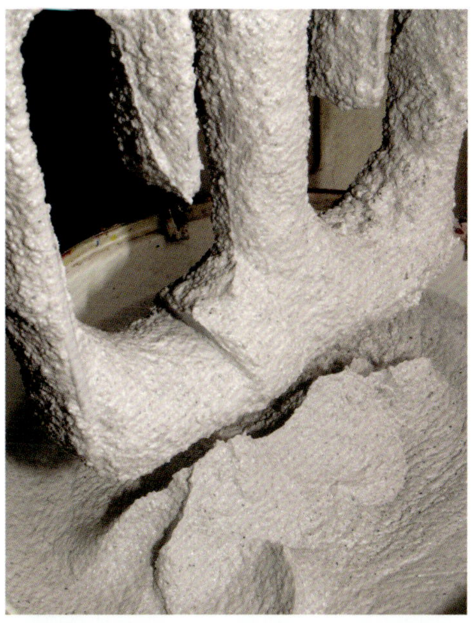

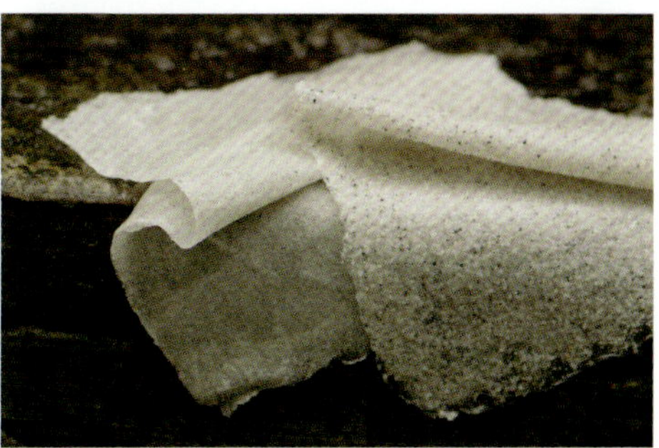

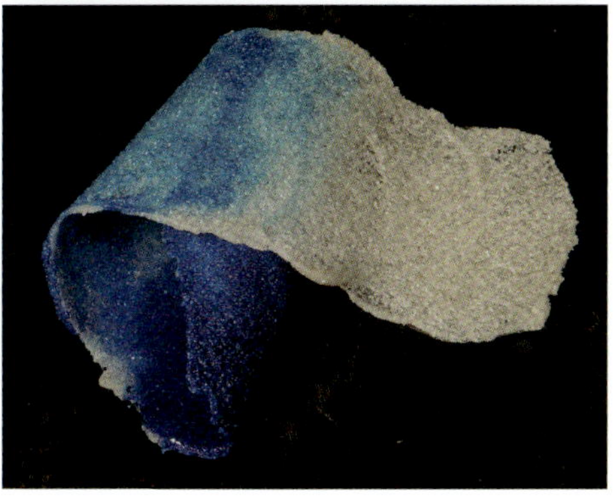

FROM TOP TO BOTTOM Nepheline gel in a mixer. • Sheets of coarse and extra coarse nepheline gel. • A curl of nepheline gel in a clear-to-blue tinted gradation. The translucency of this medium dilutes but does not dampen the brightness of a color.

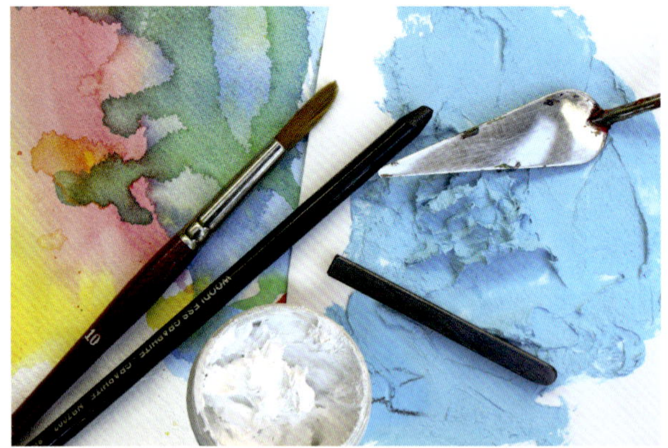

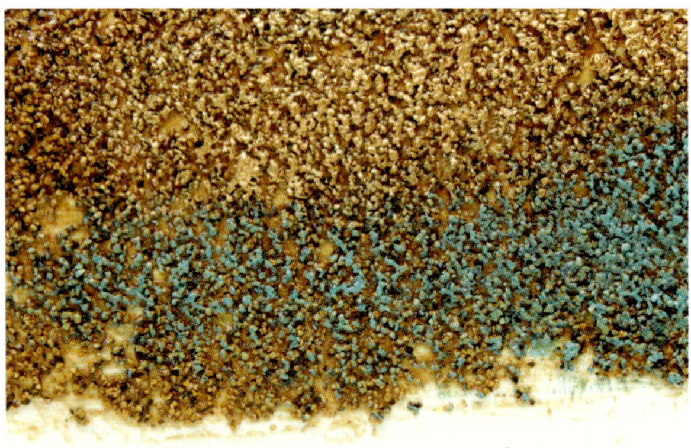

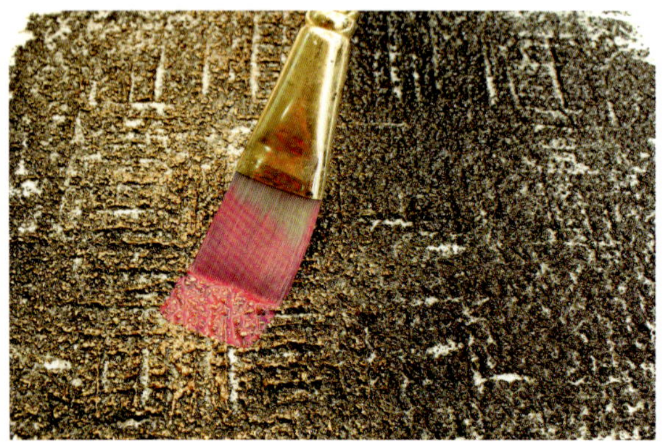

ABOVE, TOP TO BOTTOM Fiber paste—shown here in smooth and rough, untinted and tinted applications—absorbs water-based paints like paper and is toothy enough for drawing on with dry media. • Dry-brushing of cobalt teal over reharvested walnut-shell medium. • Adding a thin wash of quinacridone magenta over and through the textured medium.

Some primary uses for textured mediums are these:

- *Blending color.* Lightly granulated mediums provide a rough surface whose toothiness is ideal for blending color. The transitions are not as gradual as on a smoother surface, but the lightly broken color can create an optical blend.
- *Producing bas relief.* Even unintentionally, these mediums produce bas relief, a raised surface for the light to catch, grabbing the viewer's attention.
- *Mixing and applying color.* Various overpainting techniques yield different results depending on the application, volume of paint, and transparency of color. A thin layer of color, especially one that has been watered down, will sink into the textural crevices, leaving less color on the uppermost surface. Conversely, a thick layer of paint will stay crested on the top of the texture without descending farther. Opaque color layers will obscure the color beneath without changing the texture, creating a uniform color field with an irregular surface. A transparent layer will create a broken color appearance; by allowing the color beneath to manifest itself, it adds visual texture to the tactile texture.

Mediums with textural additives are quite different from one another, so you must take the time to get to know them to determine how you want to use and accentuate them.

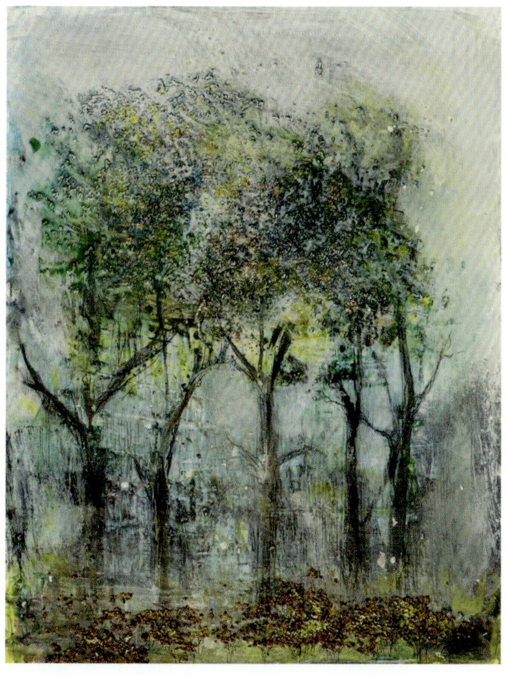

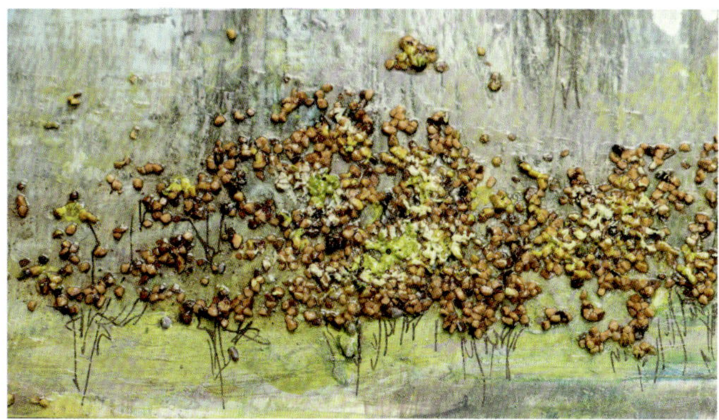

LEFT A quick sketch, with details added in waterproof fine liner. Granular mediums in the trees suggest foliage, and recycled crushed walnut-shell gel suggests shrubbery. Graphite rubbed into the light textures adds depth.

ABOVE In this detail from the sketch at left, walnut-shell medium masquerades as low-lying shrubbery.

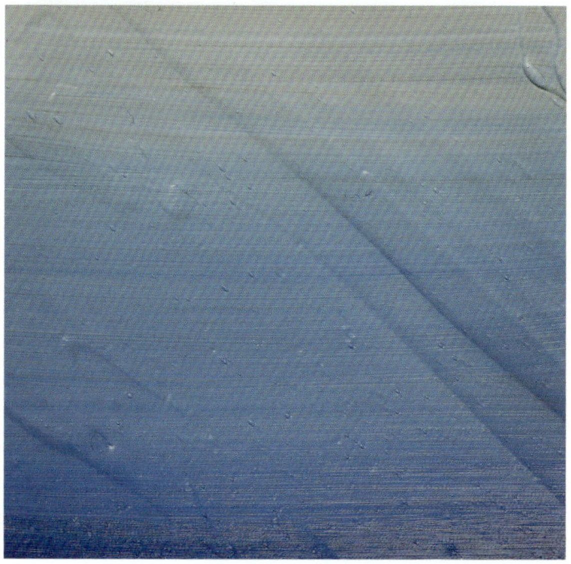

ABOVE, LEFT Blending color on a smooth paint surface results in a streaky blend.

ABOVE, RIGHT Dry media ground provides just the right, uniform level of tooth to allow for exceptionally smooth color blends.

LEFT A granular medium shows a blend that is part broken color and part smooth gradation.

There are many aggregate-filled gels on the market; each acrylic manufacturer has its own versions and selections. Because gel medium provides such a friendly environment for various inert materials, this part of the acrylic mediums catalogue is continually growing in diversity. For the table below, I've chosen a selection of Tri-Art textured gels, which is representative of the general scope of such products. Some of these bear resemblances—in appearance and handling properties—to other brands' textured gels.

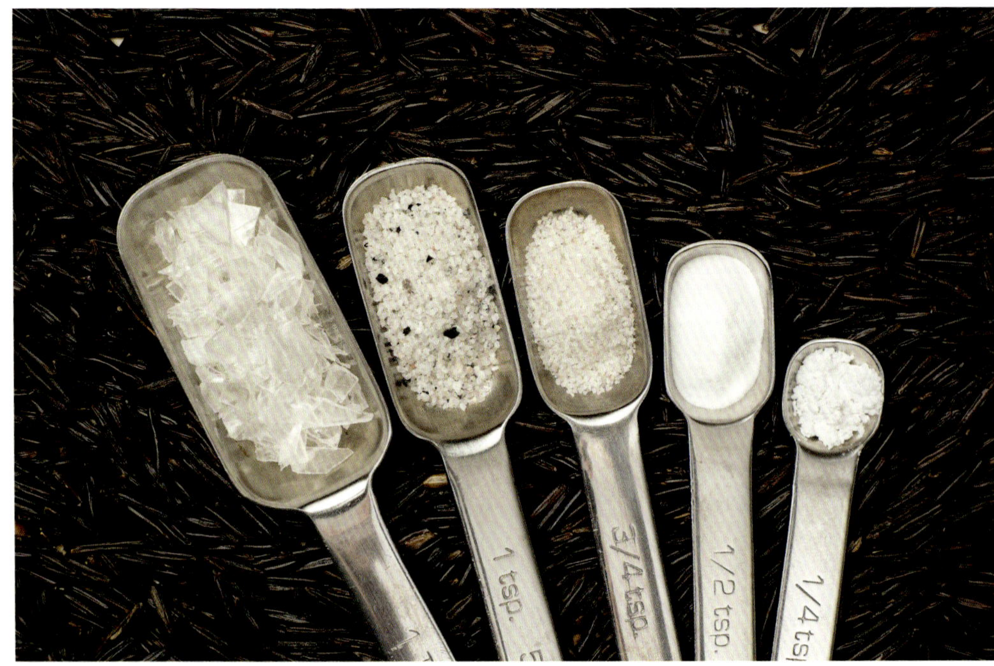

ABOVE Here are just some of the different kinds of granular particulates that give specialty gels their body and texture.

OPPOSITE A few bits of textured gels here and there are enough to suggest the elements of this landscape. Nepheline fine gel on the sky and foreground adds a little grit and color disruption. Recycled glass gel is used for the trees, and copper cinder gel at the picture's lower edge suggests shrubbery (see details, below). The pen lines, drawn in water-based ink, sweep along with the movement of the brush.

KEY CHARACTERISTICS OF OTHER TEXTURED GELS

KEY TERM	RECYCLED GLASS	PETE PLASTIC	WALNUT SHELL	COPPER CINDER	FIBER GEL
VISCOSITY	high	high	medium	high	high
RHEOLOGY	short	short	short	short	short
LUSTER	gloss	gloss	matte	gloss	semigloss
RELATIVE COVERAGE	translucent	transparent	opaque	semi-opaque	semi-opaque
TEXTURE	granular, coarse	sharp, irregular	granular, irregular, coarse	granular, irregular, fine to coarse	fibrous

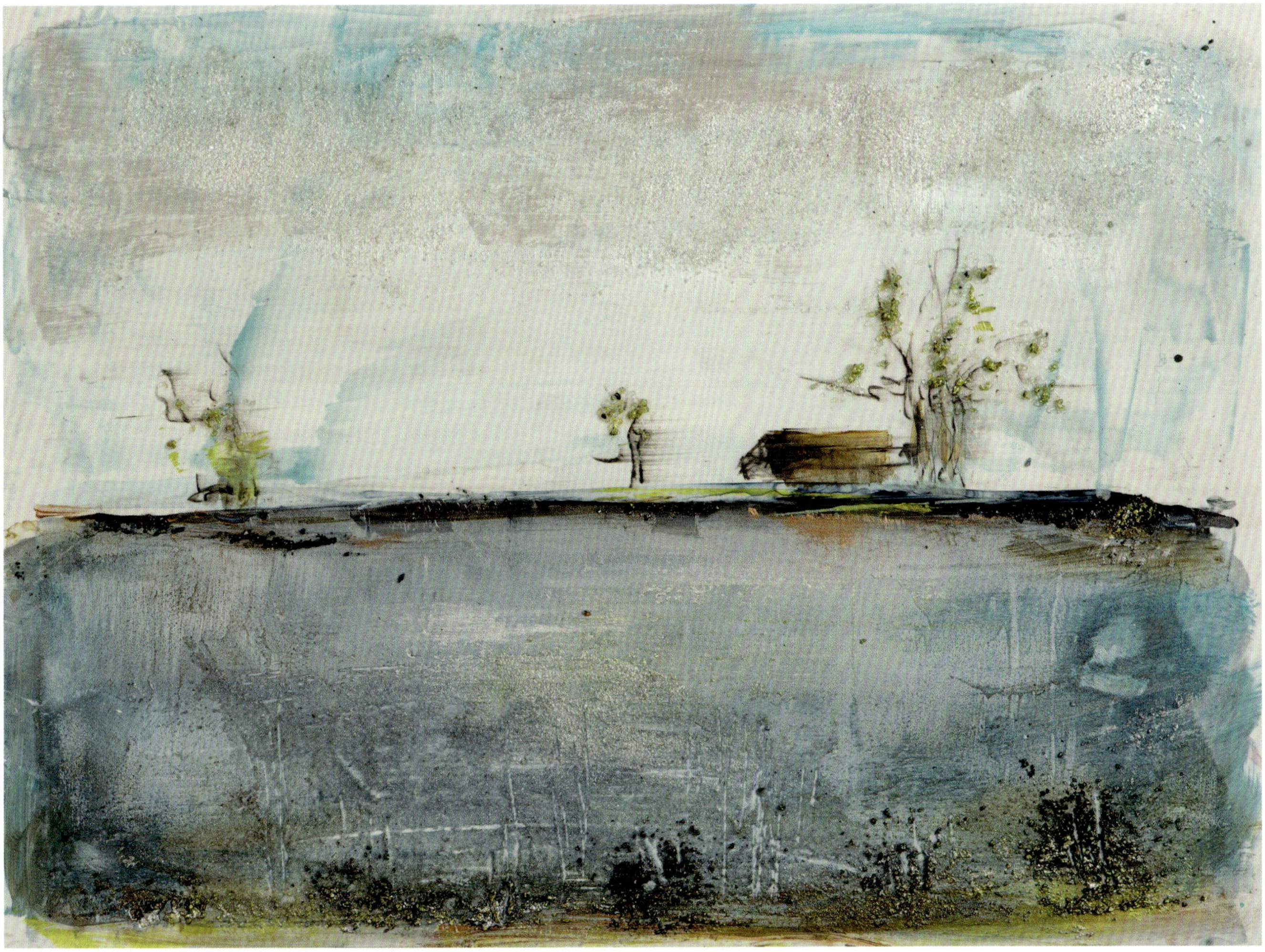

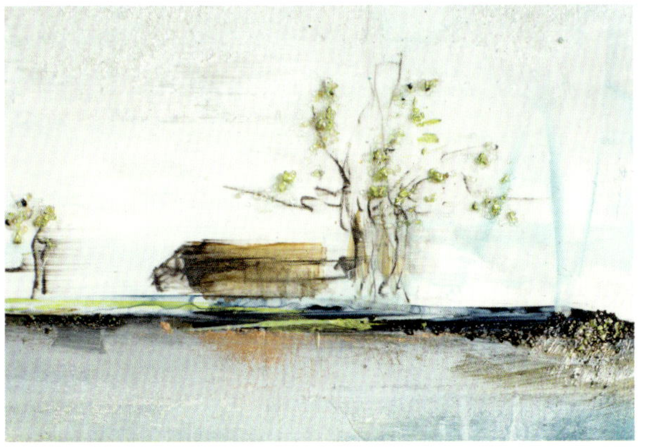

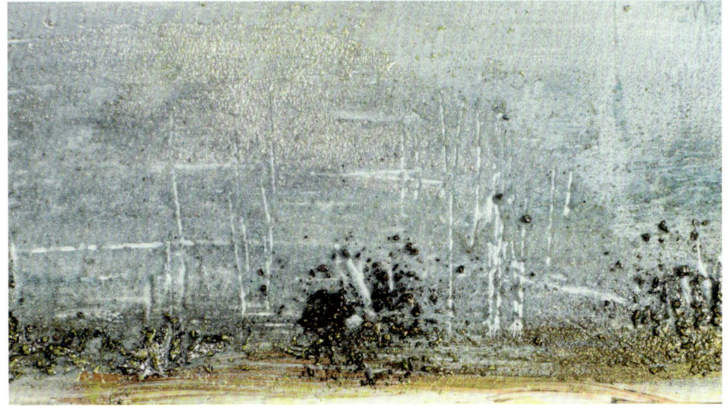

PROCESS

EXPLORING UNUSUAL TEXTURE MEDIUMS AND MIXED MEDIA

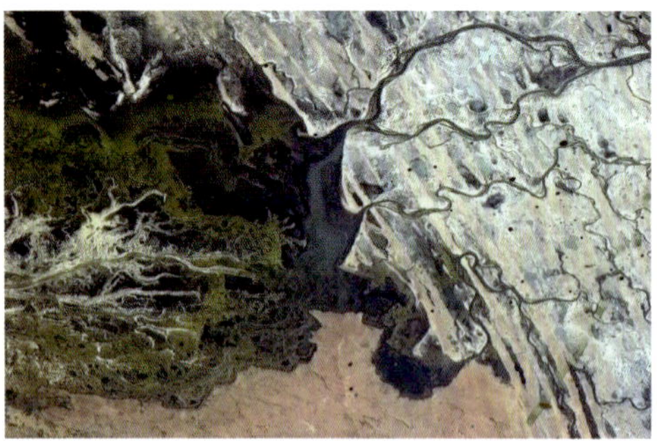

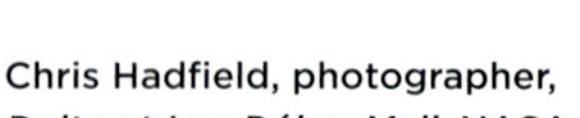

Chris Hadfield, photographer, *Delta at Lac Débo, Mali.* NASA.

Wood based her painting *Delta* on a photograph that astronaut Chris Hadfield took from space of an actual river delta in the African nation of Mali.

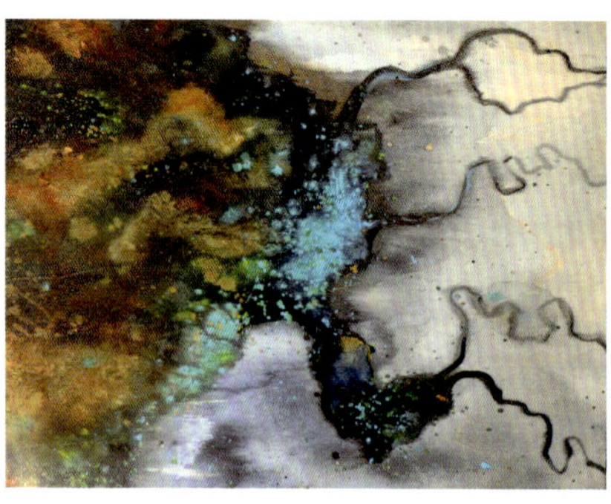

1
In the first stage, Wood laid down washes, building glazes. Colors used for the watered-down washes and splatters on the left side of the painting were golden orange, iridescent gold pale, cobalt teal, and green gold. Colors for the light area on the right: titanium white, phthalo blue, and Payne's grey.

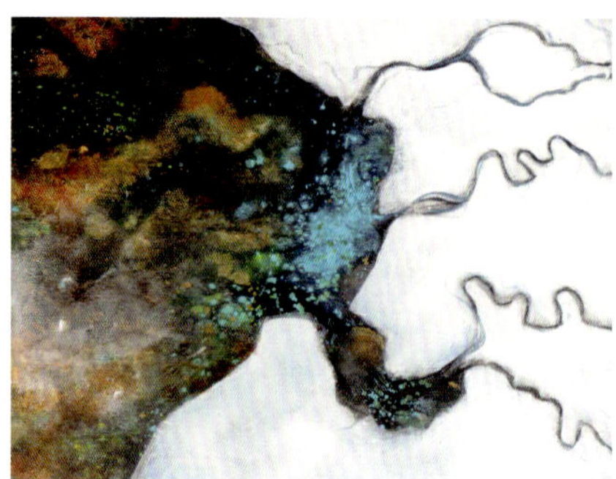

2
Then, she used zinc white with polymer glaze to establish the shape of land, building it up with a lot of very transparent layers. Dry media ground was applied all over the surface.

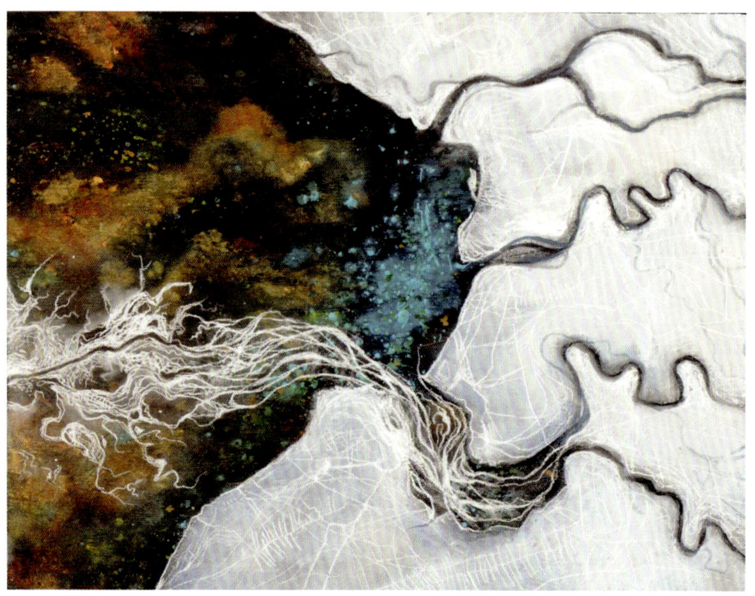

3
Wood next drew lines on the surface with white charcoal.

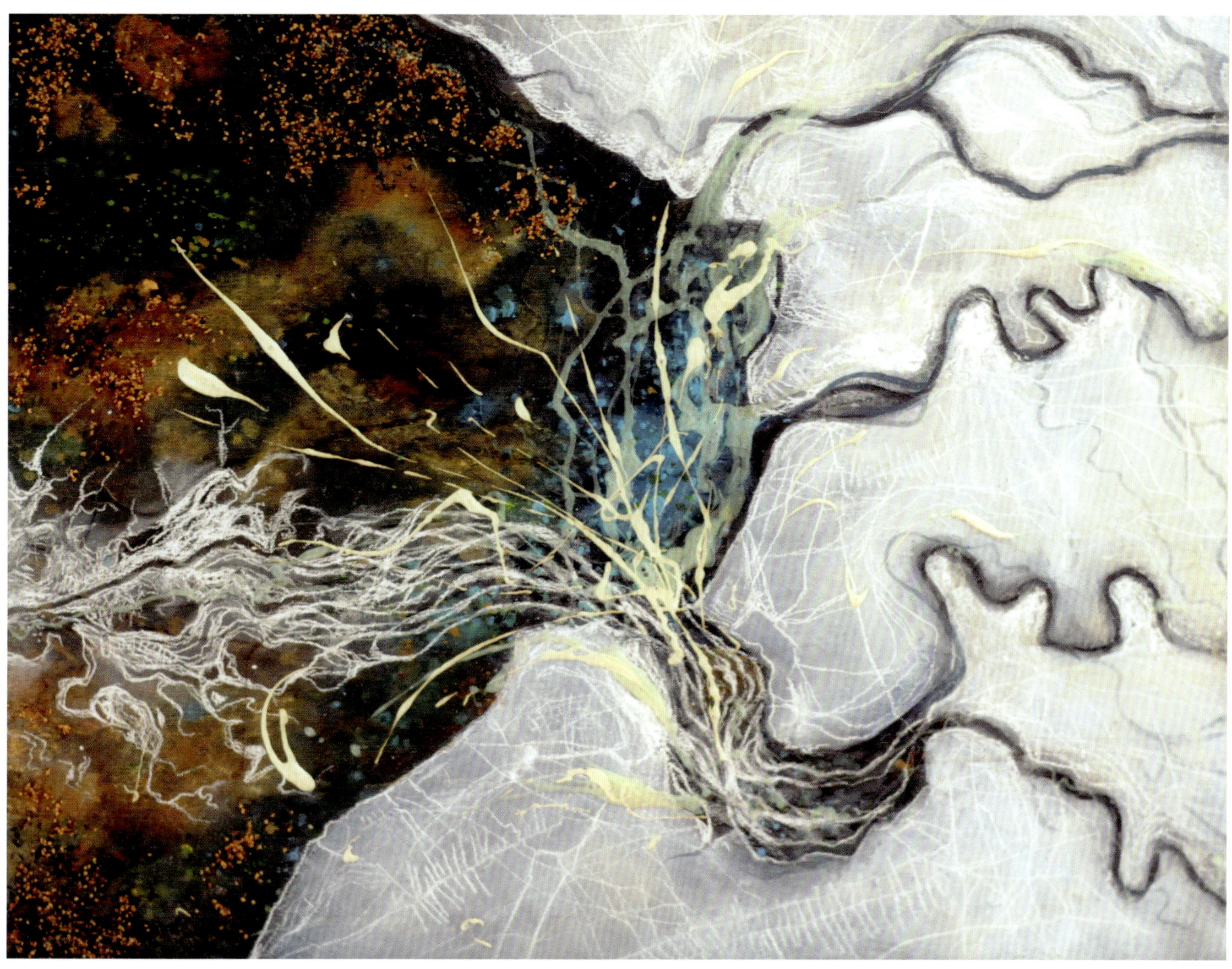

ABOVE **Sharlena Wood, *Delta*, 2017, acrylic and pastel on uncradled birch panel, 11 x 14 inches (28 x 35.5 cm). Collection of the artist.**

For *Delta*, Sharlena Wood used self-leveling gel with interference gold and self-leveling gel with interference turquoise, expressively flinging them around to connect the design. Walnut-shell and copper cinder specialty mediums were applied straight out of the jar.

About her use of mediums, artist Sharlena Wood says, "There are a handful of favorite mediums and colors that I've come to love for their unique personality traits. Dry media ground has been my best friend for a long time, as I love to draw with charcoal and soft pastels on gorgeous acrylic grounds that have been built up with a lot of thin layers of glazing. In my latest series, I'm exploring expressive abstraction of aerial earth views, drawing me toward experimental mixed media and the cosmic evolution of acrylic mediums. Recycled mediums with earth elements satisfy the texture I am craving to represent, and then things get especially cosmic when gold leaf and interference colors come into play! Every work demands and inspires a new combination, and my excitement and curiosity are at peak level when playing with and reacting to complex collisions of media."

MEDIUMS FOR SPECIFIC USES

Multiple uses for multiple mediums—that's been the story so far, the common thread that binds the mediums family. But there are a number of mediums conceived of and manufactured for specific purposes. Every maker offers its own collection of special-use mediums. This section covers some of them.

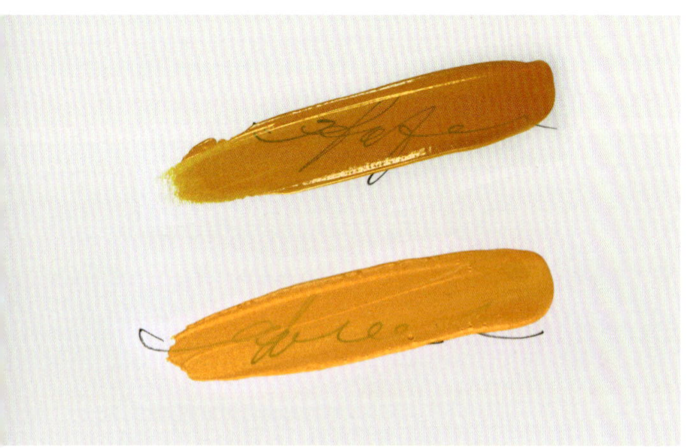

ABOVE Opacifying mediums will blanch the color somewhat, reducing chroma only slightly, while making it significantly more opaque. At top is the pure color (isoindolinone yellow); at bottom, the color with opacifying medium added.

OPPOSITE Marie Lannoo, *Beyond Blue*, 2017, acrylic on panel, 36 inches (91.4 cm) in diameter. Collection of the artist.

At the top is *Beyond Blue* photographed under normal light; at the bottom, the painting is shown photographed in darkness. Lannoo created a custom paste using phosphorescent powder, self-leveling gel, and polymer medium. This paste was filled into the curved cut in the panel (which is, in fact, the wavelength of the color blue). Photographs by Shannon Brunner Photography.

OPACIFYING MEDIUM

This liquid medium contains hollow acrylic microspheres that trap and effectively scatter light. The traditional method of opacifying colors has been to add white, but white significantly decreases the intensity of a color, often much more than desired. Opacifying medium radically opacifies transparent and semi-opaque colors, increasing the color's relative coverage with only minimal color alteration. It will also minimize the reflectivity of interference and iridescent colors.

There is no set amount of opacifying medium that can be added to the color, as each pigment exhibits its own degree of opacity. Opacifying mediums show their effect only when totally dry. This medium only affects the color(s) it is mixed with directly and has no effect on layers added over the dried film. You should test varying percentages of the medium with your chosen transparent color before using it.

UV REACTIVE MEDIUM

There is something compelling about things that glow in the dark or fluoresce under black light (ultraviolet, or UV, light). An acrylic medium that does this opens doors

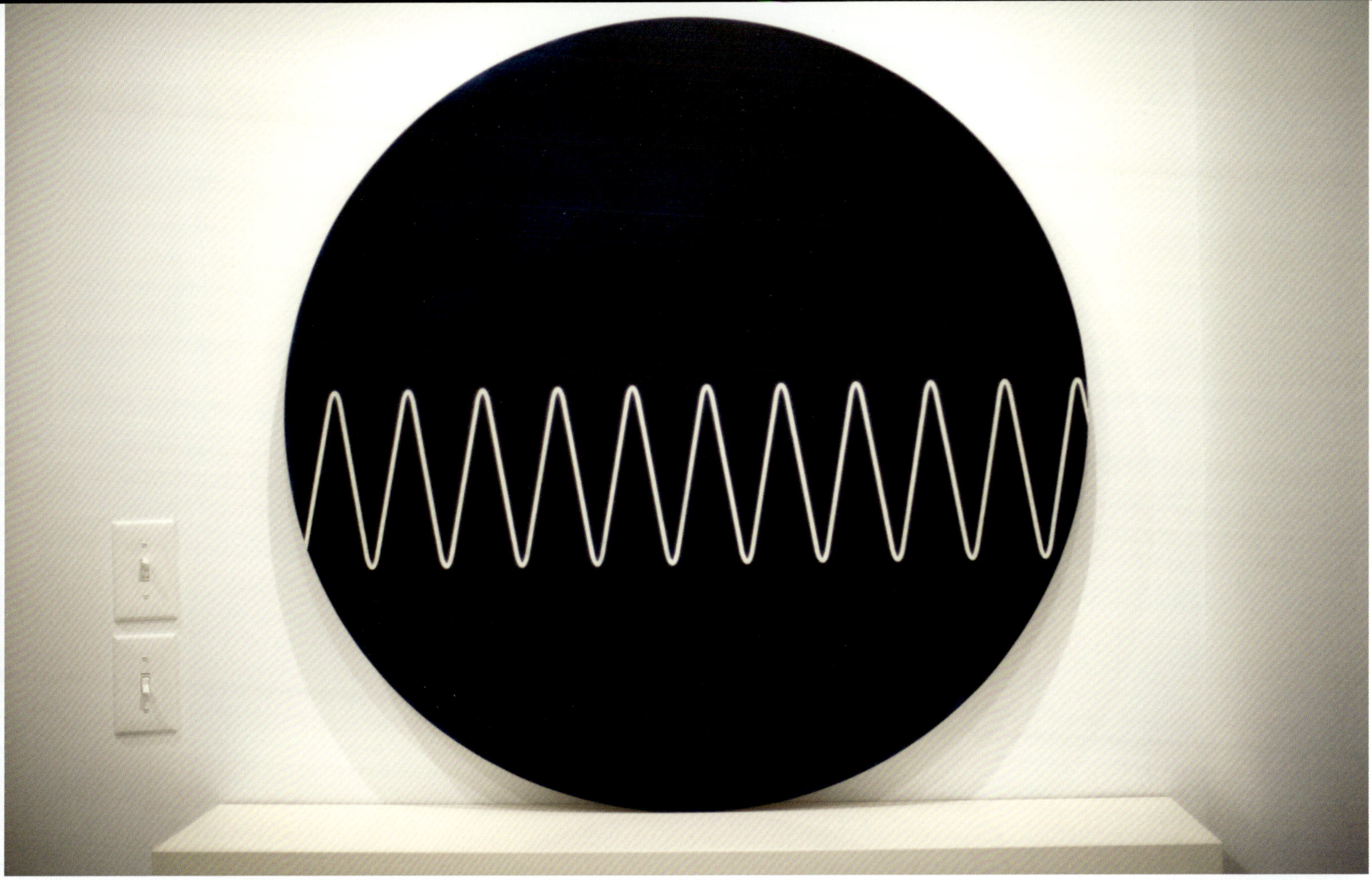

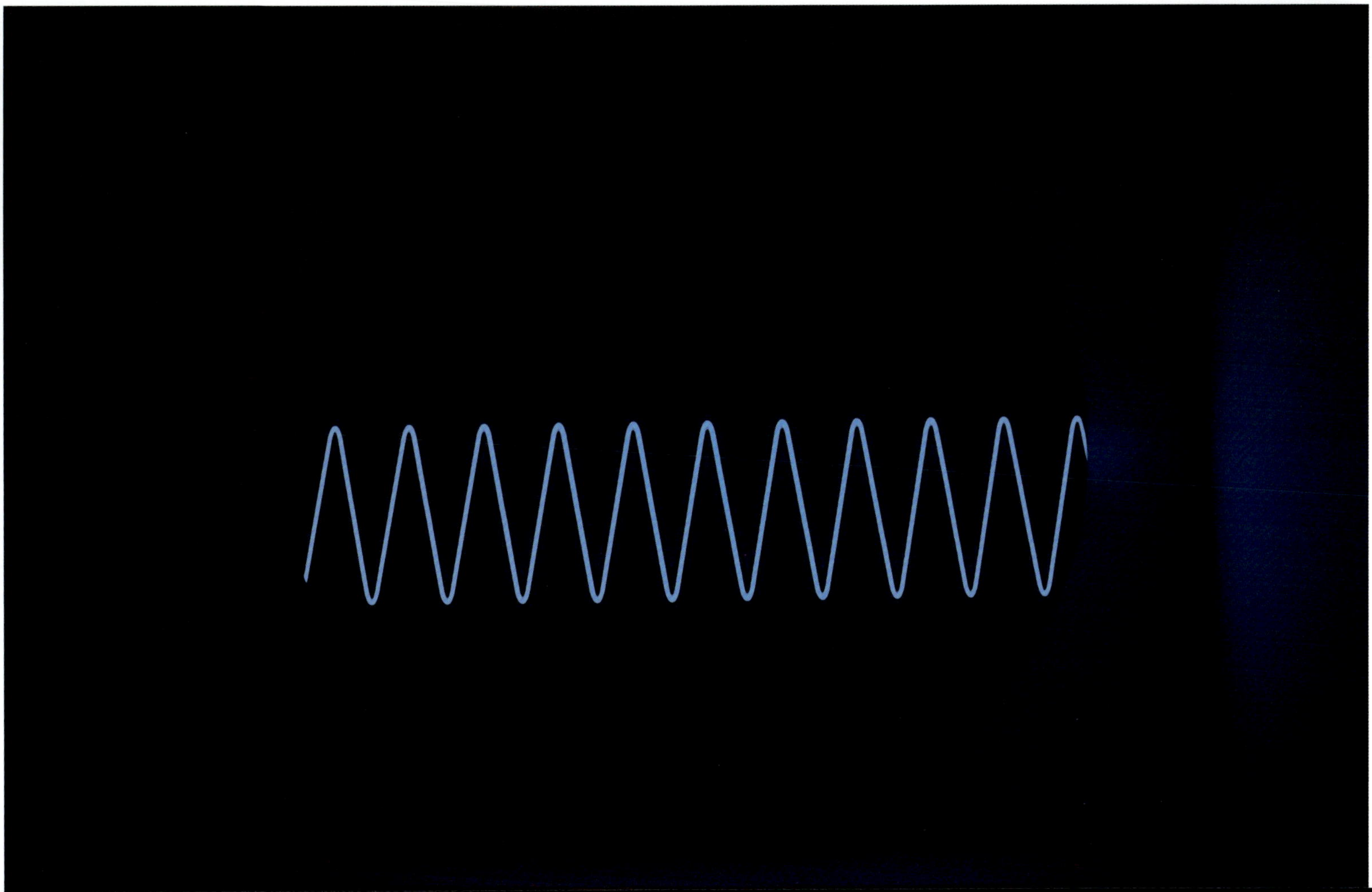

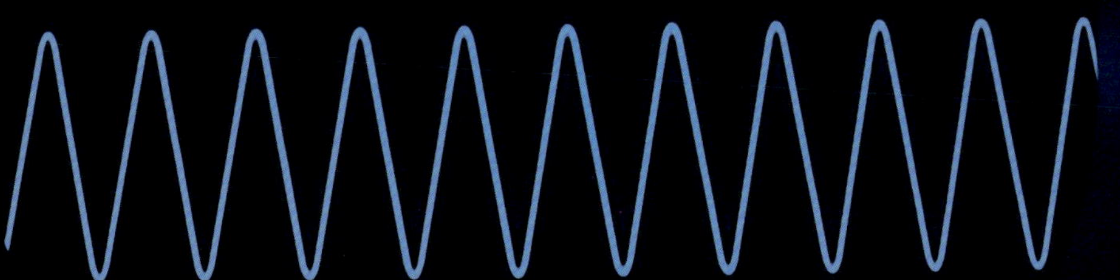

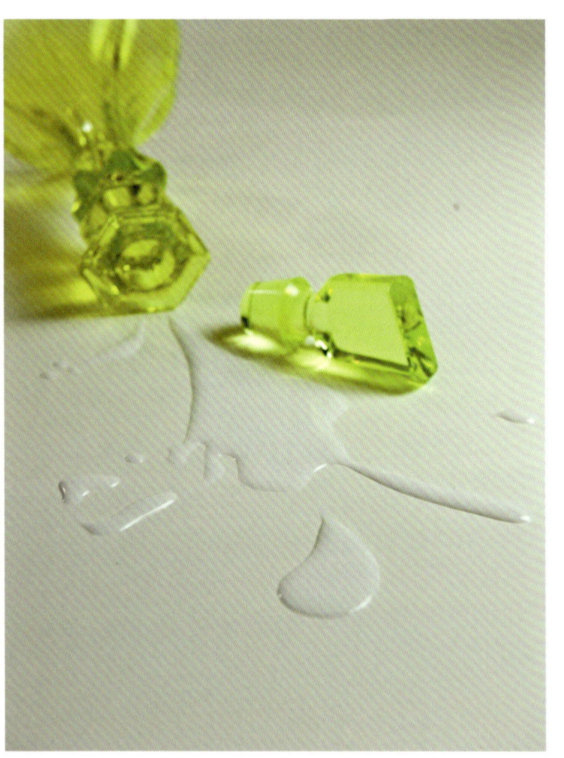
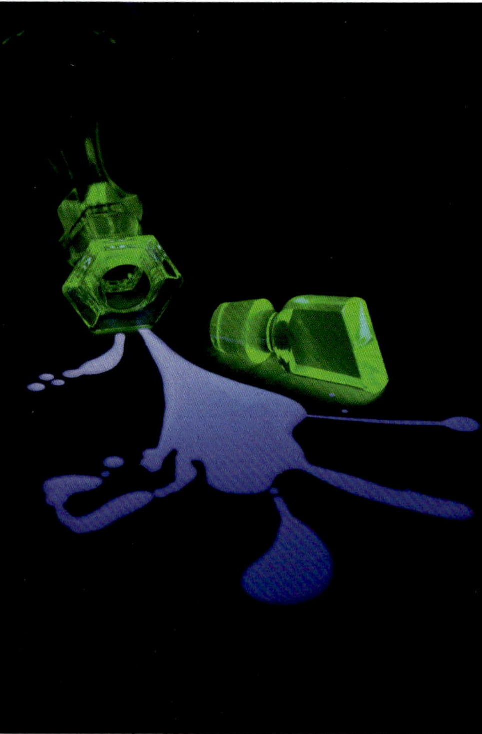

of optical possibility for black-light art installations. UV reactive medium remains virtually invisible in normal light but glows bright blue under black light. When wet, the medium has the same appearance and consistency as other liquid mediums, and it dries clear and transparent. It may be used on top of layered colors to create accents or mixed directly with transparent colors, which will then glow under black light. Adding this medium to a color, even a transparent one, will reduce its ability to fluoresce, however.

ABOVE, LEFT AND CENTER In daylight (left), the UV reactive medium is almost invisible. But under black light (center), the medium glows with an eerie blue color. The uranium-glass vessel is also UV reactive.

ABOVE, RIGHT UV reactive polymer (untinted) painted on paper and photographed under black light.

UV STABILIZING MEDIUM

As resilient as acrylic paints are, there are some that contain pigments that are not completely lightfast. What this means is that a color that starts off brightly hued will begin to fade, changing in both saturation and tonality as it is exposed to the ultraviolet (UV) component in ordinary daylight. Acrylic polymer emulsion itself offers a small amount of protection against this, and this protection can be significantly amplified by the addition of a UV light stabilizer (UVLS), which improves lightfastness and weatherability. This is the medium of choice for collage and works with paper where added protection is desired. UV stabilizing medium can be mixed with acrylic colors in any proportion or applied over color. While it does not render a surface completely invulnerable to UV-caused deterioration, it can retard the process—giving your work some added longevity and you some peace of mind.

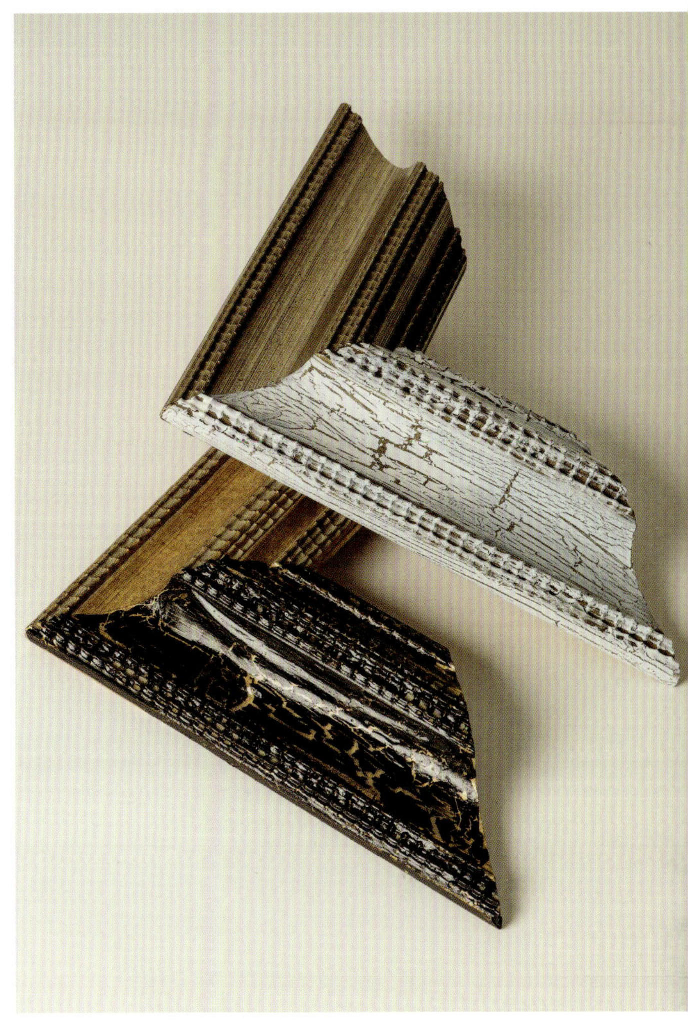

ABOVE Crackling medium makes gilded picture frames look old and weathered.

CRACKLING MEDIUM

There are times when a damaged-looking surface may actually be the effect you desire. Crackling mediums alter the paint film of acrylics as they dry, spreading and pulling it apart to produce fissures, either fine or dramatic, that reveal the underlying base color. Different brands use different materials to produce the crackling effect, so read the label instructions to ensure you achieve the best results

Artists, with their insatiable curiosity, have inevitably taken these special-use acrylic mediums beyond their intended purposes, and they will continue to do so. This keeps manufacturers and other artists on their collective creative toes, and it expands our database of methods and materials. Unusual mediums are not for everyone, but for some artists they open doors to new ideas.

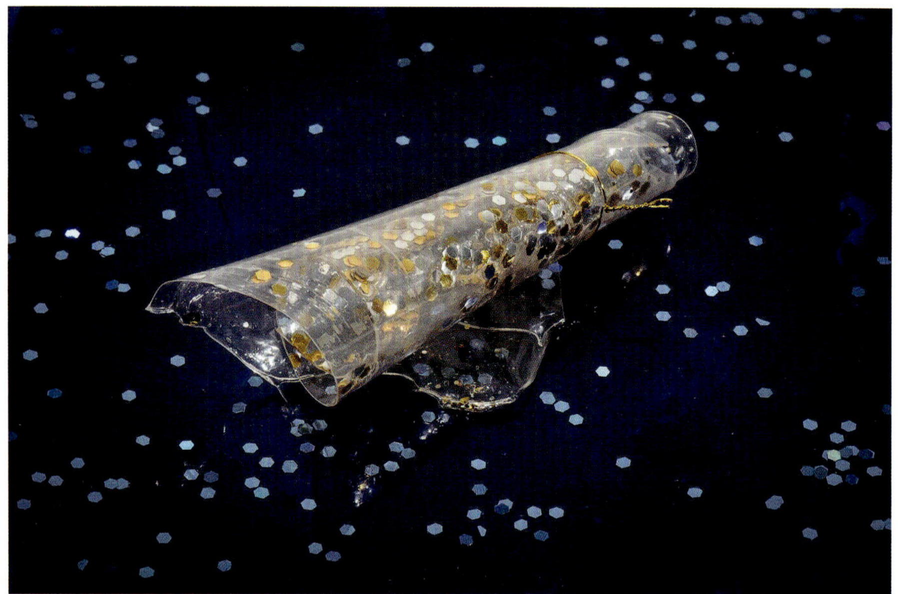

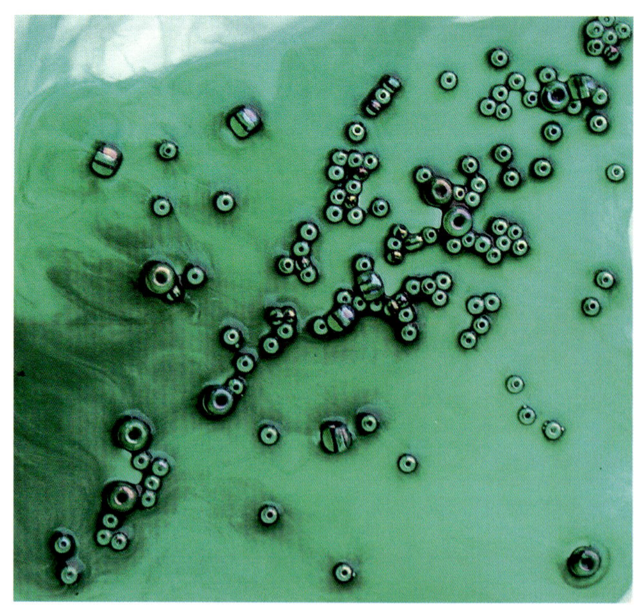

ABOVE, LEFT A rolled acrylic skin with embedded glitter elements.

ABOVE, RIGHT Glass beads resting in a thin veil of liquid medium.

Tip: A word of caution: Some metal beads and filings may at first look stable in the gel but will oxidize and discolor when they come in contact with the water in the gel. Always test a custom gel before applying it to a painting, just to be sure of good adhesion and stability.

PREPARING CUSTOM MEDIUMS

Acrylic mediums accept particles and fillers so easily that they invite you to make up your own concoctions. Just about any chemically inert material can be added to an acrylic medium to add either texture or color, provided that it is compatible with the acrylic environment. Be wary of using metals that will oxidize, organic materials that could mold, or anything that is oily, greasy, waxy, or salty. But these restrictions still leave a sizable menu of other materials from which to choose, so experiment with whatever your imagination can conjure up, and create quirky, textured gels that are distinctly your own.

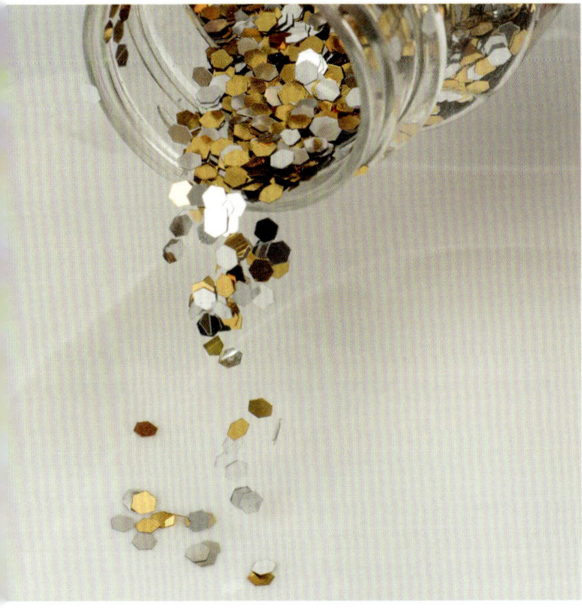

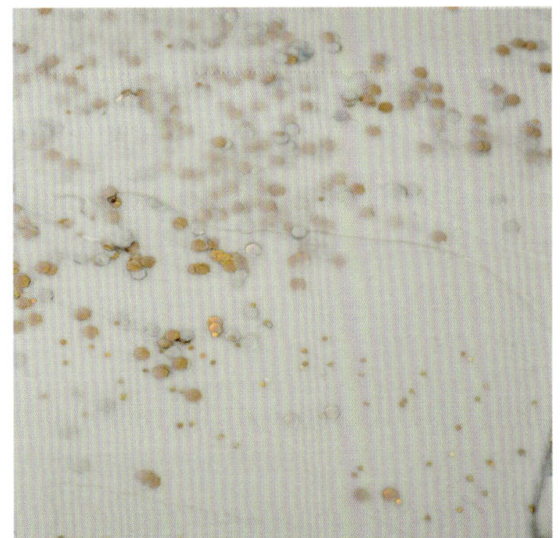

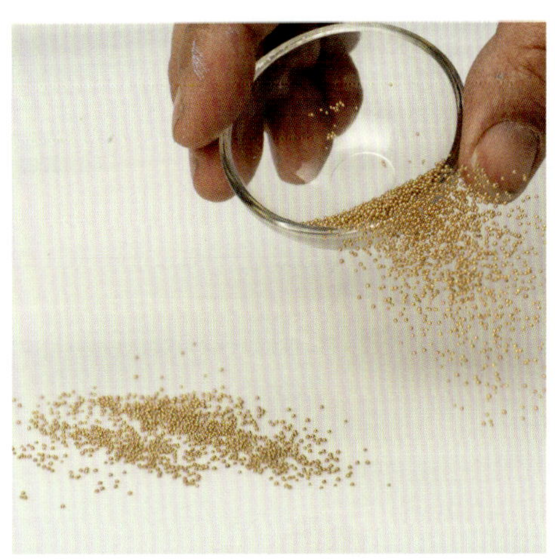

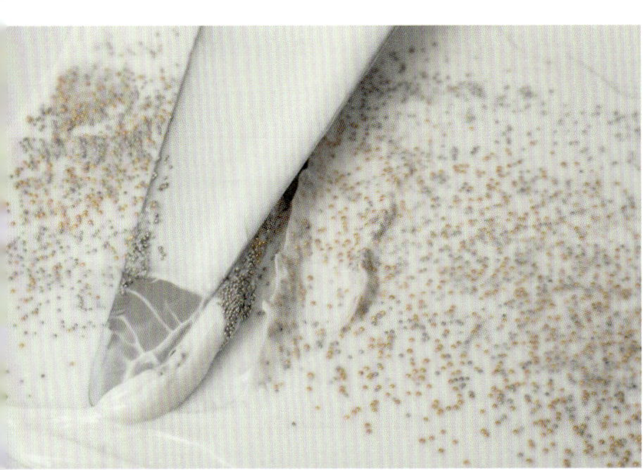

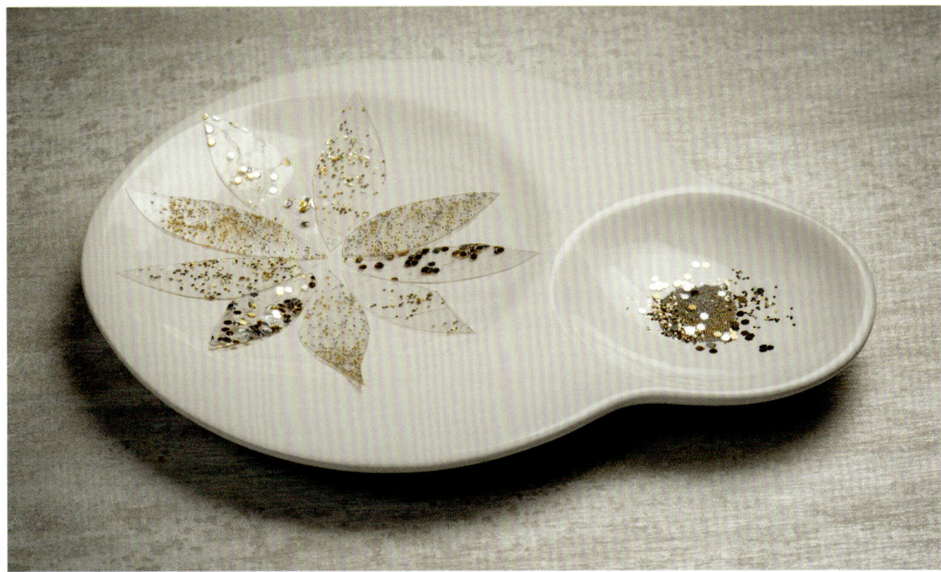

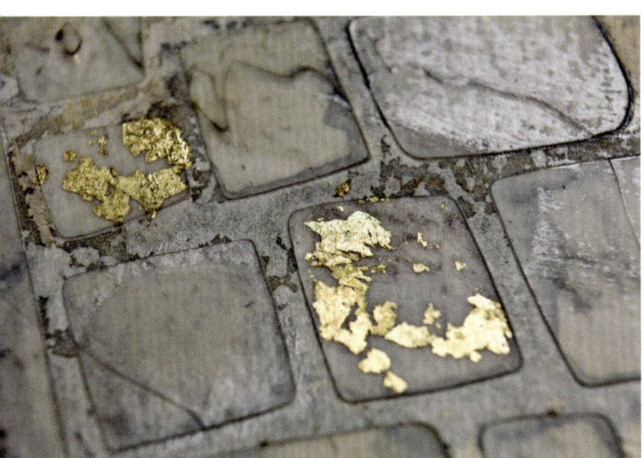

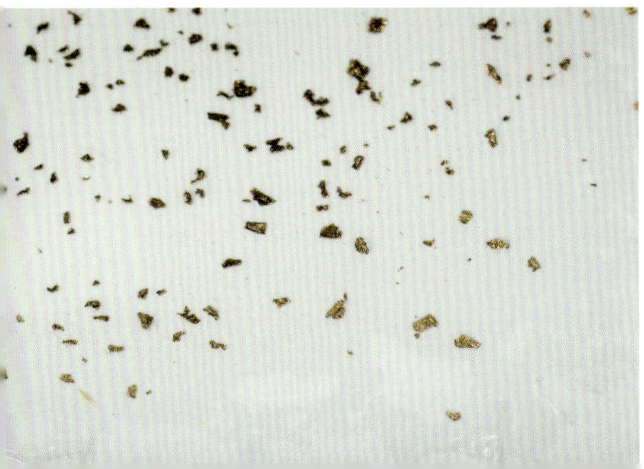

TOP ROW, LEFT TO RIGHT Sprinkling glitter into wet gel. • Embedding glitter in a thick layer of gloss gel medium. • Scattering metal beads into a waiting pool of wet gel.

CENTER ROW, LEFT AND RIGHT Folding beads into a gloss gel layer on a nonstick palette to produce a custom gel. • Petals fashioned from gloss gel medium impregnated with beads and glitter.

LEFT Here, imitation gold leaf has been lightly fused onto a stenciled acrylic surface with a touch of matte medium. A thinly applied layer of imitation gold leaf adhered with liquid medium will not have time to tarnish.

BOTTOM Imitation gold leaf, made with brass, looks great when initially mixed into wet gel, but in a thick layer, the water in the medium will tarnish the brass, turning it green as the medium dries.

Tip: Working on a surface that has been partially saturated with acrylic retarder is perfect for plein air painters. When the surface is prepared ahead of time, there's no need to carry a bottle of retarder to the site or to spend precious painting time mixing paint and retarder. Also, there's no danger that you'll use too much retarder, so the evaporation process can occur unimpeded by an excess of glycol in the paint. For a good result, brush a single layer of undiluted acrylic retarder onto your support and allow it to dry completely. Retarder is also available as a gel, which is particularly effective for plein air painting.

ADDITIVES

What's the difference between an acrylic medium and an additive? Acrylic mediums are made up of acrylic resin in varying proportions with other materials or thickeners, while additives do not contain any acrylic binder at all. They are meant to be used in conjunction with acrylic materials (colors or mediums) to facilitate a process or effect some sort of change. Also, the majority of additives are volatiles, meaning that they will completely evaporate from the paint film during curing. When they are done doing their job, they do not become part of the paint film—they simply evaporate away.

RETARDING AGENTS

For those accustomed to the long working time of oil paints or the resolubility of watercolors, the speed at which acrylic paints dry can at first be daunting to deal with. The acrylic binder coalesces and dries through evaporation of the volatiles within the wet paint film. As this takes place, the paint becomes less malleable, eventually turning into a flexible, water-resistant film. Mastery over this aspect of acrylic painting can be accomplished through certain painting methods but also by using an additive called *acrylic retarder* to control it. (Retarder changes the chemistry of the acrylic binder and slows, or retards, the drying process for a limited amount of time; it is sometimes called "slow drying aid" or another, similar name.)

An innocuous substance composed primarily of propylene glycol, retarder can be quite effective, but it must be used wisely. If you add too little, the paint will not be affected; too much, and it will not cure properly. In attempting to make acrylic paints behave more like oils, painters have a tendency to be a tad overzealous in their use of retarder. This urge must be suppressed! Overuse of retarding agents can render the paint film less water resistant and reduce adhesion between layers. It can also make the paint film weak and easier to damage. And acrylic paint that is overloaded with retarder will dry unevenly, could potentially be reactivated by new paint layers, and, in extreme cases, may fail to cure at all. Each manufacturer has its own ideal proportion for mixing retarder with other products, so it is prudent to read the package instructions prior to use.

Because retarder increases the open, or working, time of acrylic paints to a small extent, it allows for smoother blending and can decrease unwanted brush marks. This added time is more noticeable in thinner films, but the added time afforded by retarder varies according to paint format. A high-viscosity paint will hold onto volatiles slightly longer than a liquid acrylic. Higher-viscosity paints take longer to dry because they are dense and do not level out. Their mass can retain the water content more effectively, particularly in thicker films, than the mass of a lower-viscosity paint whose film is thinner, allowing for faster and more uniform water evaporation.

Relative humidity and temperature will also influence retarder's effectiveness. The more water there is in the air, the more slowly acrylic paint will dry, and the opposite is true in arid conditions. Cooler temperatures will slow down drying time, whereas heat will speed it up. Retarder is therefore more useful in hot, dry conditions.

FLOW RELEASE

Flow release—also called flow aid or flow extender—is a low-viscosity acrylic retarder. It's also potent, keeping paint "open" for hours rather than mere minutes. Flow release is aimed especially at artists using airbrush acrylics and acrylic inks, since it facilitates the flow of paint through airbrushes,

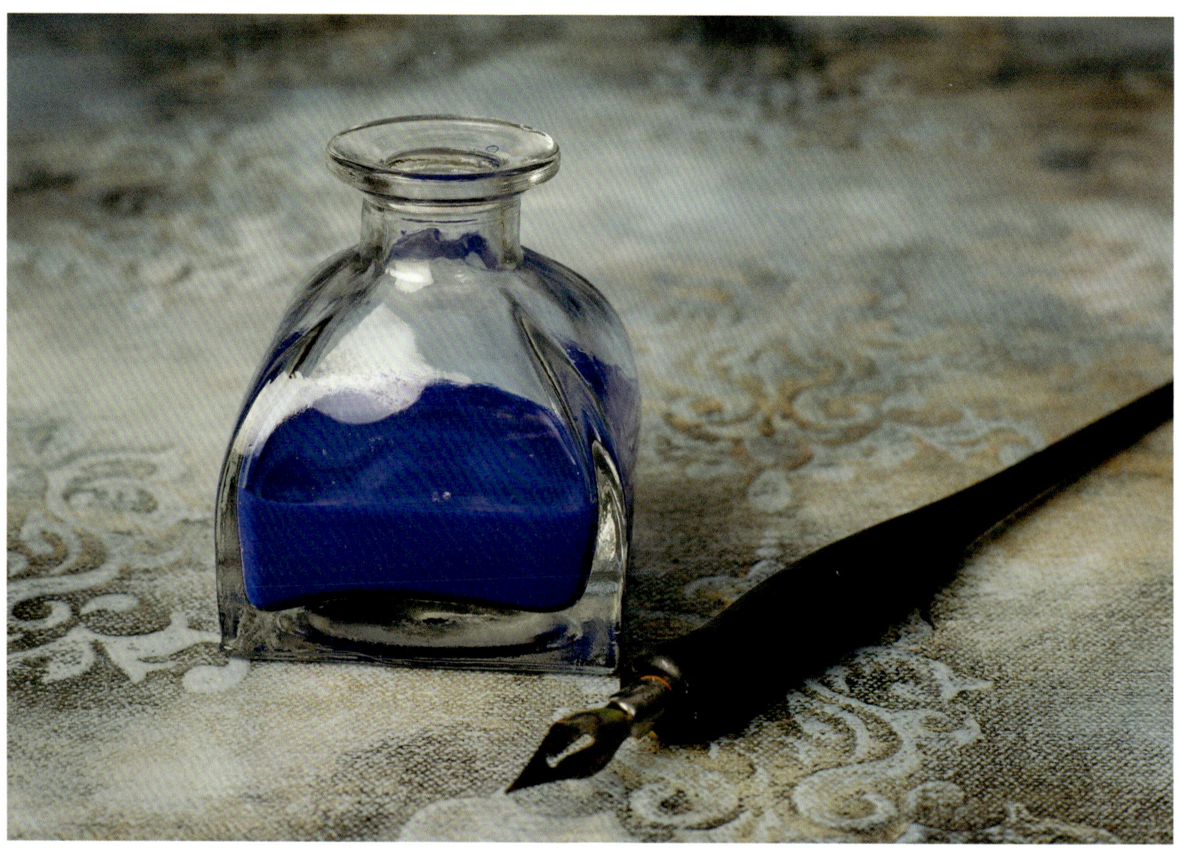

ABOVE With the addition of a flow release agent, a liquid acrylic becomes an ink whose color saturation, transparency, and density you can control.

pens, and other precision tools. A fine line or mist of paint blended with this additive will dry in a relatively short period of time because the volume of paint on the support is quite small. But if paint mixed with flow release is applied more liberally, as with a brush or spreading tool, the result will be a wet, sticky film that will not be workable but will continue to attract dust and be susceptible to damage for an extended period of time. This is a case in which the additive should be used *only* for the tools and processes for which it is intended.

When used with an airbrush, flow release significantly decreases the possibility that paint will clog at the tip of the gun. The addition of flow release to acrylic inks and airbrush acrylics makes penning, brushing, and spraying smoother and easier, without causing beading or crawling or compromising their chroma or longevity. The required proportion of flow release to color greatly depends on the tool being used and on the humidity level in the studio at the time of application. Colors to which flow release has been added will remain wet for long periods if stored in airtight containers; if separation occurs, a gentle stir will remix.

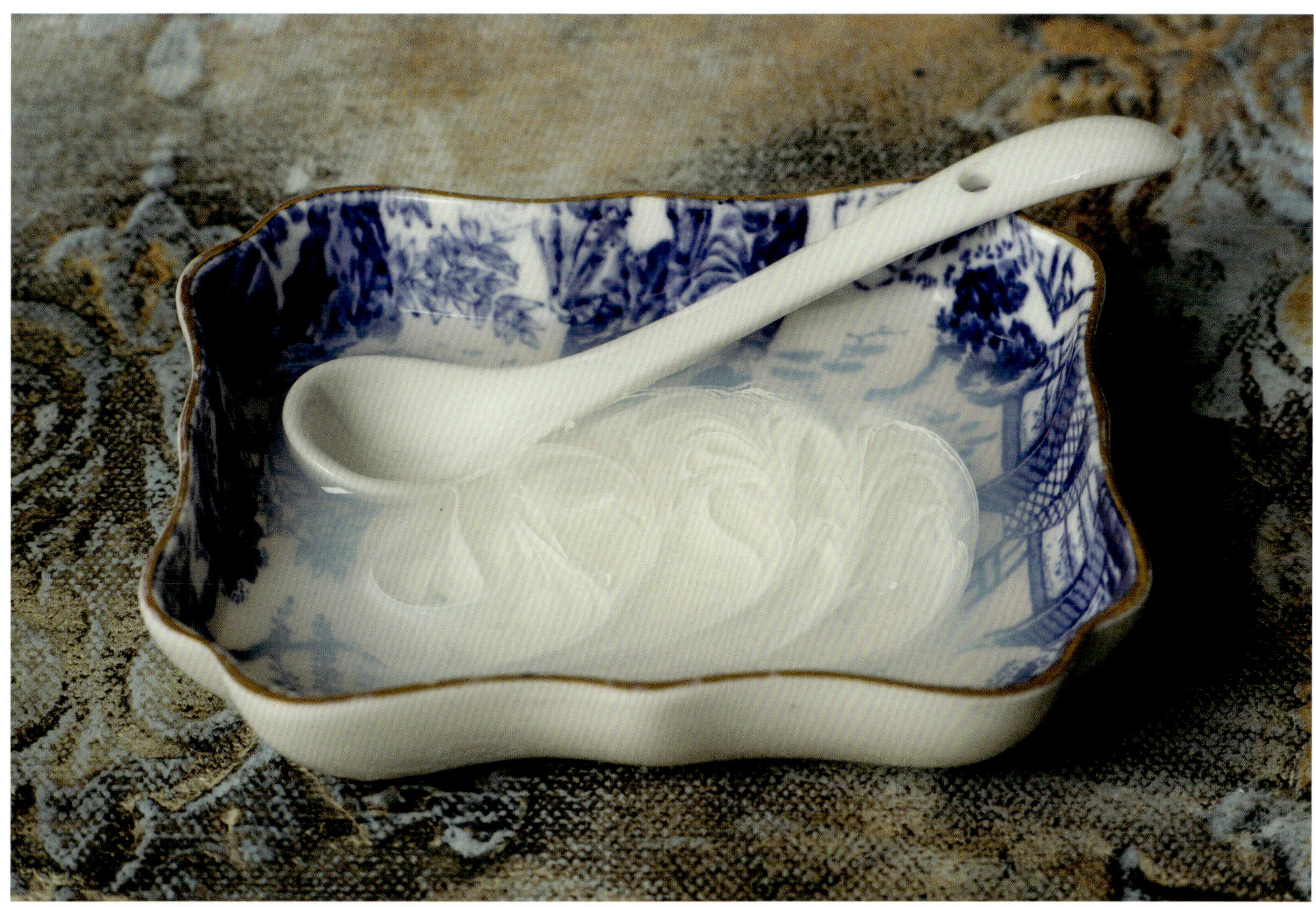

FINISHING MEDIUMS

Virtually any medium can be a finishing medium, because most clear-drying mediums can be used as the final layer of a painting. The role of a finishing medium is to achieve a particular surface appearance—not, as is often assumed, to protect the painting. Laying down a final layer of medium across the surface of a finished painting lets you unify the surface luster, adding depth and increasing color chroma.

There are many mediums created specifically for finishing, and it can be tricky deciphering the differences between them. The majority have a very low viscosity and a medium rheology, meaning they can be easily spread over large areas without leaving ridges or brush marks. They range in luster from gloss to matte, and are all transparent.

ABOVE Zinc white floating in gloss top coat. This will make a lightly translucent finishing layer.

FINAL FINISHES VERSUS VARNISHES

Any clear-drying medium can be used to finish off a painting, but this does not mean it is a varnish. A traditional varnish is a clear-drying top coat derived from some type of tree sap and thinned with solvent. Traditional varnishes are also designed to be removable in order to "clean" a painting. Because atmospheric dirt attaches itself to the varnish instead of directly to the painted surface, the painting can be refreshed by removing the varnish and then applying a new coat.

Resin varnishes of this sort, however, are far too brittle for the acrylic surface. Even when fully cured, acrylics remain very flexible and can move as the substrate shrinks or swells with seasonal changes. Sap-derived resins are too hard to accommodate this movement, making them susceptible to crackling, breakage, and eventual delamination from the paint surface.

Any top layer added to a surface painted with acrylics—particularly on a flexible or semiflexible support—has to be able to bond fully with the surface and to move with it as it changes. Unsurprisingly, the materials that fuse to acrylics most effectively are other acrylics. Therefore, an acrylic top coat or final finish is the medium of choice for this purpose.

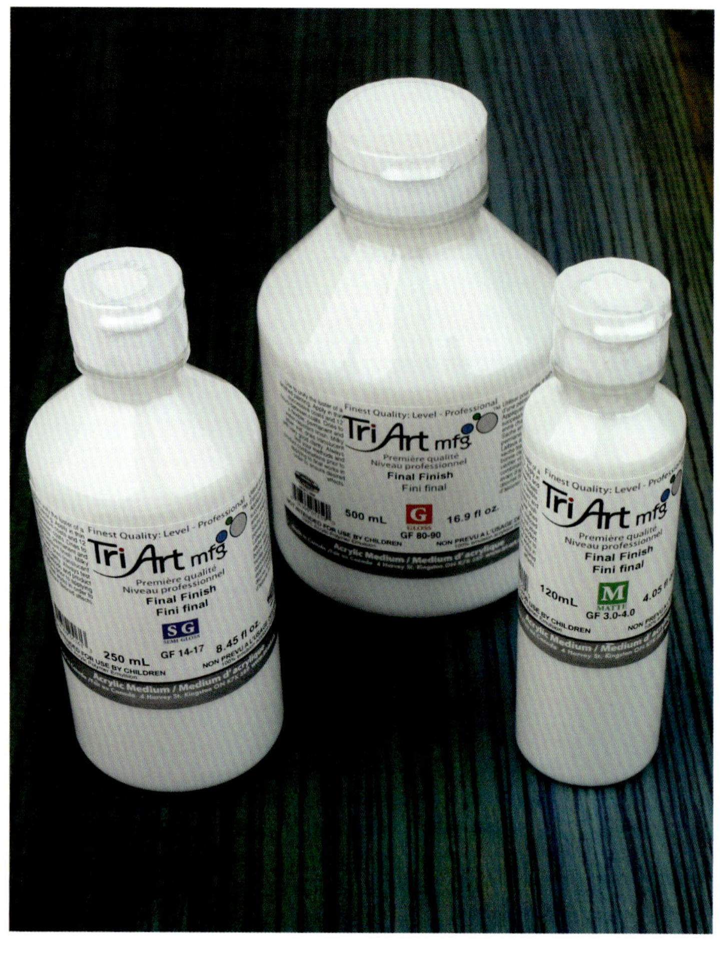

ABOVE Final finish mediums (the name differs by brand) are thin acrylic mediums that leave no discernable brush marks and serve to alter the surface luster of the painting.

ACRYLIC FINISHING MEDIUMS

To reiterate: any clear-drying medium can be used as a final layer. Be aware, however, that many mediums will impart additional textural aspects to your painting that you may not want. If you want to change the luster of a finished painting but do not want to add any new brushstrokes or other textures, then use a medium that is specifically designed for that purpose.

Final finish mediums (sometimes called acrylic varnishes) are self-leveling liquid mediums specifically designed to unify the luster of a finished painting. There are also some brands of acrylic finishing mediums in aerosol format. Clear drying, such mediums can be applied in successive thin layers to provide a gloss, semigloss, or matte surface. Liquid finishing mediums can also be sprayed through an airbrush or spray gun. Whatever your application method, allow for overnight drying between coats.

TOP COATS

I refer to finishing mediums that do not share the same acrylic base as acrylic colors as *top coats*. The majority of these are not removable. Their chemical bases differ: some are based on mineral spirits; others on urethane; and others have a tougher acrylic polymer emulsion base (also referred to as acrylic polymer resin) or are hybrids of these materials. Keep in mind that there are many different grades, hardnesses, and formats of acrylic resin. The particular acrylic resin that is used for top coats generally has a lower flexibility and superior transparency and tensile strength than that of the acrylic resin used in colors and painting mediums. Top coat resins dry to a lower tack finish (uncoated acrylic paintings retain a lightly tacky, or gummy, surface) and can be tinted (lightly) with liquid acrylic colors. Top coats are available in soft or hard formats, with the latter recommended for use only on rigid surfaces. Some top coats also contain an ultraviolet light stabilizer (UVLS) to boost lightfastness.

Urethane-based top coats and urethane-acrylic hybrids have similar handling properties to pure acrylic top coats but will dry to a harder, less flexible finish and are more appropriate for coating rigid surfaces. Mineral spirit–based acrylic varnishes are comparable to these but are more fluid and are removable with solvents. These types of finishes have a very low tack when fully cured and are less permeable to water than acrylic finishes.

CHAPTER 3

UNDERSTANDING TOOLS AND THEIR EFFECTS

THE ARRAY OF AVAILABLE PAINTING TOOLS IS VAST. Choosing which to use depends on your process and, in some cases, your materials. Some basic tools, however, suit a variety of purposes and provide a good utilitarian foundation.

Many tools are designed specifically for the acrylic medium, and there are very good reasons for choosing these. For example, brushes with synthetic bristles such as Taklon and Interlon are perfectly suited to acrylics. They are resilient and soft, do not suffer from prolonged exposure to water, maintain their shape well, and require no conditioning. Although brushes that are labeled as being for acrylics tend to have long handles, synthetic watercolor brushes, with shorter handles, provide greater control for detail work.

Choose tools designed specifically for acrylics.

OPPOSITE Still life with fan brushes. Not just for blending, fan brushes can produce distinctive brushstrokes and sharp textures.

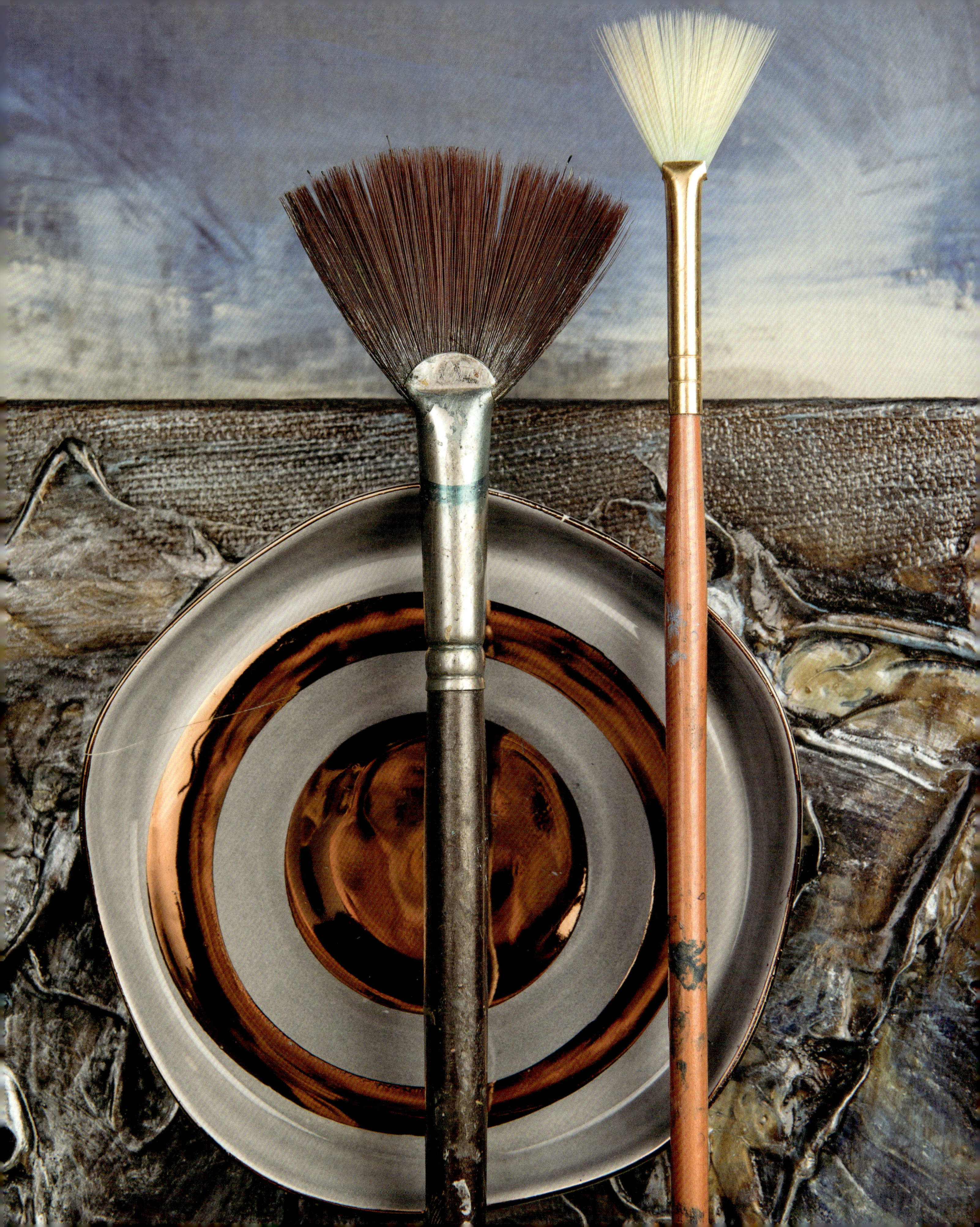

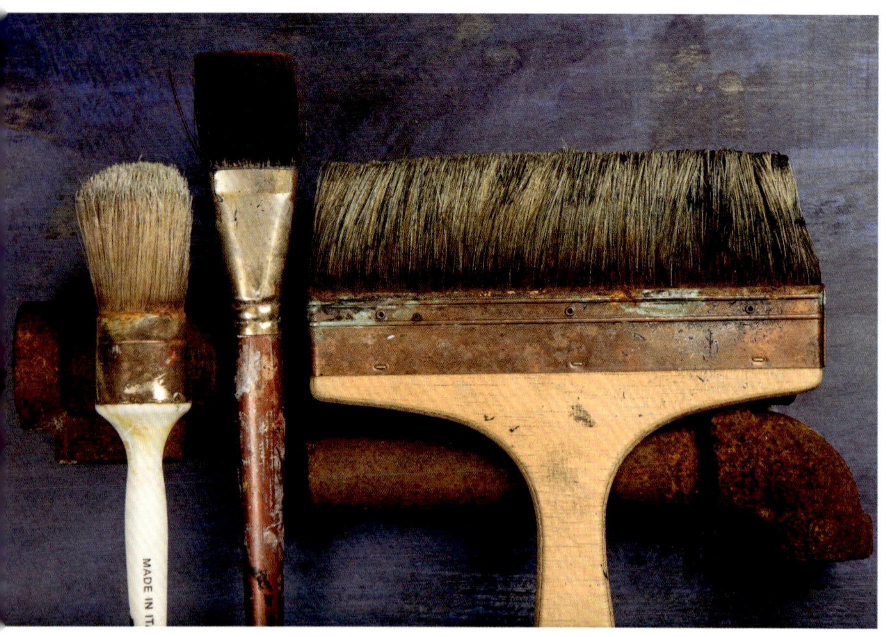

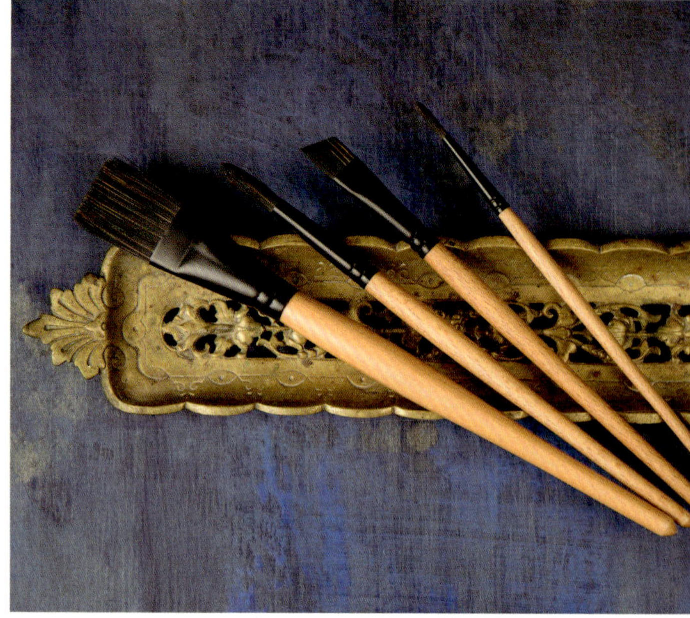

BRUSHES AND BEYOND

These days, there is a lot of excitement about acrylic-painting tools—not surprising, given how many new products are on the market. Tools in new shapes and sizes and made with new materials interact with paints and mediums to produce dramatic and interesting textures, giving artists whole new ways to paint.

For one thing, palette knives are getting bigger. This is a boon to painters who favor larger supports, and as well to those who use great amounts of mediums. When buying one of these larger palette knives, check for flexibility and structural integrity. Pick one that holds its flat edge, bouncing back to a perfectly level state after every flex. (A range of shapes in the traditional, smaller size remains available.)

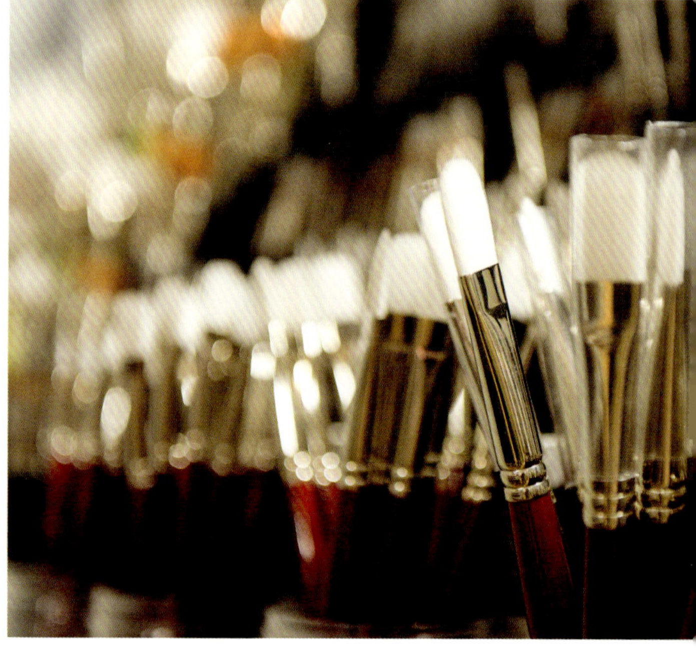

TOP, LEFT The size of your tools is limited only by your imagination (and perhaps the size of your canvas).

TOP, RIGHT Synthetic-hair brushes with resilient, springy bristles.

ABOVE Taklon brushes.

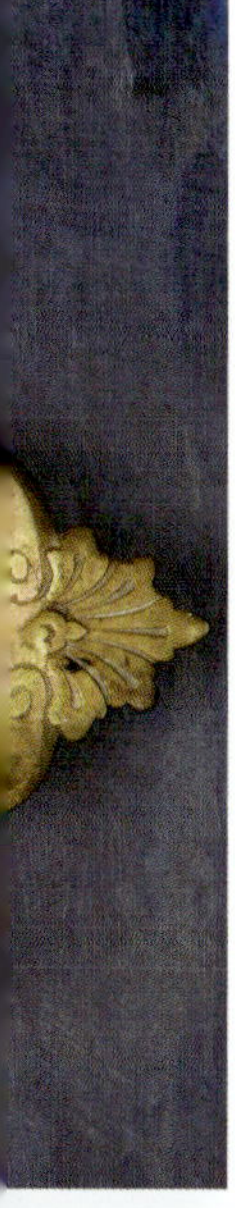

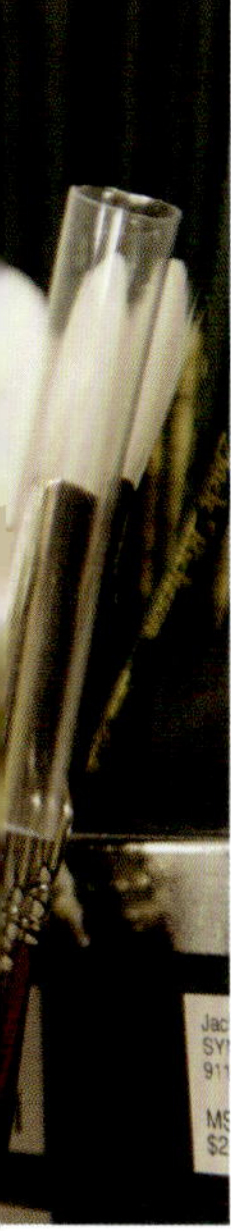

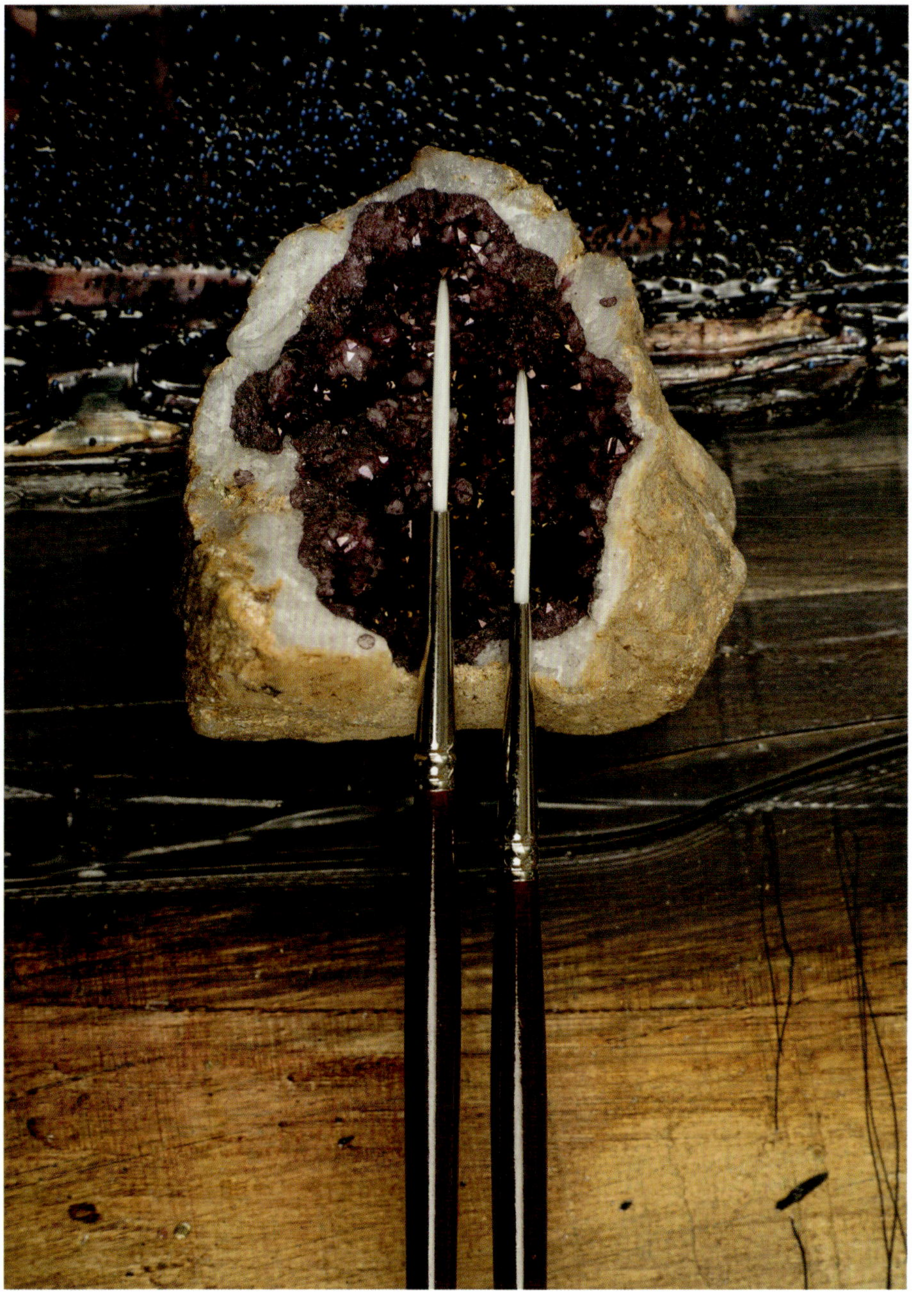

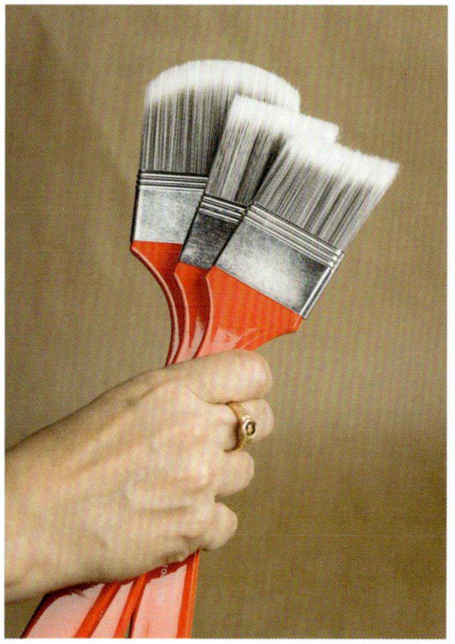

LEFT Delicate synthetic-bristle watercolor liners.

ABOVE Large-format synthetic-hair brushes for dramatic impasto techniques.

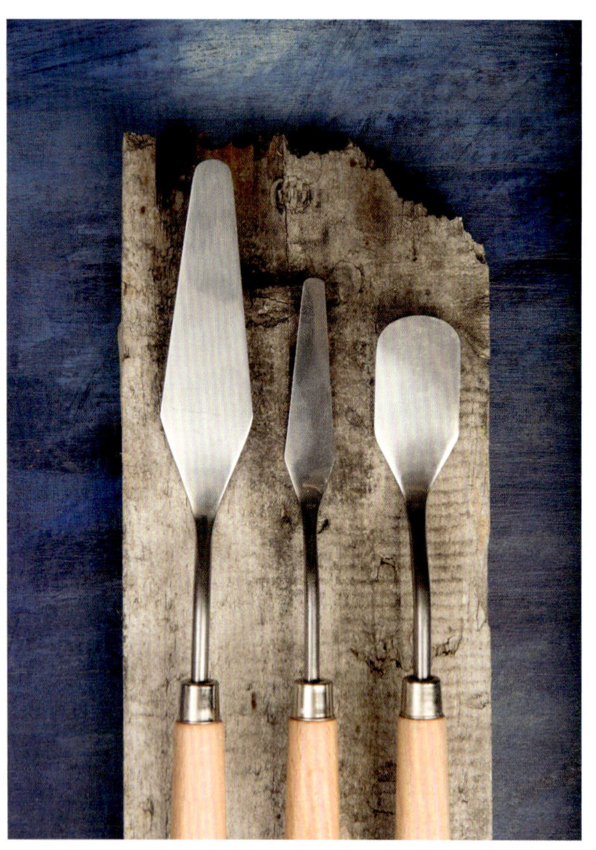

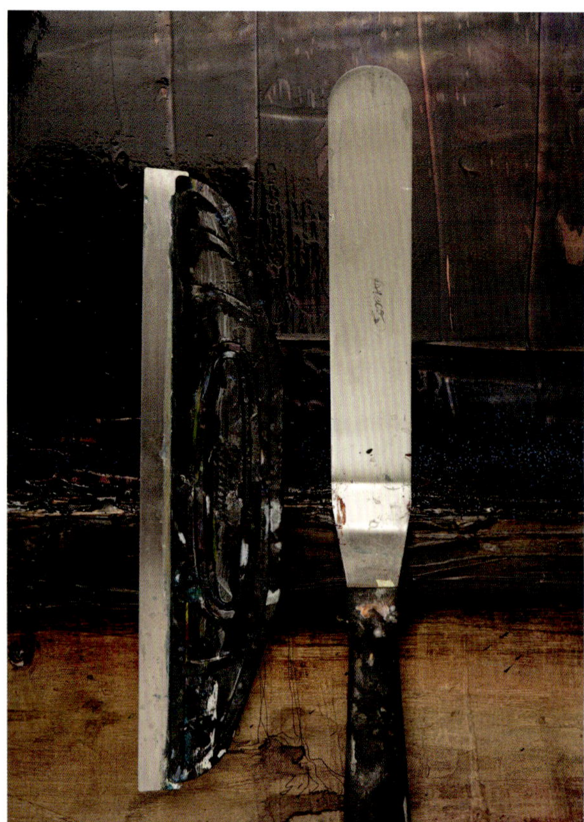

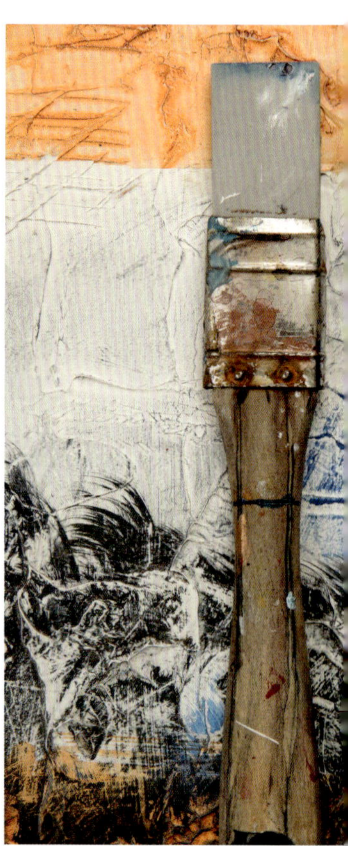

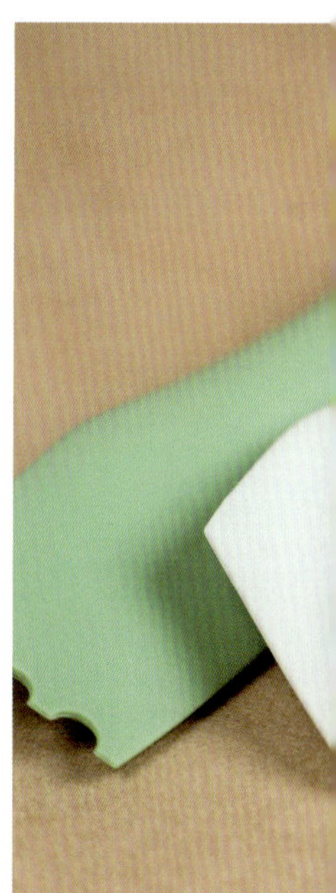

One of the fastest-growing tool trends is the use of spreading tools and synthetic (usually silicone) texturizing tools designed specifically for painters. Color shapers are plastic and silicone-tipped spreaders that are indispensable for creating smooth swaths of paint, producing furrows and linear textures, scraping away layers, and spreading thin glazes of color. Because dried paint peels cleanly off their surface, these tools maintain their edge and shape for a long time, outlasting most brushes.

These nontraditional tools can take your work in a new direction—spreading, pouring, staining, knifing, pulling, extruding, and so on. Using these tools with more substantial quantities of medium can produce more dynamic and nuanced textures than can be achieved with a brush. They can therefore be essential in adding variety and drama to your surfaces.

Moving away from intricate brushwork leaves detail and precision out of the creative equation. There are many

ABOVE, LEFT Traditional palette knives come in all sorts of shapes and sizes. They're sharp, pointy, and strong but flexible. Use them to dispense, mix, and spread paint and to make marks.

ABOVE, CENTER A large palette knife with straight edge and offset handle.

ABOVE, RIGHT Color shapers have become indispensable tools for today's acrylic painters.

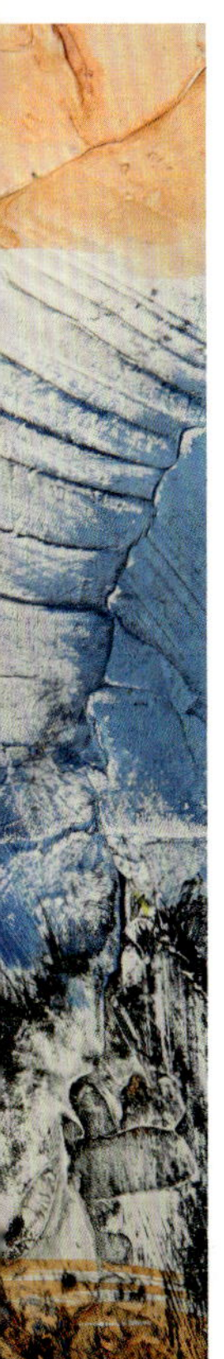

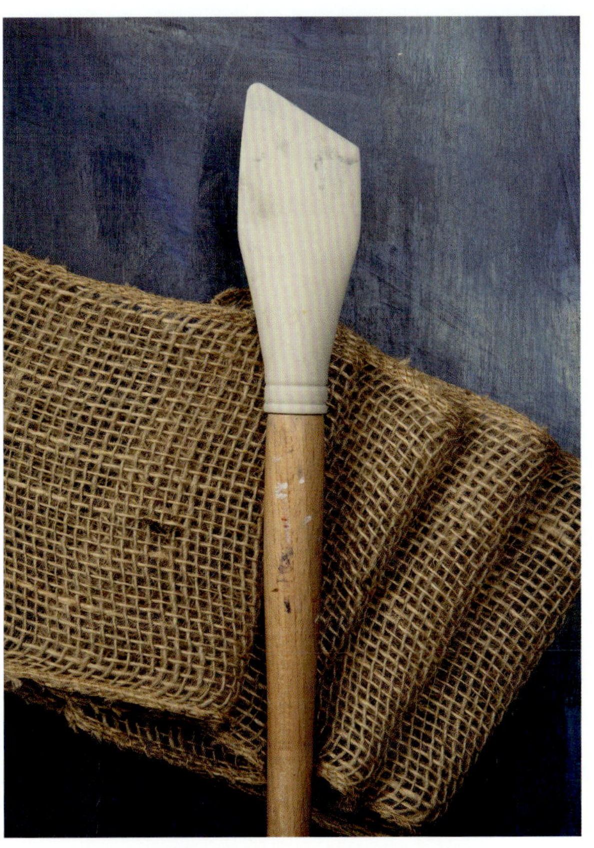

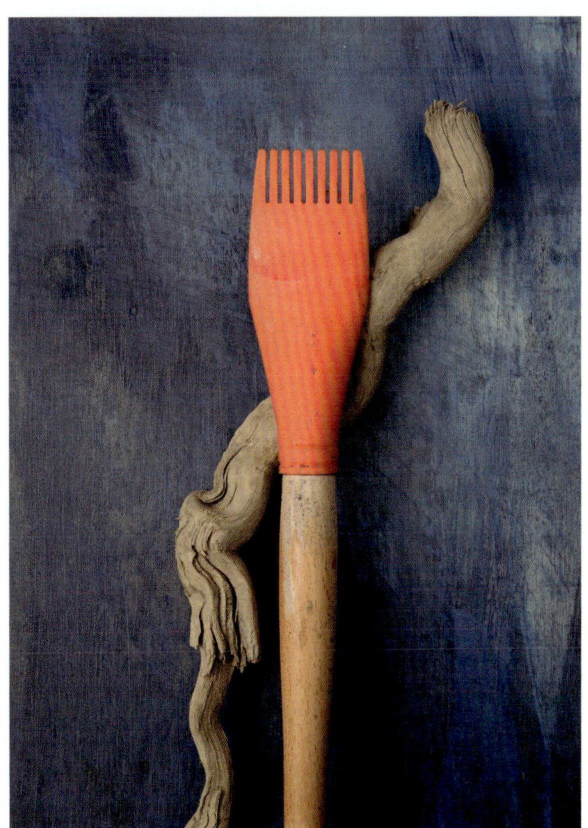

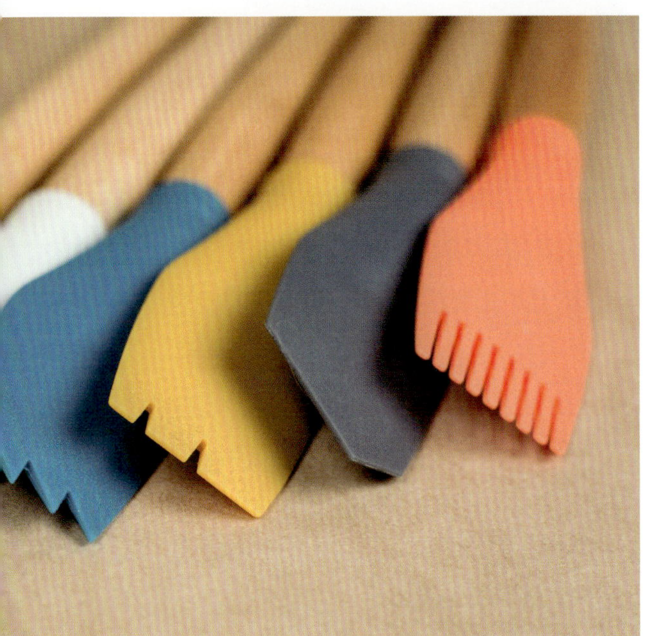

more incidences of accidental mark-making and irregular texture, and there's always the possibility of a "subtle catastrophe," which can be a great way of loosening you up and breaking you out of a creative slump. Happy accidents often mark the beginning of a fresh new approach.

The accidental obscuring and revealing that occurs can bring a depth to the painted surface. In this kind of process, the material shines more than the artist's hand, making its own signature and asserting its presence in a way that doesn't happen in studied hand-brushed work. The freedom given by the action frees the painter from the crutch of exactitude. Painting with this measure of unpredictability regarding the outcome, however, is not for the control-driven artist nor for those wanting impeccable precision in the execution of their craft.

TOP, LEFT A small, wedge-shaped paint manipulator.

TOP, RIGHT This Catalyst color shaper's teeth don't bite; they swirl and create ridges.

BOTTOM The Catalyst tools made by the Princeton Artist Brush Co. are texturizers at your fingertips.

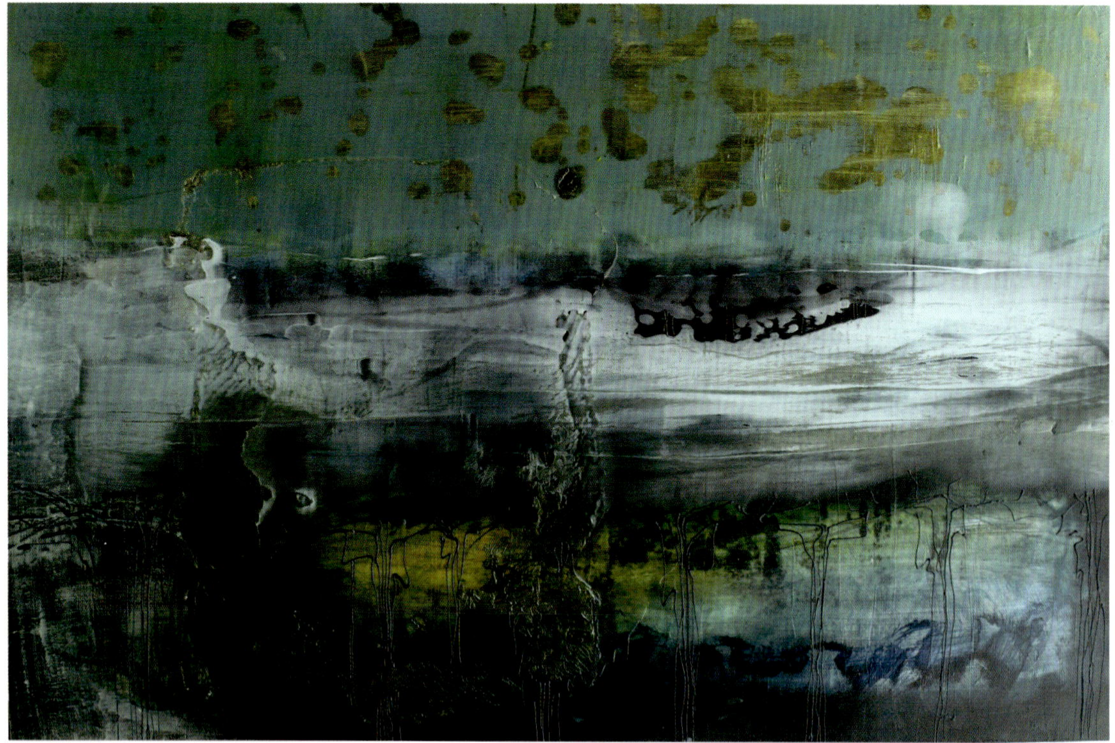

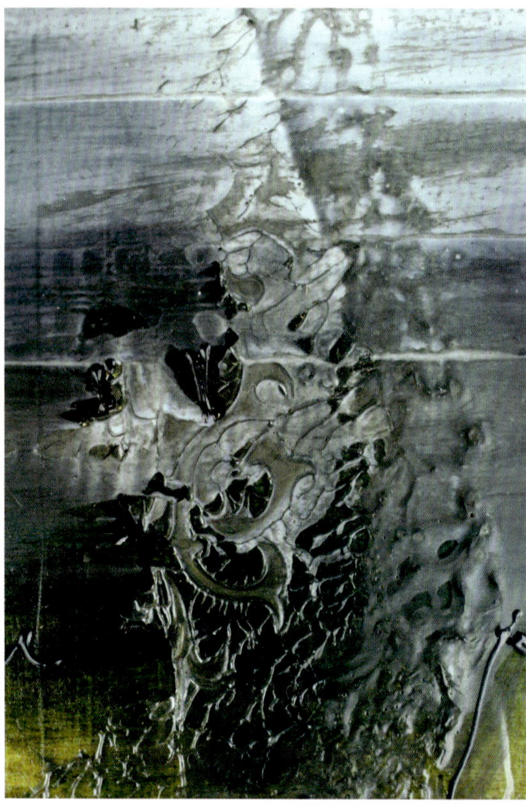

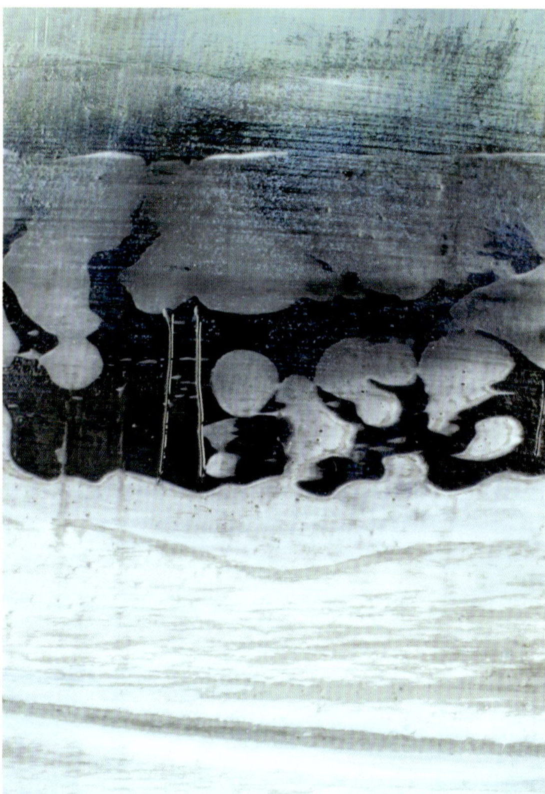

TOP **Rhéni Tauchid, *Leftovers,* 2015, acrylic on panel 36 x 24 inches (91 x 61 cm). Private collection.**

Mediums, colors, and techniques collided to form this abstract composition. Like so many of my paintings, this one evolved from my trying one technique after another.

ABOVE, LEFT In this detail from *Leftovers*, a remnant of gel wiped away from a stencil left on a nonstick palette was collaged onto the paint surface to provide a new texture.

ABOVE, RIGHT Another detail from *Leftovers*. Here, matte medium and dry media ground combined to cast a semi-opaque veil.

CHOOSING THE RIGHT TOOLS

Thick mediums—some chunky and granular, others slick and gooey—require special tools to successfully manipulate them. Not every tool is right for every purpose. When you understand these myriad materials' different characters, you'll be able to match each to the best tool. That said, there are a number of things you should always consider when choosing a tool for acrylics:

- *Water permeability.* If you tend to leave your tools sitting in water for hours or days after a painting session, don't use wooden-handled tools. Brush ferrules can trap water, causing the adhesive to weaken, the wooden handle to swell, and untreated metal to rust.
- *Ease of cleaning.* Certain synthetic tools, like HDPE plastic palette knives and silicone-tipped color shapers, are a wonder when it comes to cleaning. Even days-old dried paint can be peeled off of them fairly effortlessly. Synthetic-bristle brushes may stain a little, but because they are not absorbent, they are easier to clean than natural-hair brushes.
- *Resistance to solvents.* When acrylic paint or medium gets caked onto a tool, you may need something stronger than water to remove it. Choosing tools that will not be adversely affected by solvents is prudent.
- *Flexibility.* This is not a mandatory consideration, but flexibility will have an effect on how your paint is manipulated. A very soft bristled brush, for example, will move liquid consistencies with more ease and precision than a coarse, stiff-haired one, while a springy, stronger bristle will more effectively manipulate thick paints and mediums.
- *Non-reactivity.* Choose tools that are friendly to water-based acrylics. Metals that oxidize will rust.
- *Durability over time.* There are too many disposable tools in our world. Acrylic paints are easy to clean off of many surfaces, and if you take care to routinely maintain your tools, they can last a lifetime.

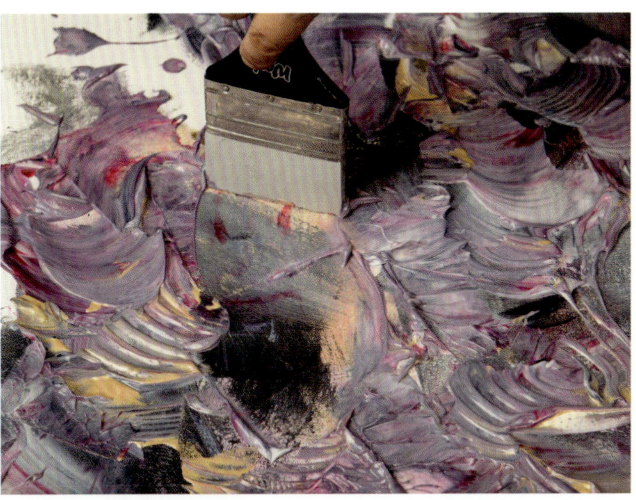

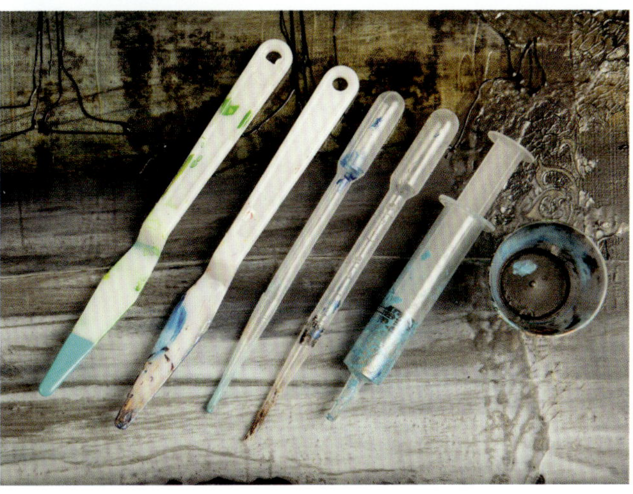

TOP Pushing away a wet gel layer with a color shaper reveals the original marks beneath.

BOTTOM Plastic tools for plastic paints: flexible, durable, and easily washable.

TOOLS AND TEXTURES

Each kind of tool produces a specific textural effect. Here, images of tools are paired with images of the textures they produce.

A soft, absorbent goat-hair brush produces washes with soft brushstrokes.

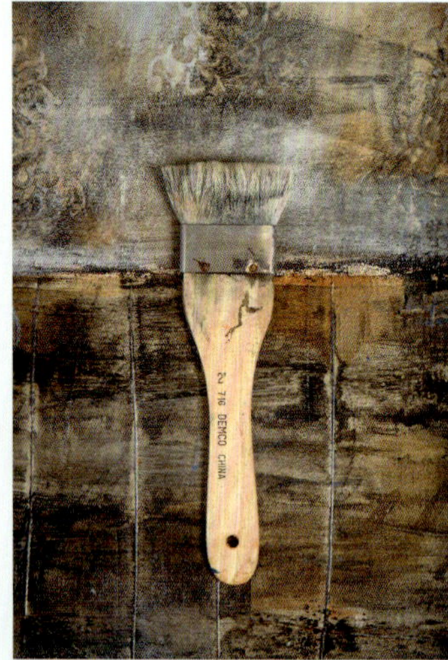

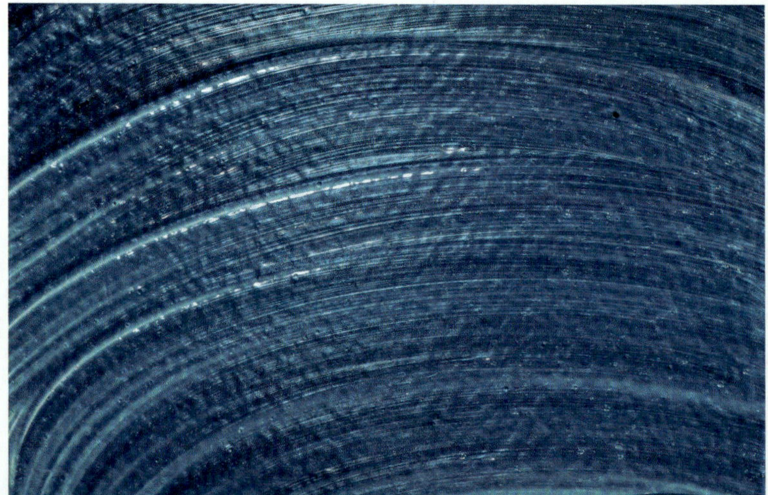

A hog's bristle mop brush can hold a big quantity of wet paint or produce very wide swaths of drybrushing and distressing. This type of bristle stands up to a lot of wear and tear.

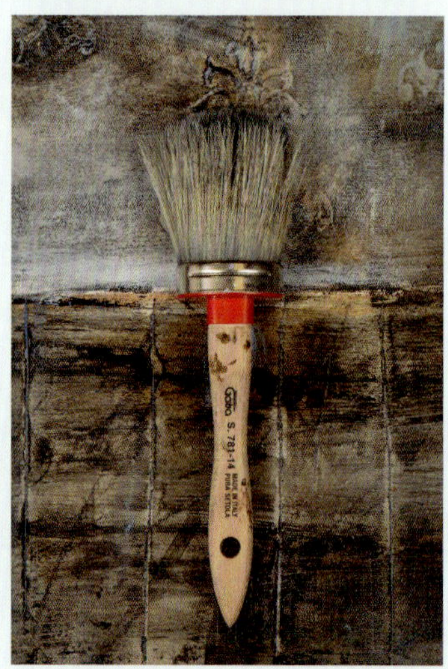

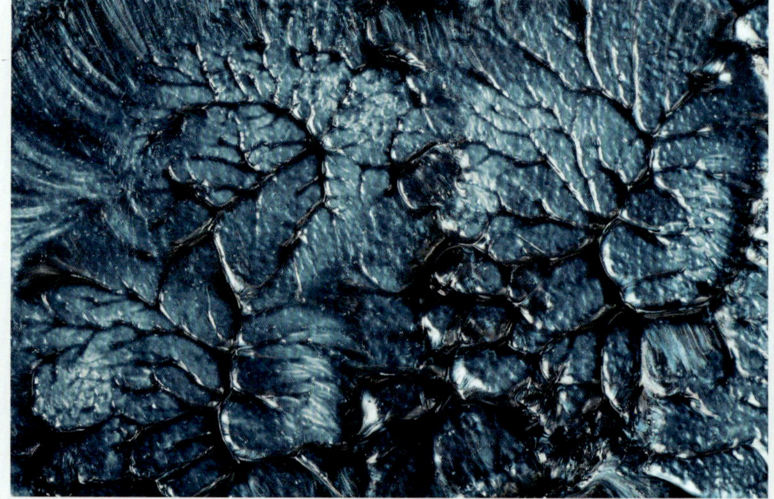

This impasto brush with tightly packed, short synthetic bristles creates wide sweeps.

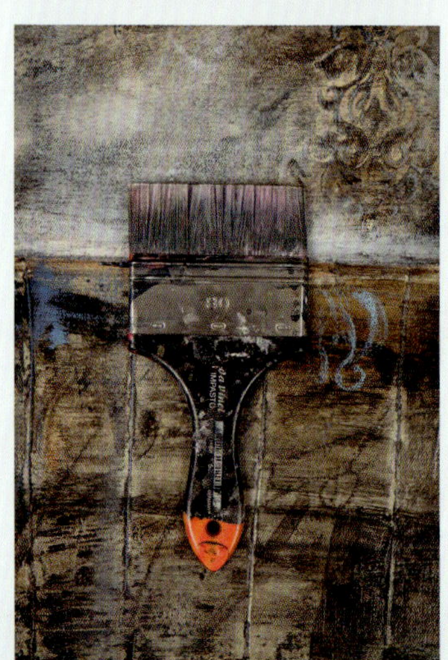

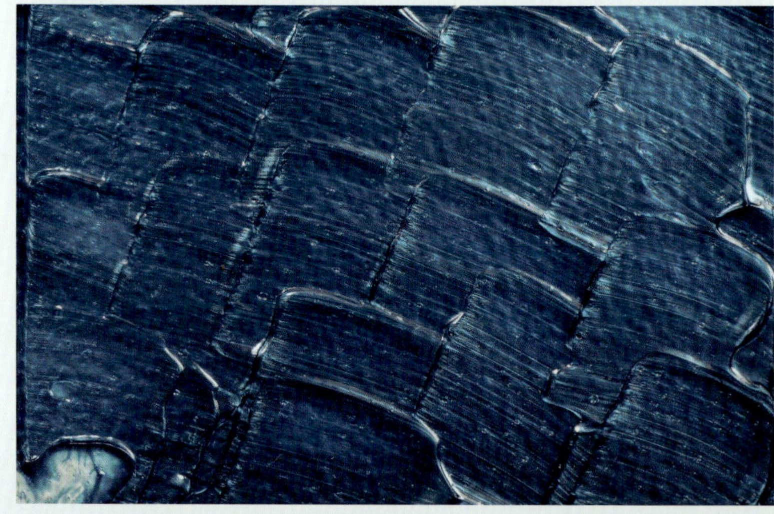

A fan brush and its distinctive marks.

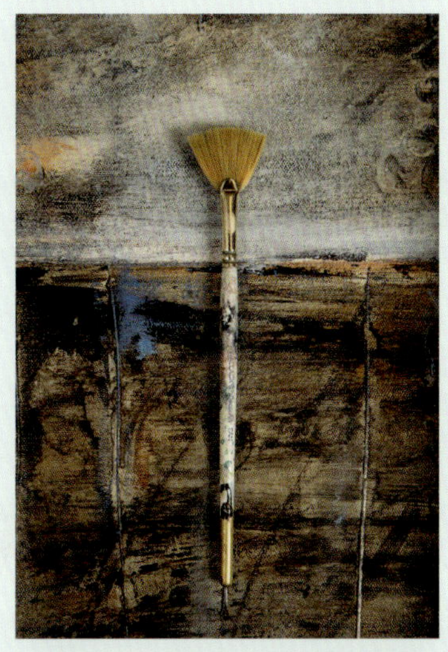

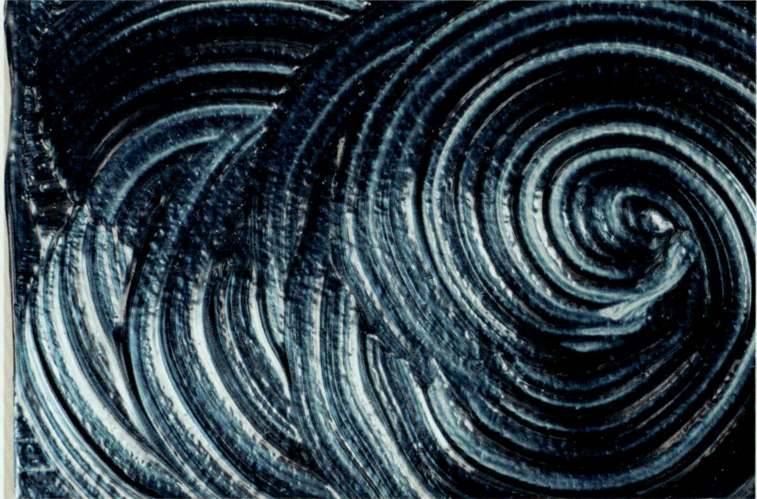

Do not underestimate the practicality and robustness of the plastic palette knife.

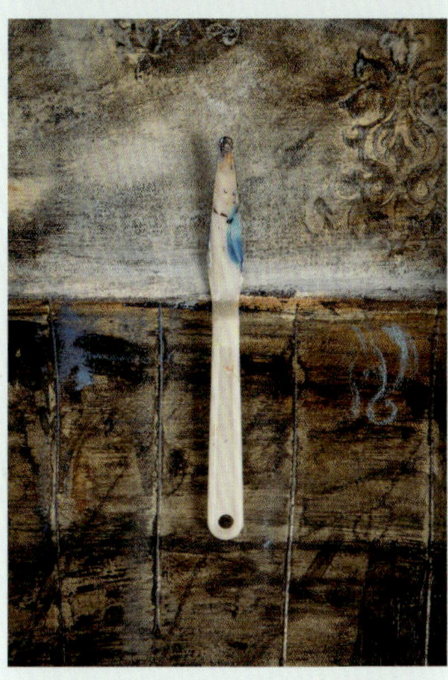

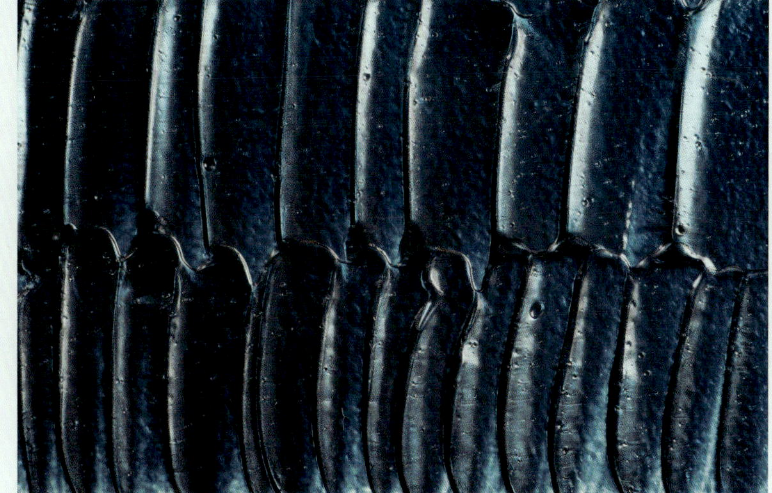

Supple, thin, and strong, the classic metal palette knife creates a loose impasto effect.

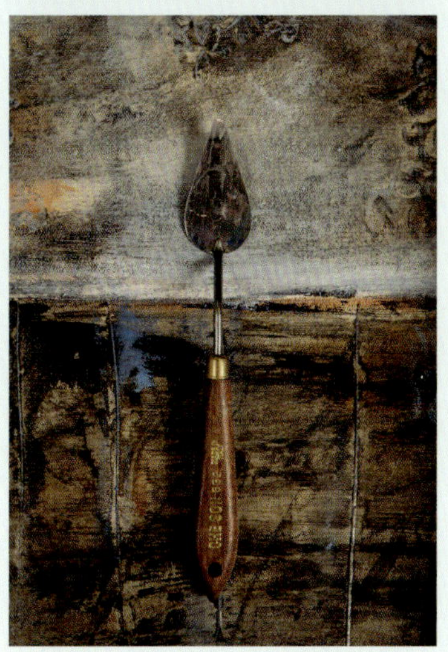

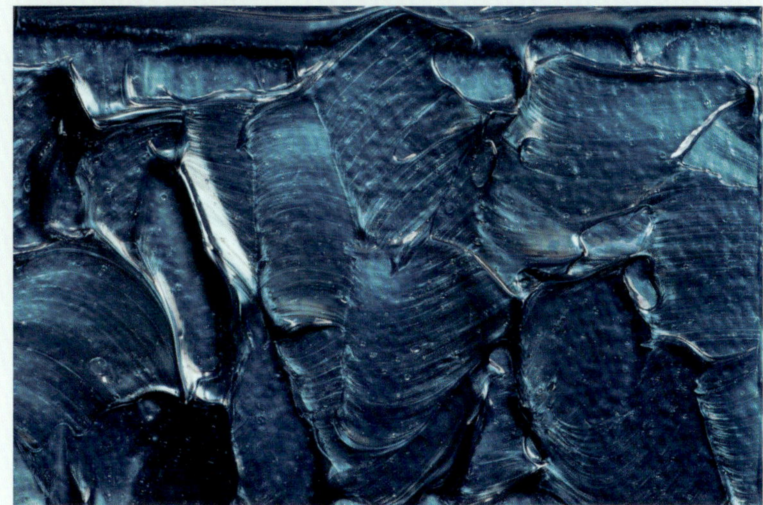

The offset handle of a larger palette knife makes it easy to move large quantities of paint around.

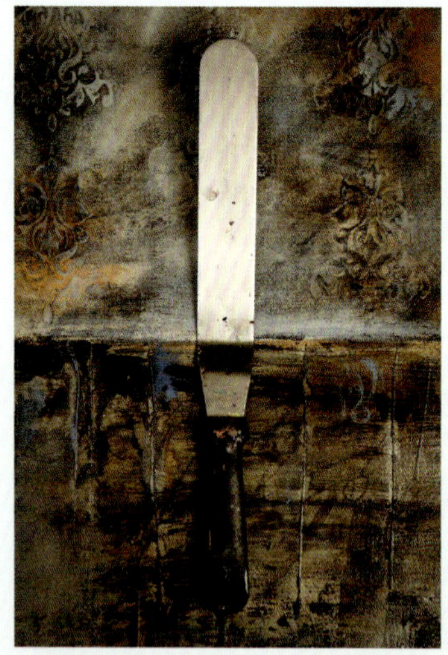

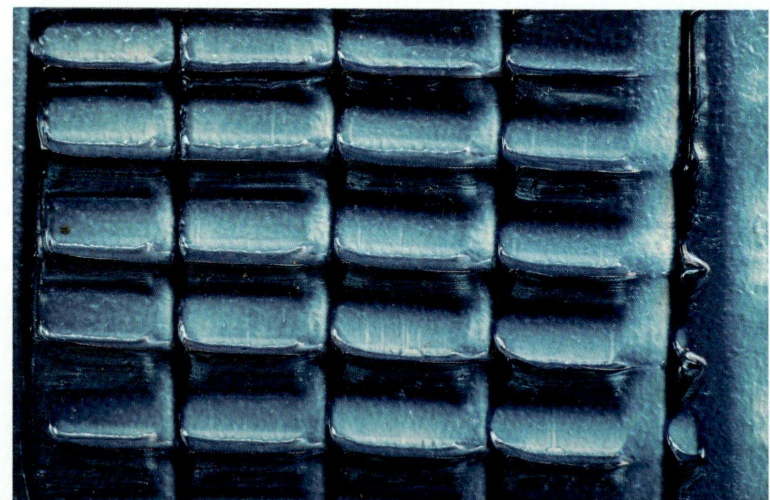

Small color shapers of varying hardnesses and shapes produce their own distinctive kinds of marks.

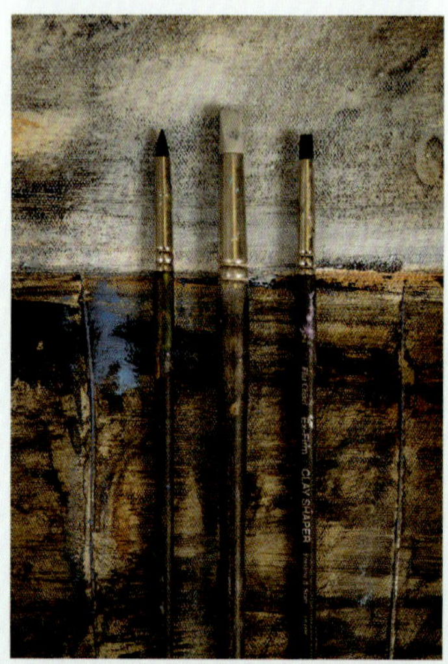

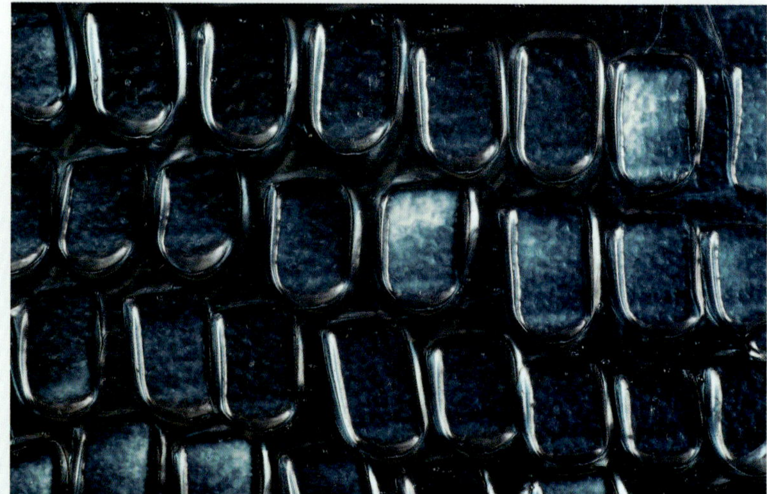

An extra-wide color shaper and the texture it creates.

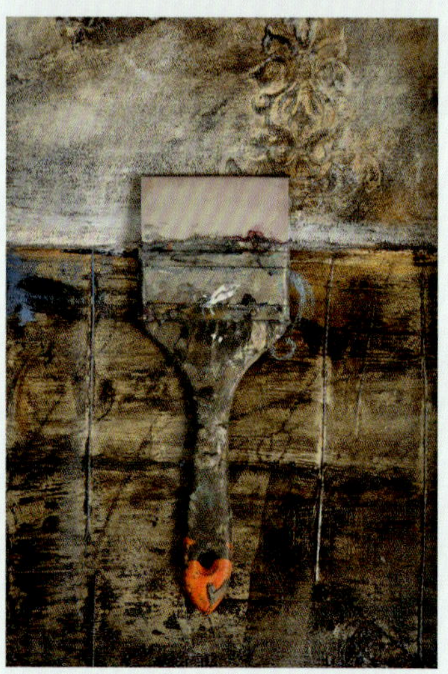

Good tools don't have to cost very much. Bamboo skewers, always close at hand in my studio, make good drawing tools.

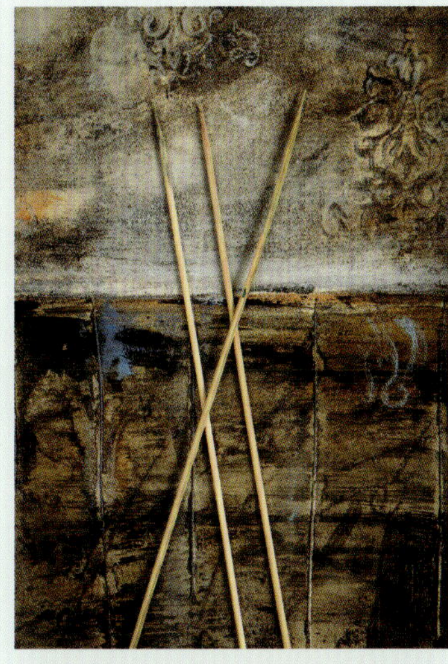

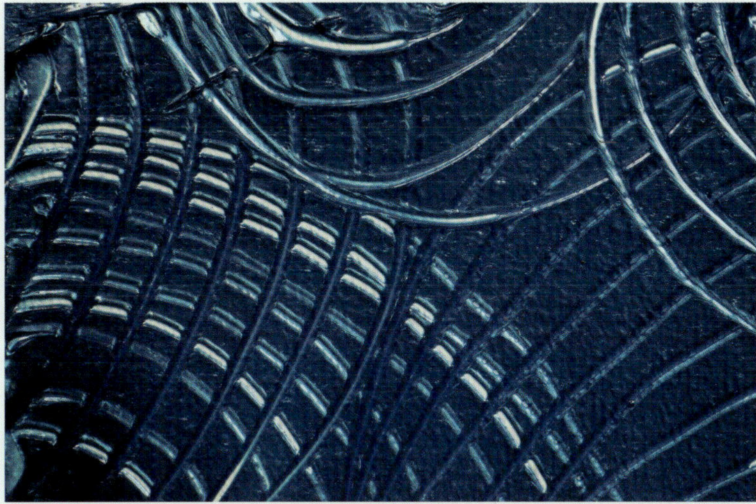

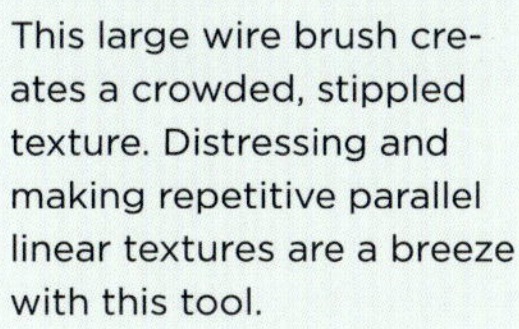

This large wire brush creates a crowded, stippled texture. Distressing and making repetitive parallel linear textures are a breeze with this tool.

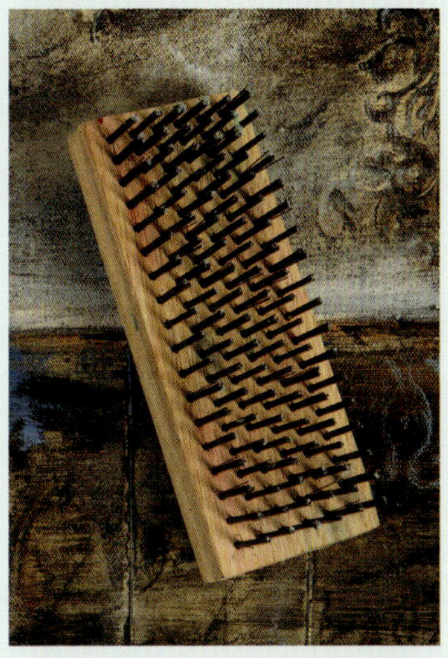

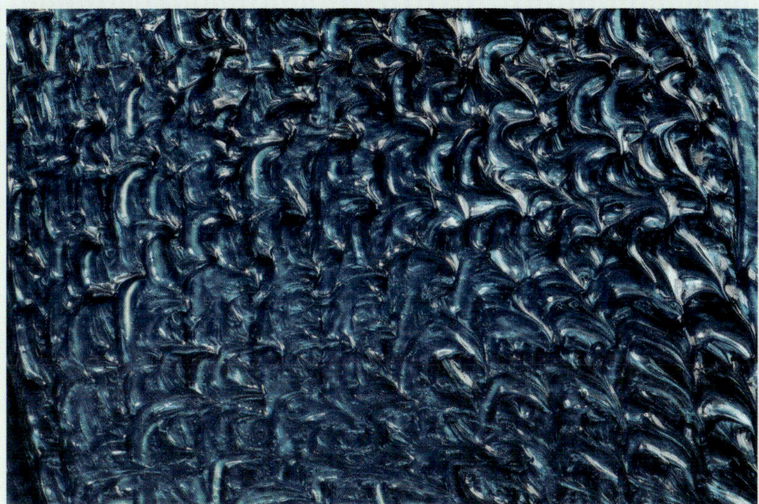

The green Catalyst wedge tool and its marks.

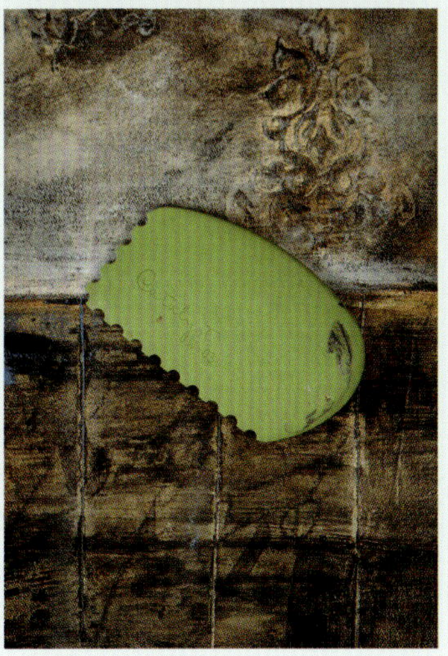

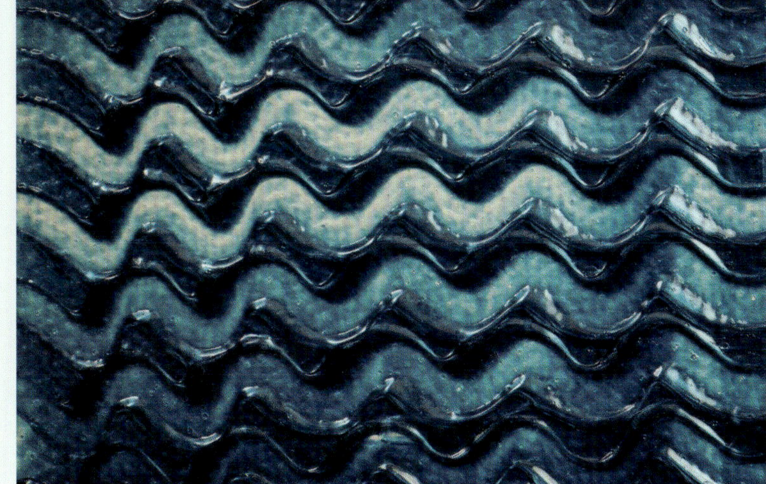

Catalyst painting tools. Texturizers at your fingertips.

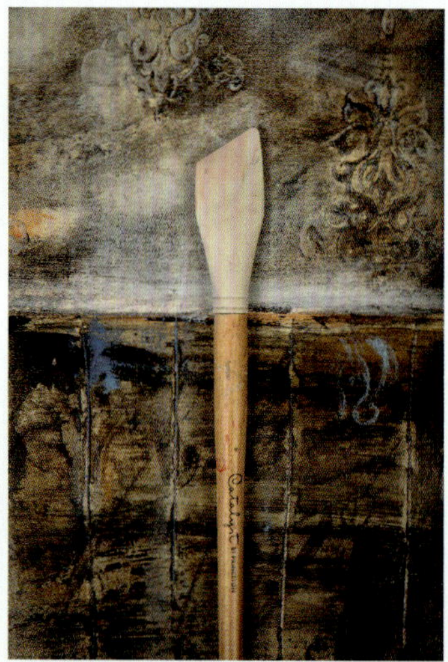

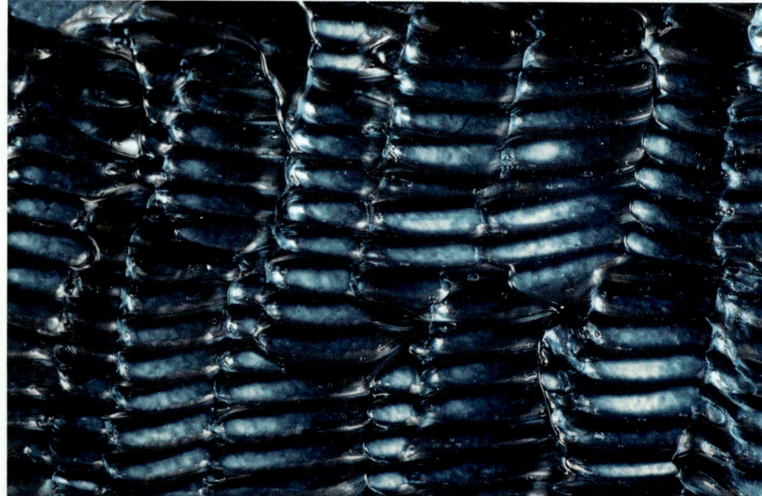

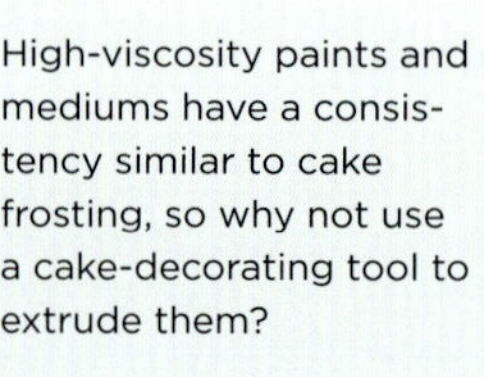

High-viscosity paints and mediums have a consistency similar to cake frosting, so why not use a cake-decorating tool to extrude them?

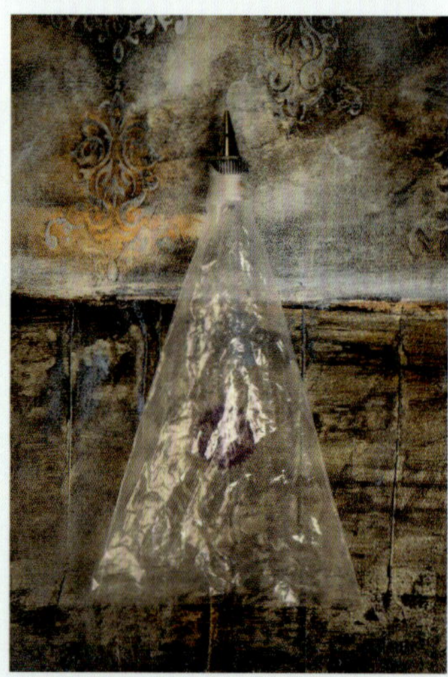

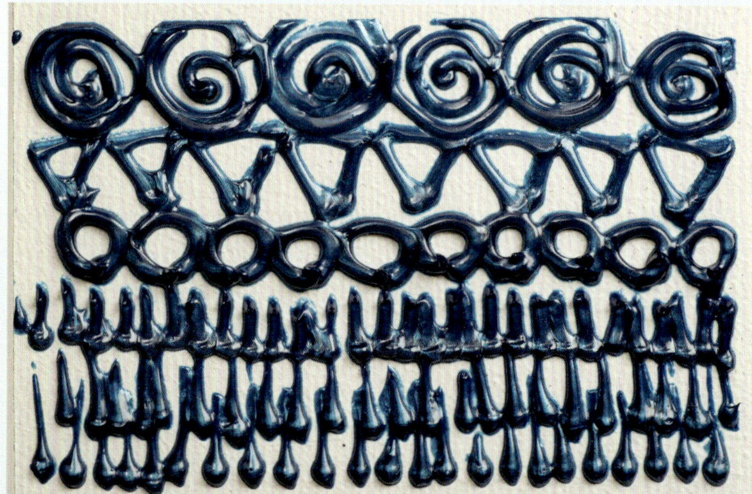

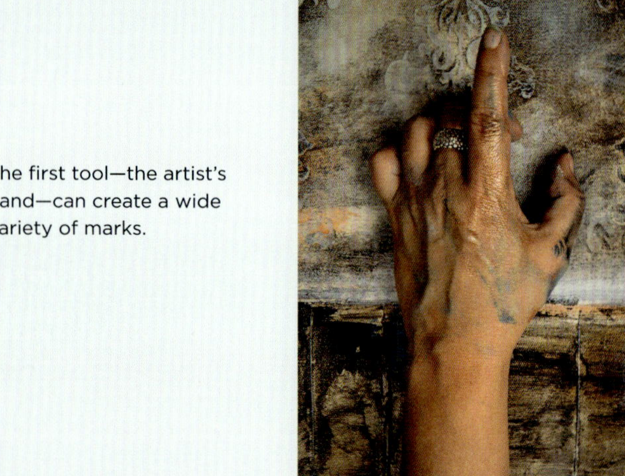

The first tool—the artist's hand—can create a wide variety of marks.

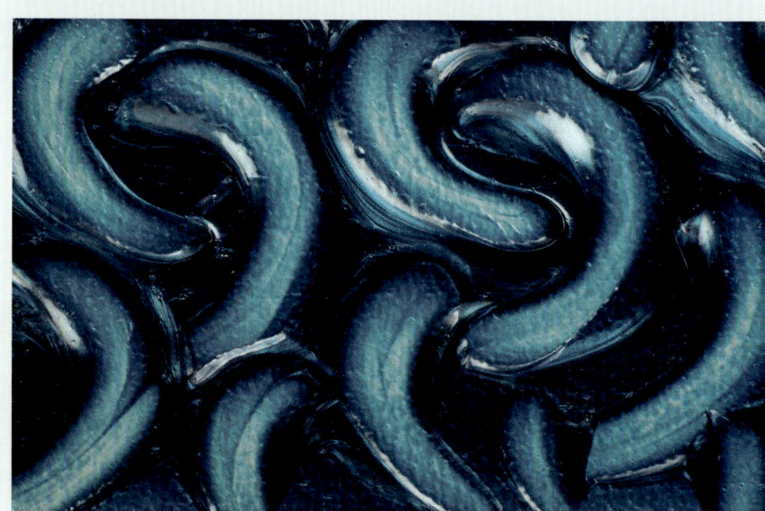

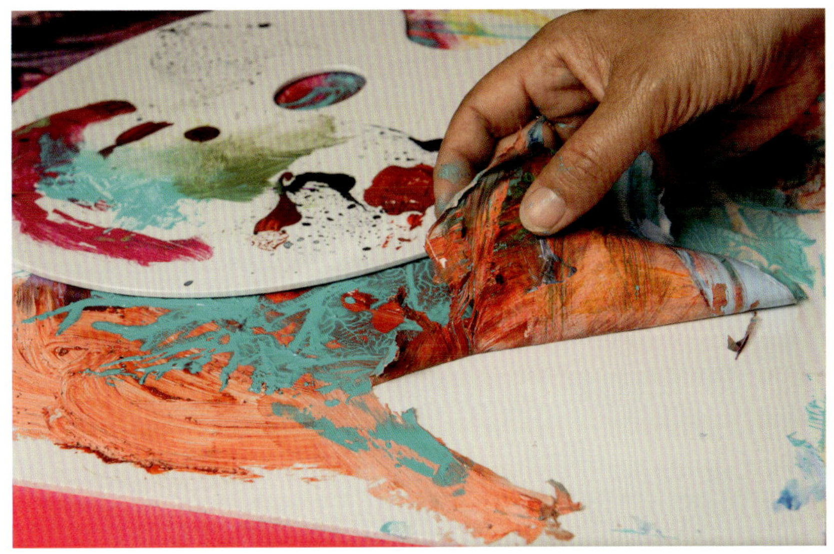

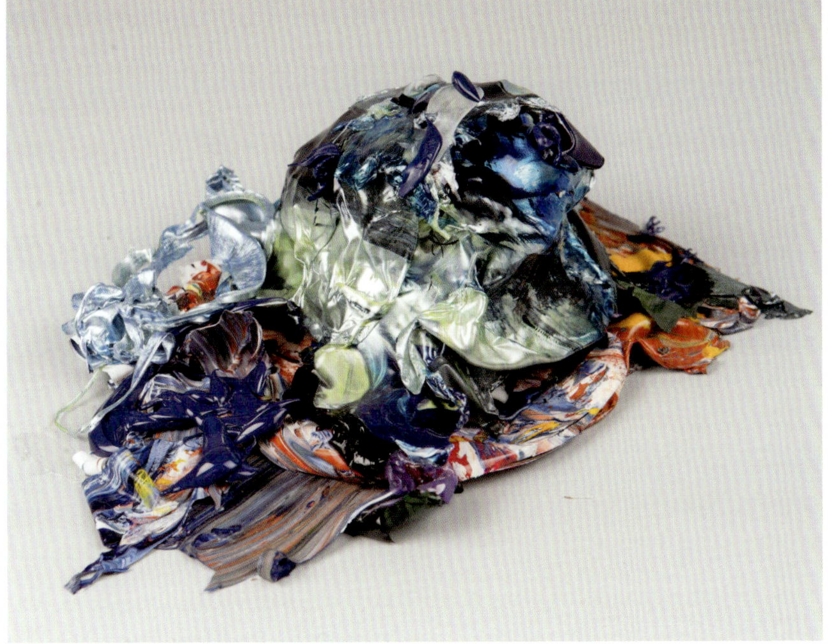

PALETTES

The tool that you likely will use most often during your painting process is your palette. There are many different types of palettes, from thick glass slabs to flimsy, disposable paper palettes. But for me the truly indispensable palette is the nonstick palette, which can be reused, will not stain or warp, and does not have to be washed with water. A nonstick palette might be the sustainable answer to all your palette needs. Even thin layers of paint and medium can be cleanly and easily peeled away. These remnants, or acrylic "skins," can then be used as collage, textural, or sculptural elements.

TOP Nonstick palettes are an essential part of the acrylic painter's tool kit.

BOTTOM Leftovers peeled from my palette take on a sculptural form.

CHAPTER 4

CHOOSING A SUPPORT

A VAST ARRAY OF MATERIALS CAN BE USED AS SUPPORTS, or substrates, for acrylic painting. Some are familiar and often used; others are relatively untested. But because acrylics are the "freshmen" of paint formats, we do not know with a high degree of certainty how any support will weather the ravages of time and the environment. So there is no guarantee that even "safe," familiar supports will endure over the long haul. Following this logic, we can feel free to try out some nontraditional supports—and some more "speculative" combinations of supports and media. Part of art-making is about breaking through the walls and stepping out into the unknown. But if you are keen to take the road less traveled, there are some basic survival rules of which you should be mindful. Test each support you try for adhesion, do your research, and always be aware that problems may arise.

Break through the walls! Step out into the unknown!

OPPOSITE **Heather Midori Yamada, *Shifting II* (detail), 2015, liquid acrylics, acrylic gel medium, and pumice gel with Japanese washi papers on canvas, 30 x 60 inches (76 x 152 cm). Collection of the artist.**

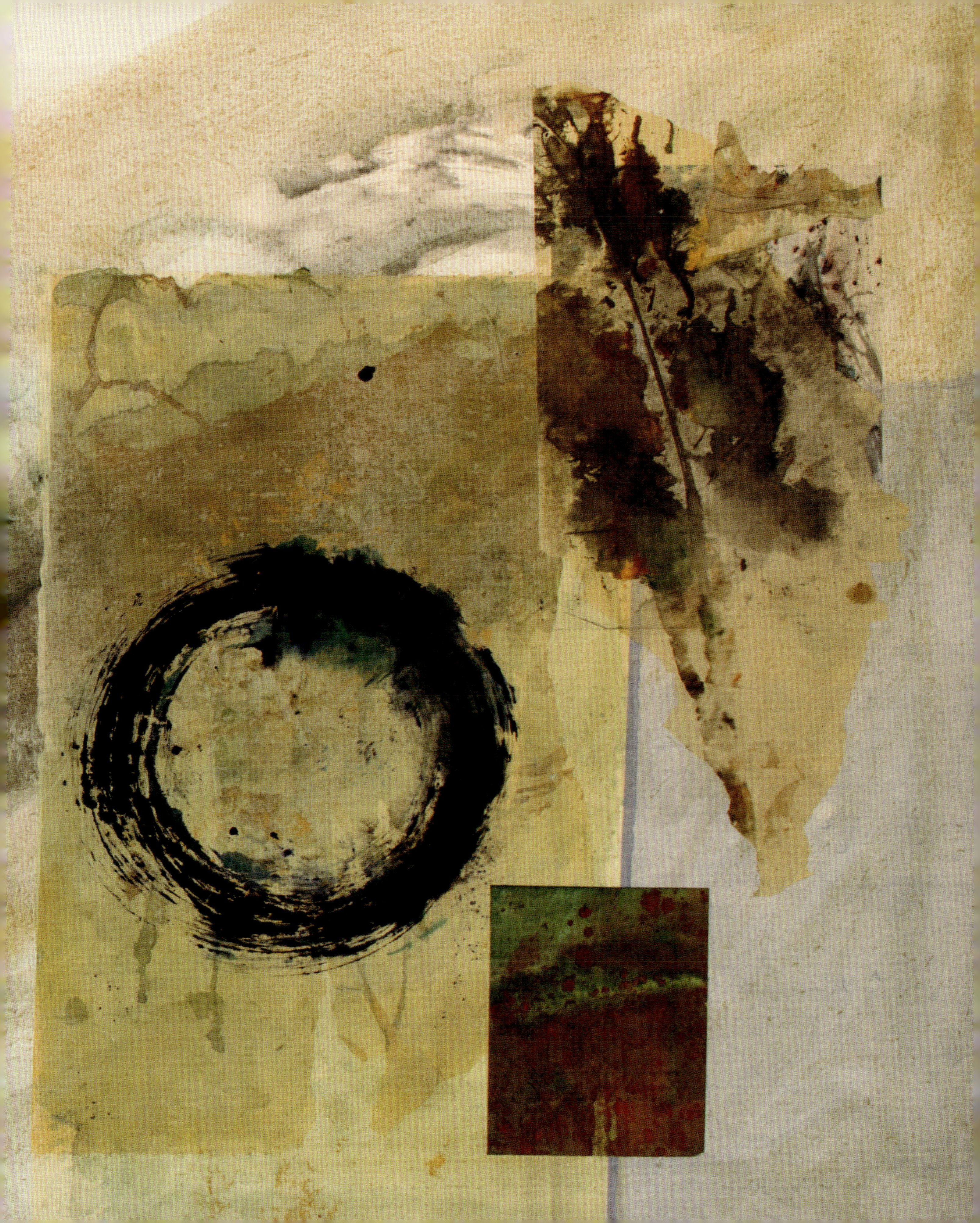

TYPES OF SUPPORTS

Regardless of the support your choose, it is essential that you do due diligence and learn whatever you can about it. This is particularly important for nontraditional supports. For what purpose was the material originally made? What is it composed of? What types of coatings, if any, are usually applied to it? What is its estimated longevity under indoor and outdoor conditions? Knowing its original purpose may be key for a successful crossover to the acrylic painting world—and to understanding what some of the potential incompatibility issues may be. For most kinds of supports, there's a variety of levels of quality and a range of prices.

Tip: Good support choices are not necessarily based on cost, but more often than not there's a price to be paid for poor ones. Make good choices!

TYPES OF SUPPORTS

Acrylic paints stick to pretty much anything (except oily, greasy, or waxy surfaces), so there is a large array of support choices. Note that the more porous the support is, the better the bond between it and the acrylic materials will be.

- Canvas and linen (pre-stretched and primed or raw—requiring you to build the stretcher and apply the gesso)
- Canvas board
- Canvas "paper"
- Other fabrics
- Wood panels (particle board [Masonite], birch, maple, plywood, or bamboo; cradled or uncradled)
- Paper (cotton, wood fiber, rice or other plant fiber; sized or unsized; hot-, cold-, or rough-pressed)
- Synthetic paper (Yupo, TerraSkin)
- Aluminum (coated or brushed)
- Plexiglas and other rigid plastics
- Fiberglass
- Mylar
- Stone
- Glass

FACTORS TO CONSIDER

Choosing the substrate is as much a matter of personal preference as it is of suitability, but here are some primary factors to consider when making your decision.

SIZE AND WEIGHT OF THE PAINTING

Consider your transportation options when choosing a support. Size matters! In my case, I can fit a 48 × 60 inch painting in my car—but bigger than that and it either gets strapped to the roof or I have to rent a van. Much as I'd like to experiment with larger pieces, I am mindful about the limitations of my studio and my car.

Even if the physical size of a painting is not an issue for you, you still have to consider weight. Weight isn't just a function of size, but also of the material the support is made from: a large canvas, for example, will weigh significantly less than a wood panel of the same dimensions. For an extremely heavy painting, you may have to take special measures to make sure it will hang securely on a wall.

APPEARANCE

Many supports become invisible once painted on. But when paint is applied thinly, scraped away, or applied in successive transparent glazes, the support may show. Canvas will reveal its weave, wood will show its grain, and aluminum will shine through.

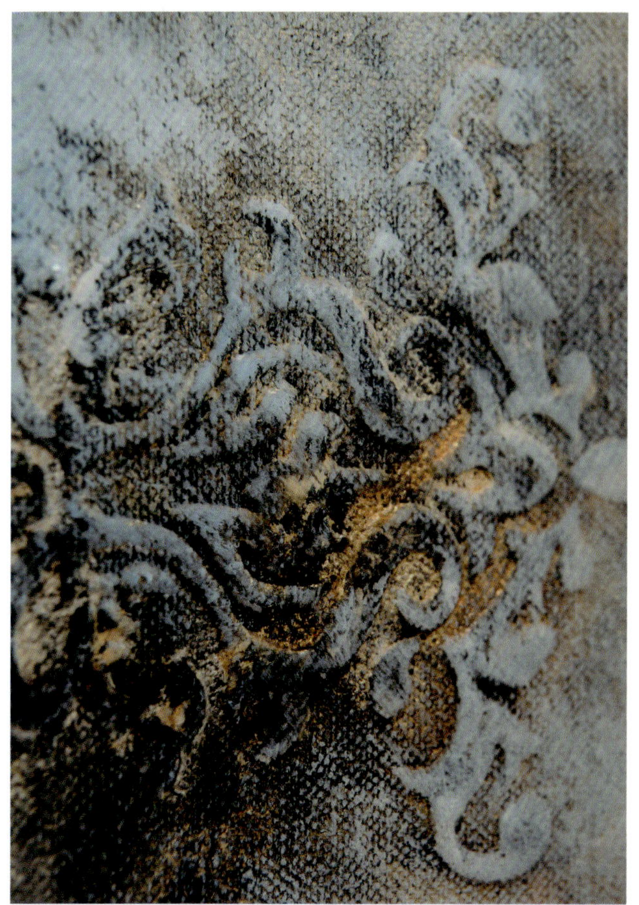

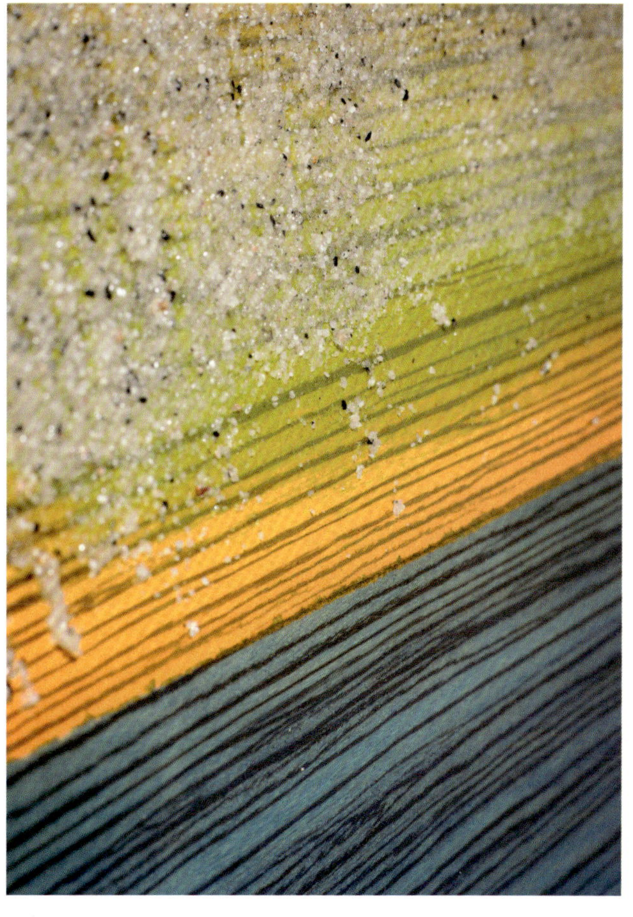

TOP From thin to thick: a water wash of color over clear gesso on a zebrawood panel plus a light dusting of nepheline gel, all revealing the beautiful grain of the wood.

BOTTOM A stenciled bas relief partially immersed in a layer of matte liquid medium and opaque colors. Despite the textures, the distinctive weave of the canvas remains visible in thinly coated areas.

TECHNIQUES USED

What are your plans for your painting? If your intention is to use a large quantity of heavy mediums or to adhere weighty collage elements, it is essential that you choose a substrate that will support them. Wood panels or tightly stretched canvas strainers are best for heavily painted pieces, not just because of their rigidity but also their porosity.

Supports that have low porosity and little tooth, like glass and aluminum, are best for relatively thin applications of paint and medium. Abrading the surface will give them more tooth and increase adhesion, but ultimately these are nonporous surfaces from which acrylics can be easily peeled or scraped away.

Tip: Acrylics will stick to just about anything, but will be much more stable on a porous, toothy surface. So give serious thought to adhesion when choosing a support.

FRAMING AND HANGING CONSIDERATIONS

Be mindful of where and how your work will be exhibited. Not every wall can support a very heavy piece, and delicate art may require special framing or mounting. It is all well and good to experiment with supports, as long as you keep a realistic eye trained on the physical constraints of the exhibition space, be it a gallery or a home.

TRANSPARENCY

If you want light to pass through your painting, you obviously shouldn't choose an opaque support. Glass and Plexiglas (or other clear, rigid plastics), either clear or frosted, are ideal for transparent painting. Be sure to test first for adhesion.

Regarding plastics, there is no limit as to what you can experiment with. Many plastics share a basic chemistry with acrylics, but there are some plastics to which acrylics will not adhere. Primers that facilitate binding between layers can solve this problem, allowing acrylics to go where they previously could not. Most of these primers are formulated for industry, not fine art, however, and it can be difficult to diagnose potential long-term compatibility issues between paint, ground, and support without detailed information on their makeup.

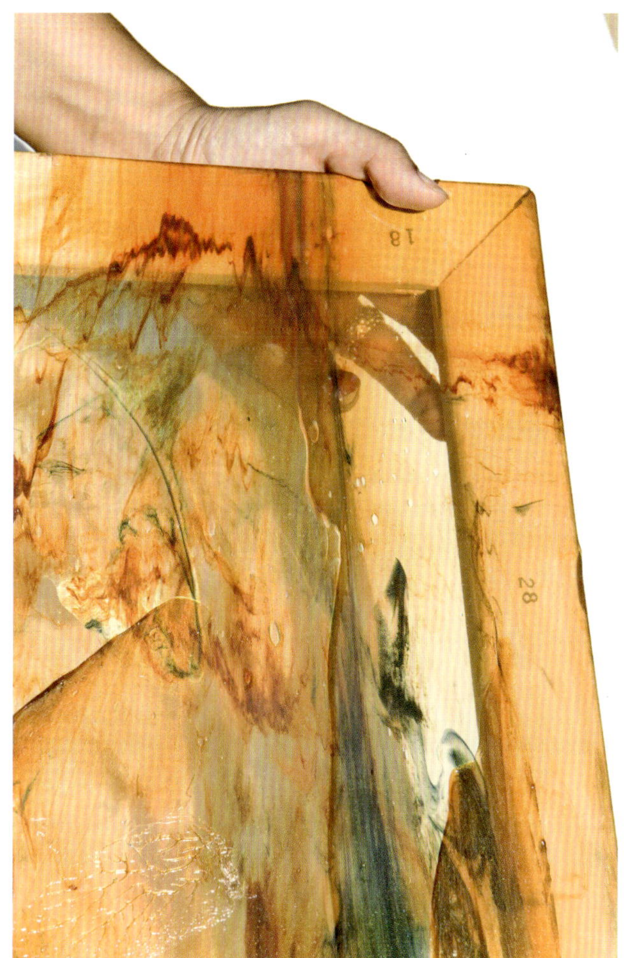

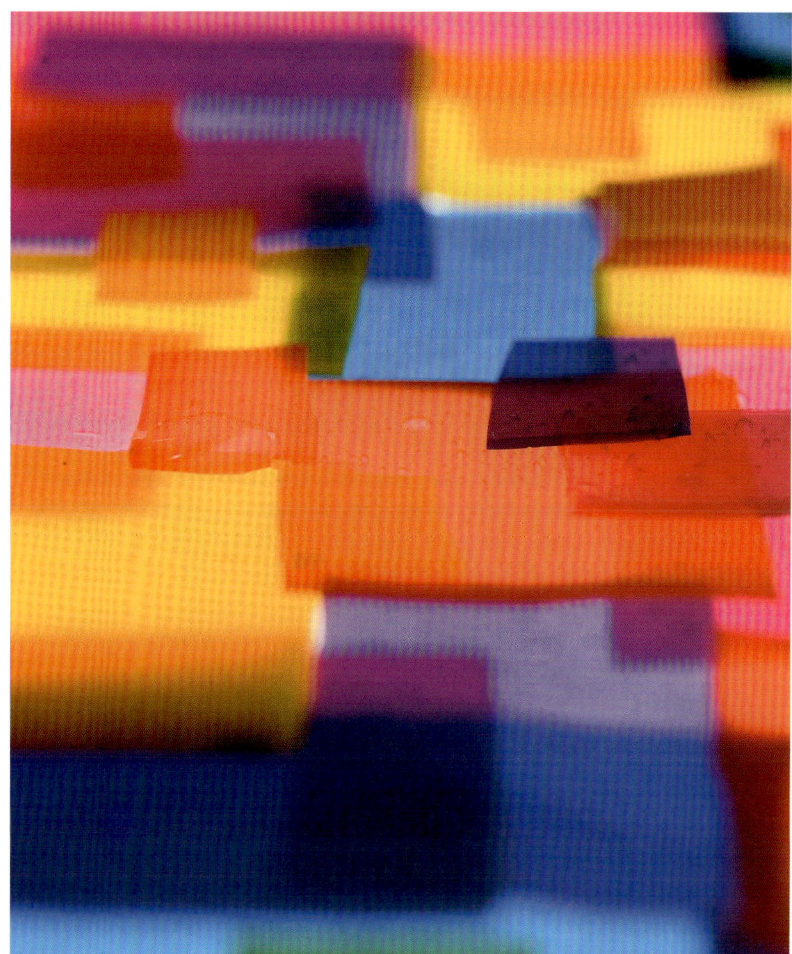

Mylar and parchment also offer varying levels of transparency. Some printing and watercolor papers (up to 140 lb) are wonderful for backlit works, as their semi-opaque surfaces diffuse the light, providing a soft, uniform glow. And acrylic paint film, all on its own, can also serve as a support.

ABOVE, LEFT This acrylic skin painting demonstrates the transparency of the paint film. The painting is composed of nothing but paint (or, in this case, mostly gel medium). The paint is the support as well as the image. This type of painting is produced by painting onto a nonstick surface; once dry, it is peeled away and stretched onto a frame.

ABOVE, RIGHT Layered rectangles of acrylic skin compose a loose mosaic on glass with background illumination.

RIGHT Purple and blue petals (acrylic skins) on Plexiglas with background illumination.

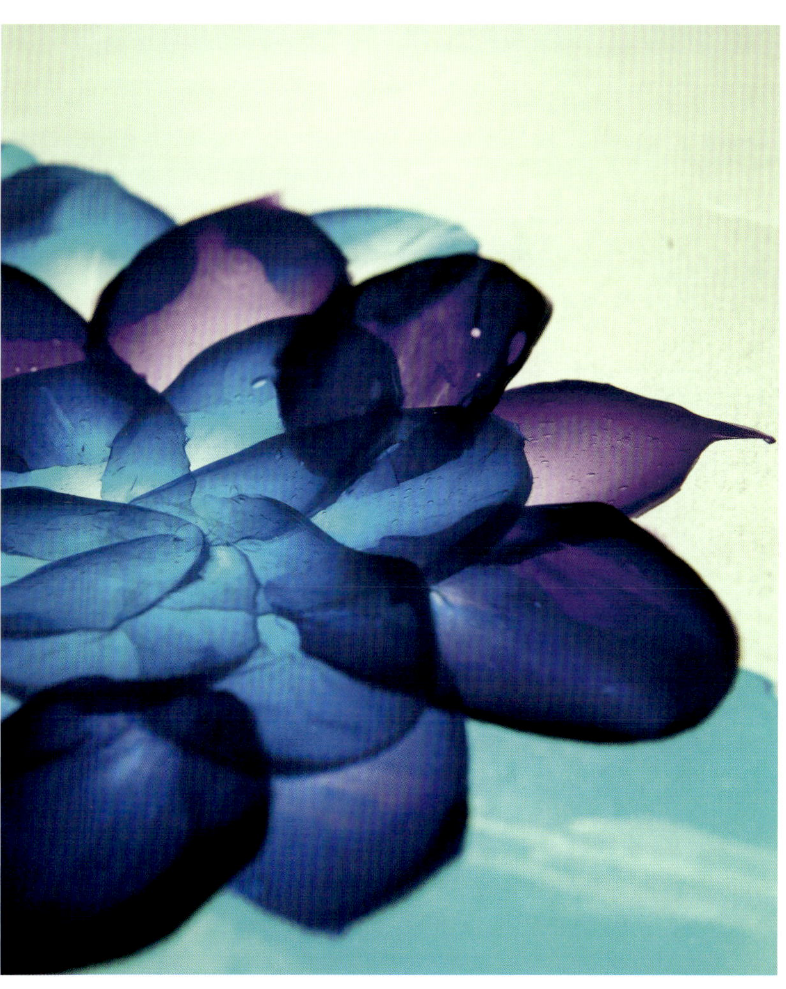

USING PAPER AS A SUPPORT FOR ACRYLICS

Some art papers are made specifically for water-based media; others are designed for printmaking. Either may be suitable for acrylics, but those made for water media will take acrylics more readily. Some printmaking papers are simply too absorbent and will not stand up to heavy water use and may tear, blotch, or absorb unevenly. To render a printmaking paper a bit more friendly to acrylics, you can lay down a light layer of undiluted liquid matte medium before beginning to paint. An acrylic gesso can also provide a suitable ground. If you do use gesso, make sure it's artist-quality gesso, not some bargain brand. Only use gesso that is labeled 100 percent acrylic.

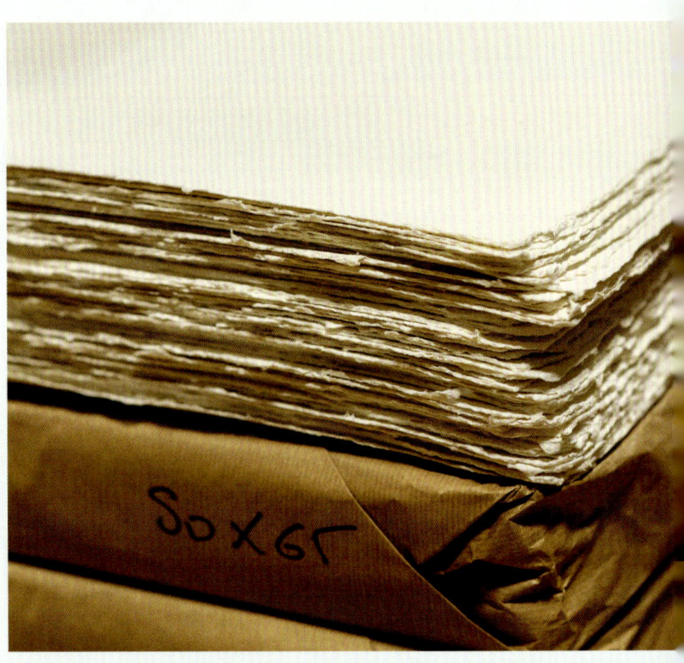

RIGHT A stack of 300 lb cold-pressed watercolor paper.

COST AND QUALITY

The cost of art supplies, including supports, is an important consideration for most artists. How much does cost matter to you? If you balk at what you think is the exorbitant price of a support or any other material, I hope you will consider this: Higher cost generally corresponds to higher quality, and the quality of materials you use has an impact on the appearance and longevity of your work. We see examples of this everywhere: Skimping on the materials used for the foundation of a house, for example, will ultimately degrade the structure, regardless of how beautifully it is decorated. The foundation will crumble, shift, or break, rendering the surface design elements worthless.

Tip: Some painting surfaces are treated with a sizing agent, which controls or inhibits the rate of water absorption. Others will soak up any amount of wetness like a sponge.

Spending more on a high-quality stretched canvas, panel, or other support can make a significant difference. Inferior supports often come with hidden problems that don't manifest themselves until long after the paint has

Tip: The longevity of your work should also be a consideration when choosing a support. Not all artworks are created with the intention that they will last forever—and most won't—but do give some thought to producing durable, well-constructed work.

dried. Canvas that is not well stretched can warp the stretcher bars, causing the canvas to sag or cave in. Wood that is not sufficiently dried and cured can warp or crack as temperatures and humidity levels change. I've even heard of cheap wooden stretchers being infested with carpenter ants and other destructive insects. Yikes!

Your materials choices should be based on purpose and performance. In reality, this is not always possible, but it's the best way to go. The bottom line is, if you can, *always* use the best materials for your particular process—the ones that are most suitable and that will give the best performance. (This is often a matter of price, but not always.) If you want your work to remain stable over the long term—an important consideration when selling a piece—then materials matter.

We primarily create art to share it, exhibit it, or sell it. If the high cost of good-quality materials is upsetting you, do keep in mind that the initial outlay for even the most expensive materials may represent just a fraction of the ultimate selling price of your artwork.

BRAND LOYALTY

Some artists stick to particular brands when purchasing materials. Brand loyalty can be based on many factors, including price, quality, perceived prestige, familiarity, practicality—even a desire to support the local economy. But for the same kinds of materials at similar price points, aren't all art-supplies brands more or less the same? The answer isn't clear. It is not possible to accurately compare every aspect of different makers' products, because most manufacturing specifics—raw materials, recipes, production processes—are proprietary. The only thing you can say for sure is that for products designed for the same purpose, the basic ingredients used to make them will be about the same.

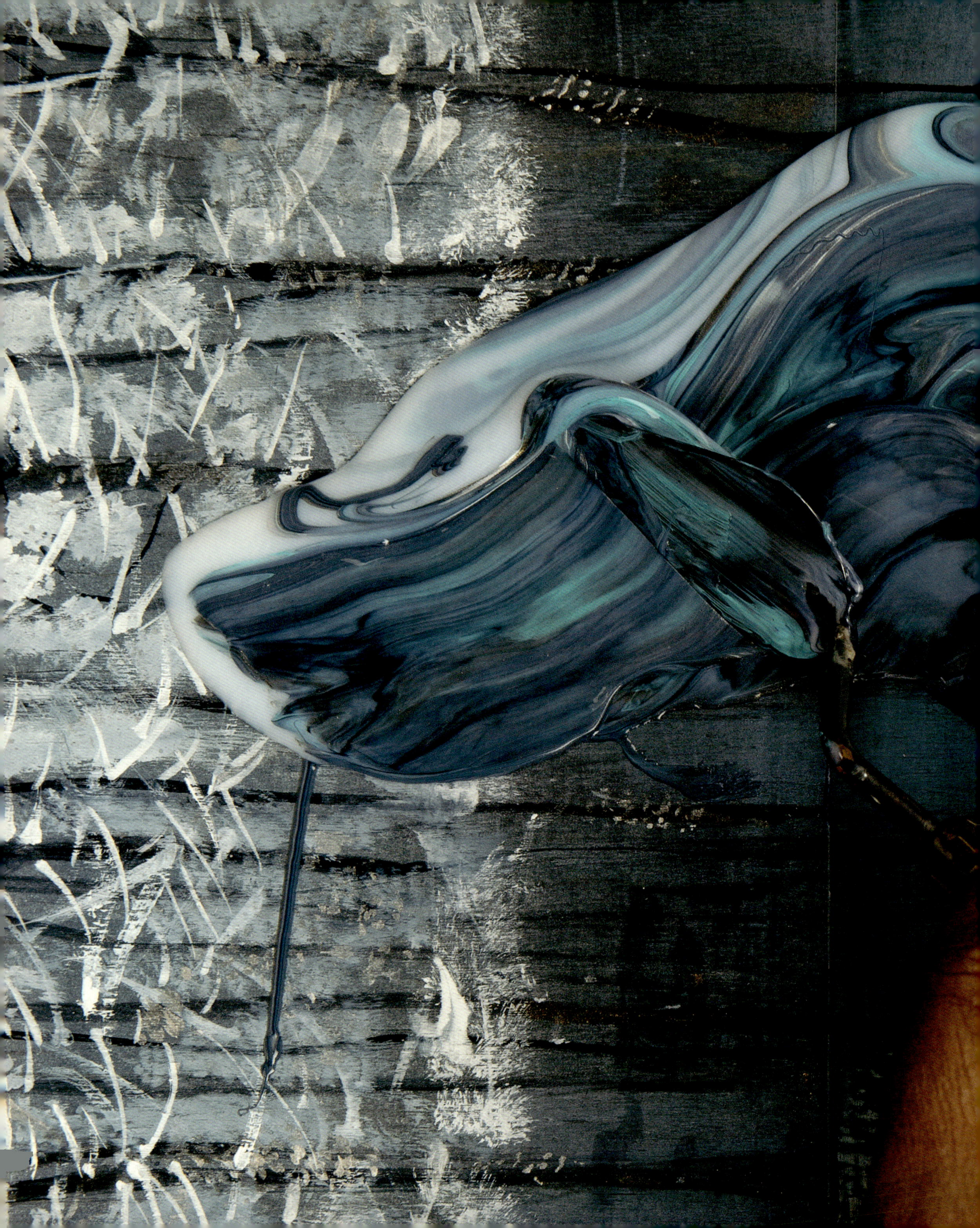

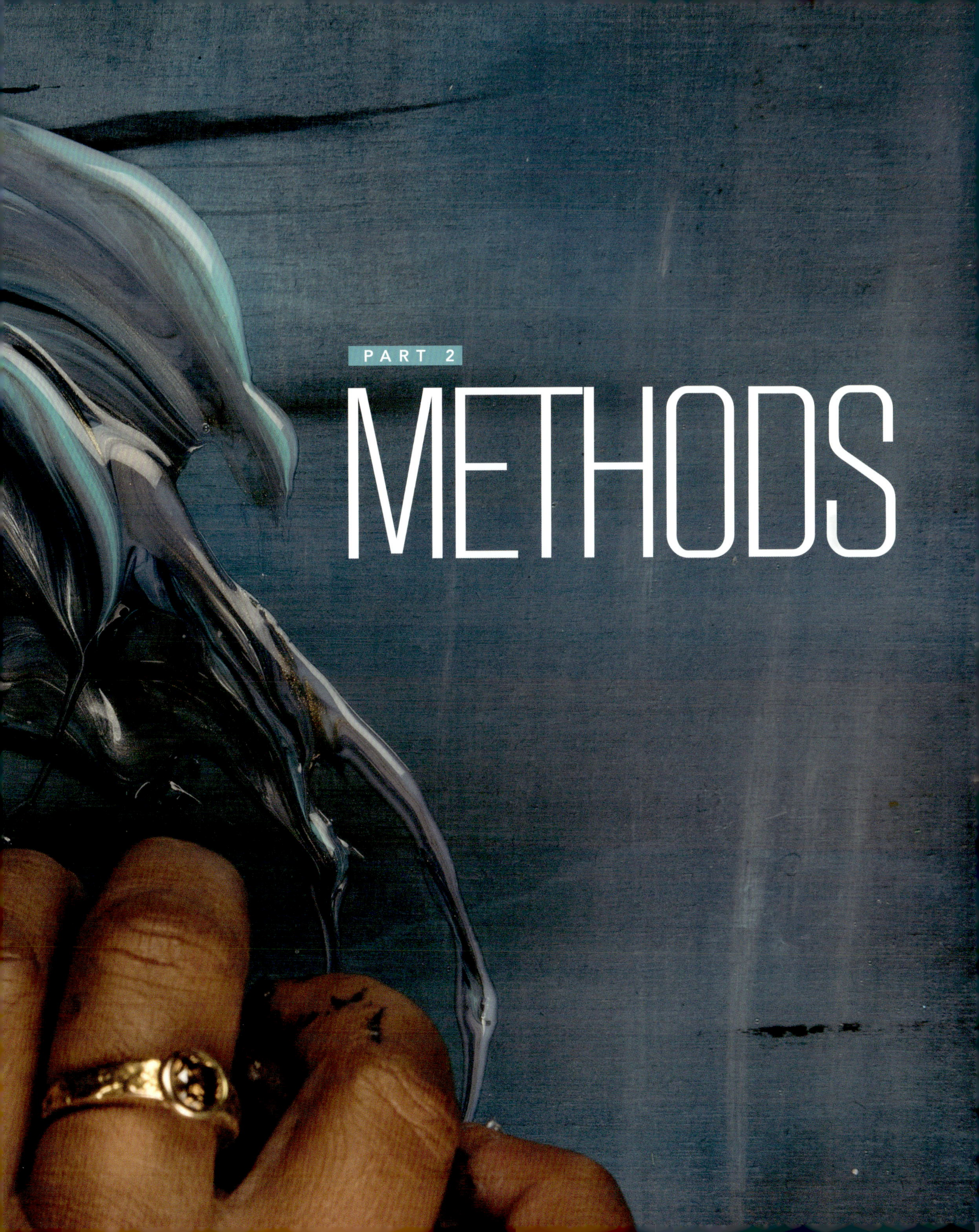

PART 2

METHODS

CHAPTER 5

INCORPORATING MEDIUMS INTO YOUR WORK

THE WORLD OF MODERN ACRYLICS has become a world of *more*. More excitement, more possibility, more fun. Acrylic mediums have not only made acrylic painting more versatile—they have made it better. They are gateways to unlocking creative potential. The ever-increasing variety of materials, however, can also make for more frequent episodes of confusion and frustration. To avoid being overwhelmed by this glut of acrylic toys, let's take a pragmatic approach and carefully map our way through their potential.

Methods, techniques, and applications are essentially recipes—combinations of chemistry and creativity. If you approach constructing a painting as a formula, you will more easily find the correct method to follow with the materials you've chosen. As every adventurous cook knows, though, recipes are guides rather than rules, and improvisation can—and usually will—lead to new discoveries.

Mediums are gateways to unlocking creative potential.

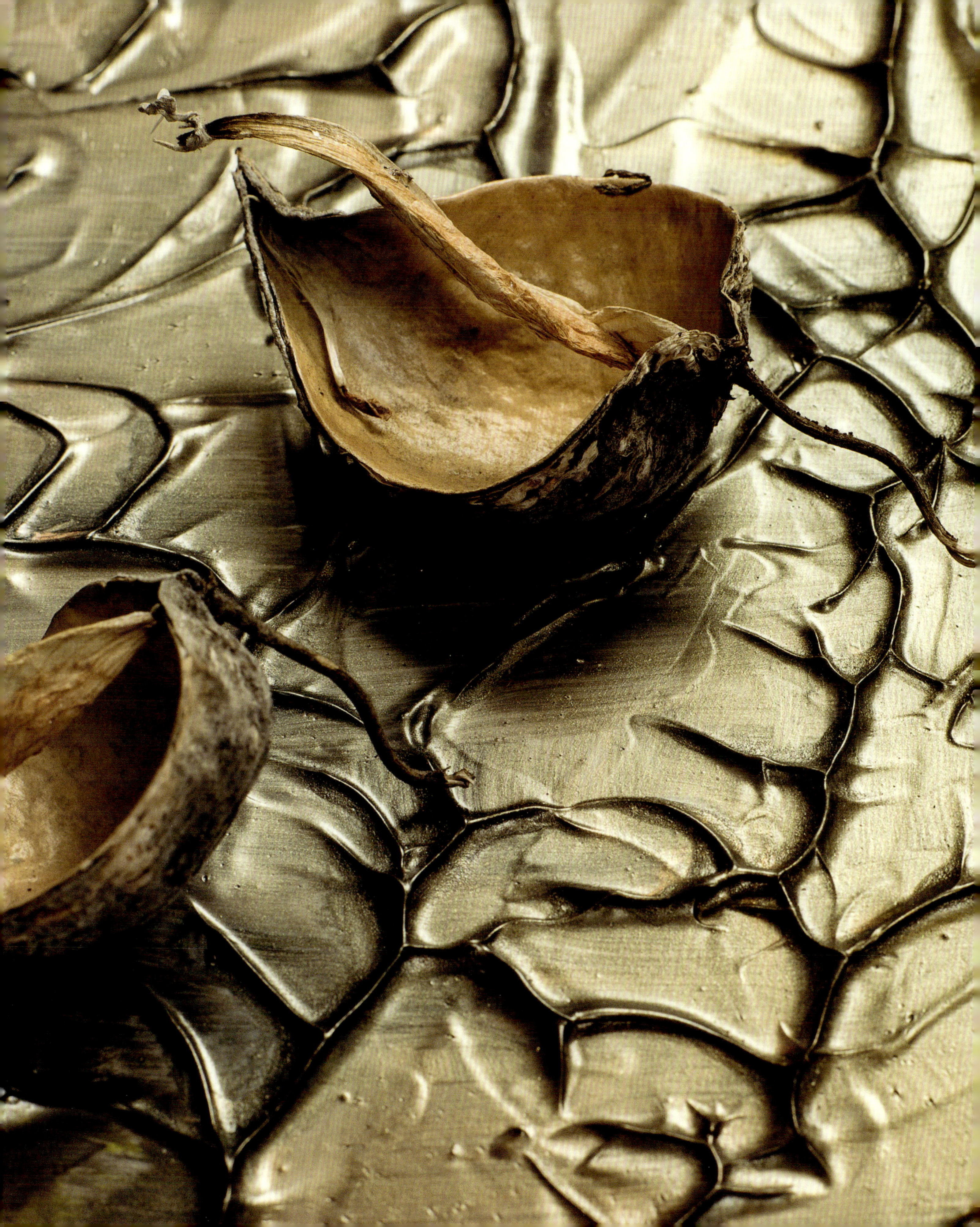

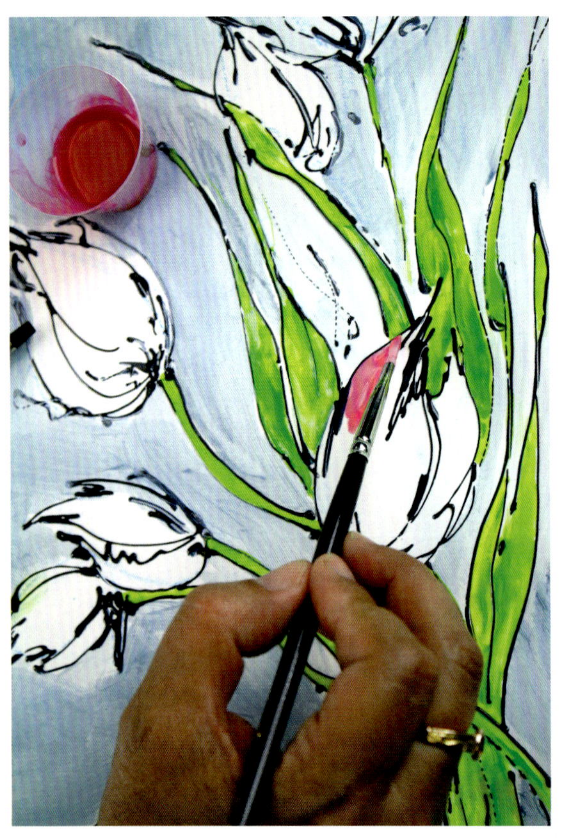

Even when rooted in some fairly solid rules, process is ultimately a personal journey, one that develops through inspiration, practice, and creativity. Because acrylic mediums are so many and so varied, it is not possible to compile a recipe book that fully covers the possibilities they offer for every individual artist, especially since there are usually no limits as to how much or how many can be used. So, instead, this chapter provides a sampling of some typical (and some novel) paint and medium combinations to inspire you to create your own. Always remember that practice, attention to foundational information, and experimentation are all crucial to achieving success.

LEFT Painting one of the tulip sketches opposite.

ONE SKETCH—MULTIPLE POSSIBILITIES

The sketch of a bunch of tulips opposite, top left, is the basis for the interpretations—using many different mediums and methods—that follow. These provide just a small sample of the myriad ways that mediums can shape your creative ideas.

OPPOSITE, TOP CENTER An *alla prima* interpretation—all paint, no mediums. Here, the acrylic colors were primarily applied with a palette knife, without dilution with water or mediums.

OPPOSITE, TOP RIGHT In this acrylic ink interpretation of the sketch, the colors were thinned to watercolor consistency using water as the medium.

OPPOSITE, CENTER ROW FROM LEFT TO RIGHT Here, the sketch was indirectly transferred onto paper; granular mediums were loosely applied, as were thin washes of acrylic color. • In this version, the background was textured with modeling paste; gel gloss and matte mediums were tinted with acrylic colors. • Here, tinted pouring medium was eased into areas delineated by extruded lines of tinted gel medium.

OPPOSITE, BOTTOM ROW FROM LEFT TO RIGHT Soft pastels were used to color the sketch on a paper coated with a fine layer of acrylic dry media ground. • Here, dry-brushed colors and light washes were applied over a liquid medium resist. • Cut and fused acrylic skins applied to a painted ground create a more three-dimensional rendition of the tulip motif.

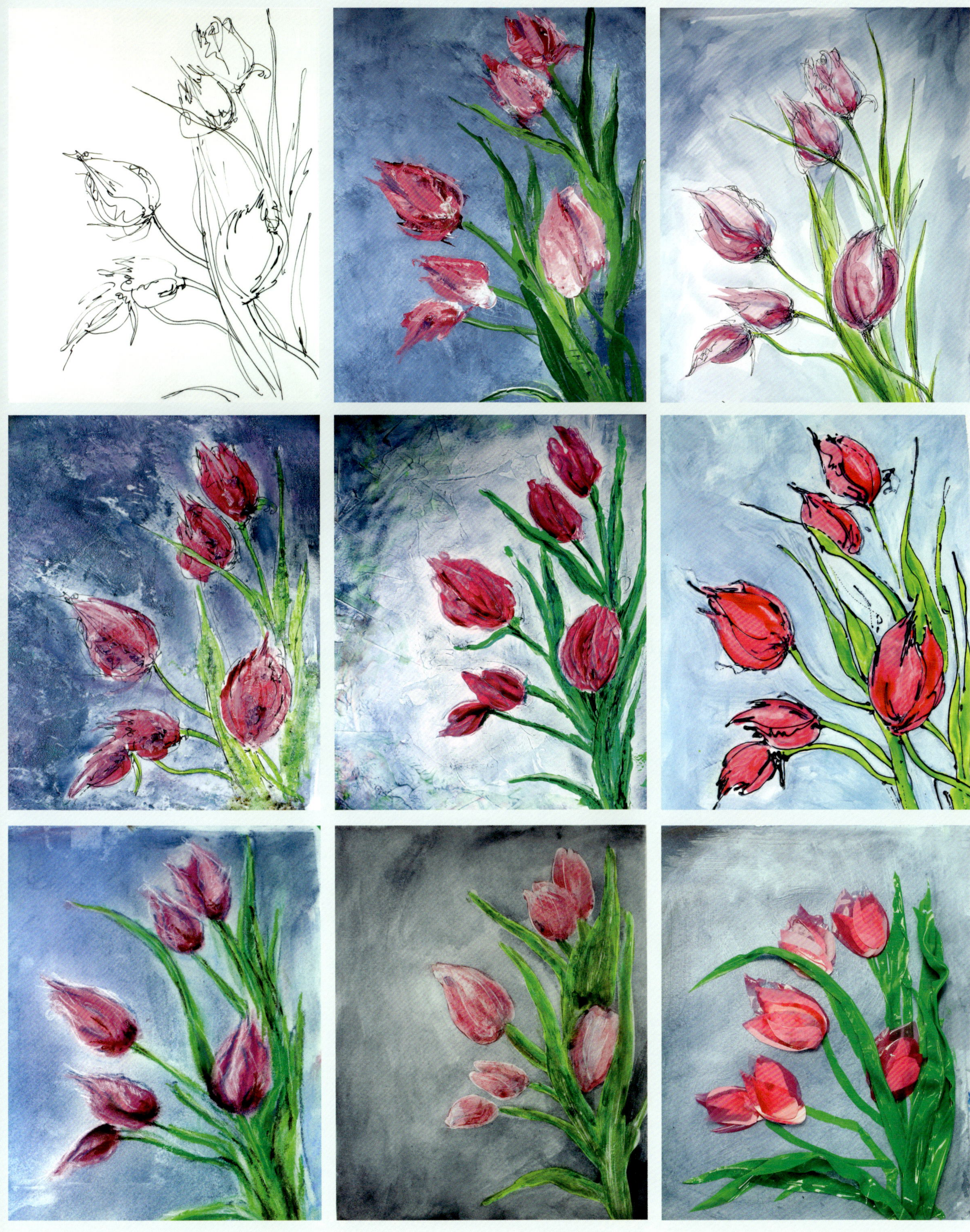

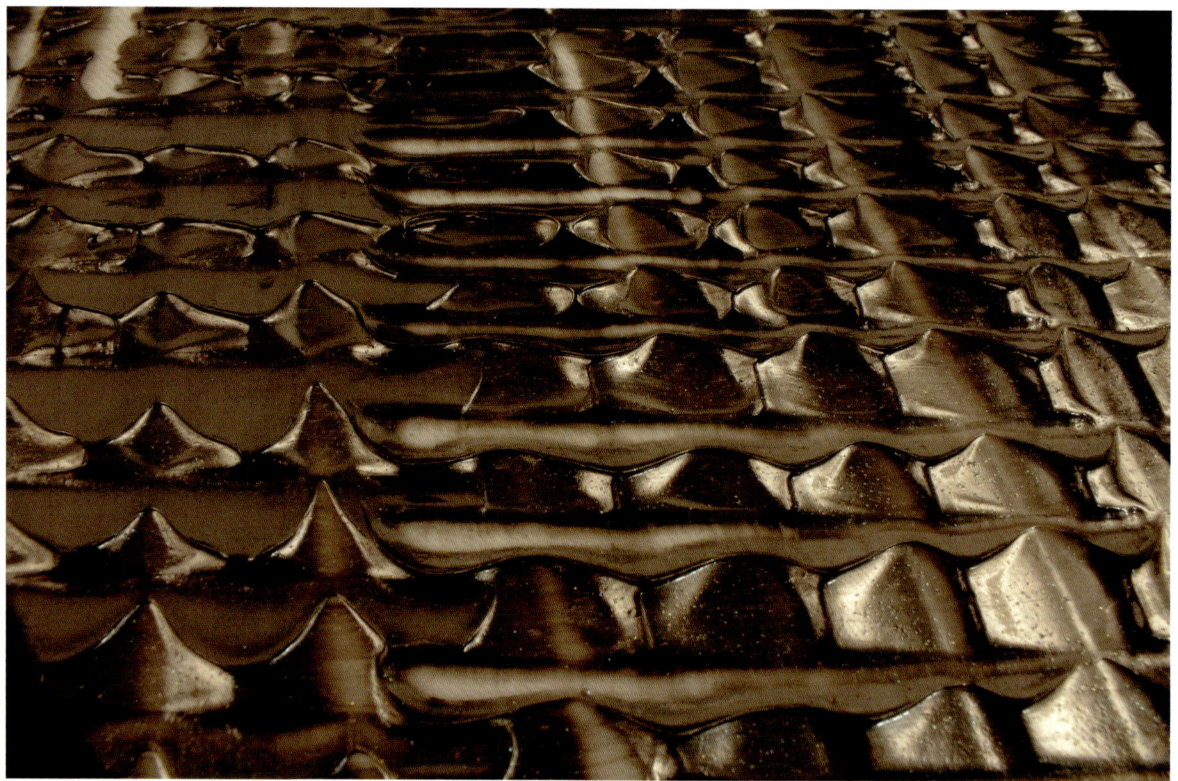

ABOVE Gloss gel tinted with carbon black, liquid mirror, and hologram pearl (spectral color) formed into sharp ridges with the side of a large-format palette knife.

SUBTLETY AND BOLDNESS

There is a common misconception that mediums are primarily used to produce abstract, expressionistic works and have little or no place in more formal or realistic painting. I hope to dispel this fallacy by showing examples of artwork, produced using mediums, that span a wide array of styles.

Mediums are capable of more than moody color fields and dripped spatters. Some, such as glazing or polymer mediums, are actually essential to achieving a high degree of subtlety in detailed compositions. Because professional-grade acrylics have a very high pigment load, diluting them with a clear, thin medium gives you greater facility when repeatedly glazing, allowing you to build rich color in a controlled, gradual manner.

In such cases, mediums are not always visually evident in a finished painting. There are telltale signs, however, that give them away: transparency, surface textures, and changes to luster. Of course, the more obviously textural mediums—those with granular particulate or the capability of holding high, tight peaks—are the easiest mediums to spot.

WATCHING PAINT DRY

Before going further, let's take a look at one of the biggest challenges you'll face when working with mediums: drying time. A real learning curve is involved, and you should observe and annotate how long the drying process takes, the factors that affect it, and the differences in appearance from wet to dry. These notes will be valuable as you progress through your education in mediums.

Acrylic colors dry quickly. But the time it takes a color to dry can change significantly when mediums are added. When any quantity of medium is added to a color, the drying time increases. This has two causes: First, adding medium to the mix means adding water, since water is a component of any medium. This increases moisture content and keeps the paint hydrated. Second, the solids (acrylic polymers) in the medium bulk up the acrylic mass, which slows down evaporation.

Be aware, however, that not all of this additional time is working time. Once the surface of the medium-enriched paint has begun to form a skin, it should be left to dry, undisturbed. Continuing to poke at it will only cause mess and frustration. The curing time for mediums is similar to that of pure colors and is dependent on the thickness of the film, the porosity of the support, and the ambient humidity.

The drying time of mediums varies. Liquid mediums will dry more rapidly than their gel counterparts, but they don't all dry at the same rate. One simple test to determine whether or not the paint film is still drying is to place your hand on the painted surface and then on the back of the support. If the paint and support feel cool, evaporation is still taking place.

Ideally, a thick layer of medium should be left to dry overnight or longer before adding new paint layers. This is of particular importance when you're working with transparent and translucent layers. If you fail to wait the appropriate amount of time, the channels through which volatiles are evaporating in the underlying film may get blocked, causing a milkiness that in some cases will not disappear.

Some mediums contain "hidden" components. For instance, some brands of glazing medium contain a larger quantity of humectant than others, which increases the open time of the paint. Such glazing mediums require longer drying time between layers to ensure proper adhesion. Adding a wet layer over uncured glazing medium with high humectant content can cause the new layer to blend with the underlying paint. These factors vary from brand to brand, so testing is necessary when you incorporate a new player into your system.

ANTICIPATING CHANGE (EXPECT THE UNEXPECTED)

The other big challenge you'll encounter when working with mediums is anticipating the visible change that occurs as a medium mixed with color dries. This can be really tricky.

Mediums are white. That is to say, they appear white in their wet state, but with the exceptions of those that contain pigment, they will dry clear. The change in color from wet to dry is due to the clarification of the paint film as the "white" water evaporates. Colors deepen, and those that are transparent show their true depth. (Note that the shift is more dramatic in economy paints because of their lower pigment concentration and higher water content.)

This change from wet to dry may not seem so dramatic when you're just adding a small glob of gel or a squirt of liquid medium to your color mix; however, the more you add, the whiter your color will appear. This can throw you off as you strive to blend your tones or to perfect a glaze. Moreover, the more medium you add, the longer the mix will take to dry, meaning you won't see the true result for minutes or hours. With a very thick application, you may not see the full change for several days.

If you are trepidatious about making the commitment to heavily dilute your color with medium, prepare a test batch of your mixture. And don't assume that a given kind of medium will be the same from brand to brand. One company's gloss gel will not be identical to another's. The ingredients and formula may be similar, but results may vary. So testing is important and necessary *whenever* you incorporate a new product into your system. Once you become accustomed both to the time it takes for mediums to dry and to the changes that occur over that span, you will find the process of working with mediums much easier.

CHANGES FROM WET TO DRY

A color mixed with a generous quantity of medium will appear very light and opaque when wet, gradually becoming deeper in tone and more transparent as it dries. Here are some examples of this transformation, showing small amounts of a color (phthalo turquoise) mixed with large amounts of various mediums.

LIQUID MEDIUM + COLOR

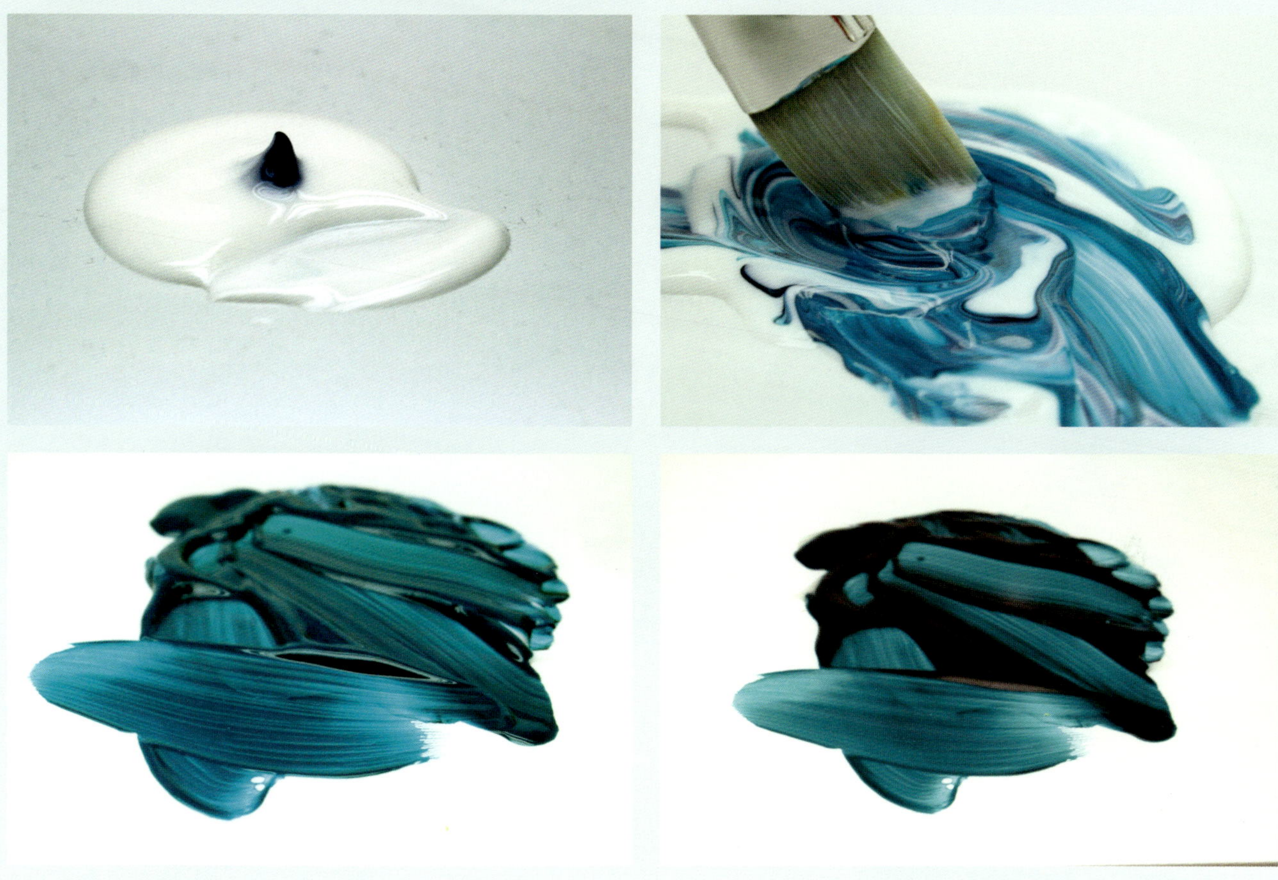

TOP, LEFT A drop of phthalo turquoise in matte polymer medium.

TOP, RIGHT The phthalo turquoise and matte polymer medium, loosely mixed.

BOTTOM, LEFT The fully blended mixture when wet.

BOTTOM, RIGHT The fully blended mixture when dry.

GEL MEDIUM + COLOR

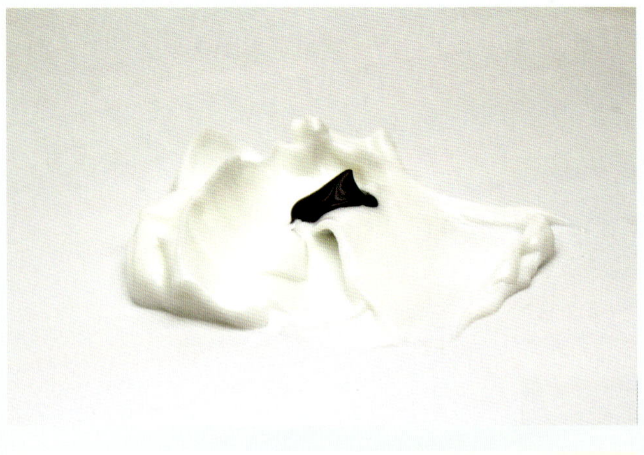

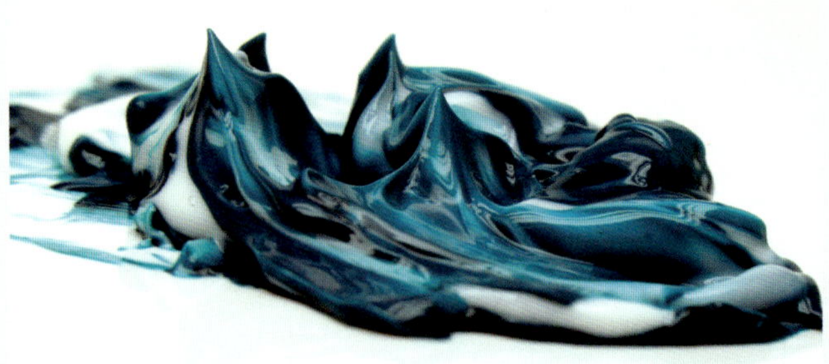

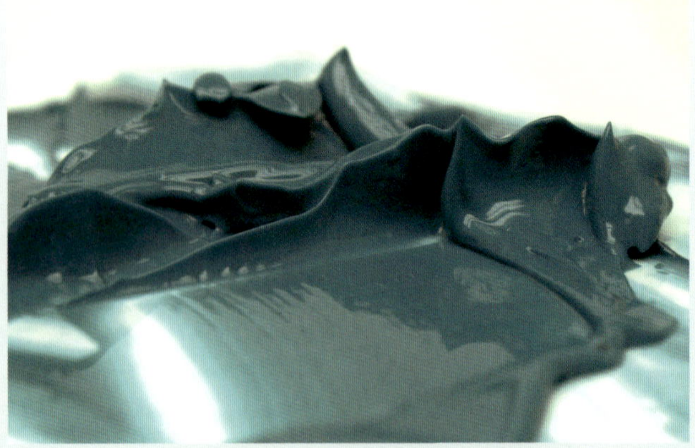

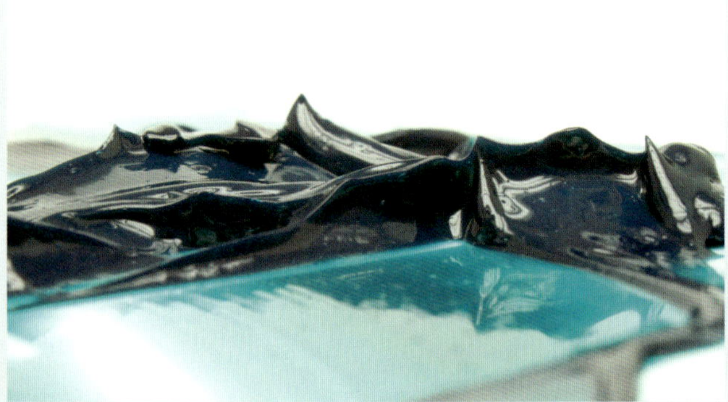

TOP, LEFT A drop of phthalo turquoise in gloss gel medium.

TOP, RIGHT The phthalo turquoise and gloss gel, loosely mixed.

BOTTOM, LEFT A large amount of medium, tinted with a single drop of this rich color, will appear quite pale when wet.

BOTTOM, RIGHT The fully blended mixture when dry.

GRANULAR MEDIUM + COLOR

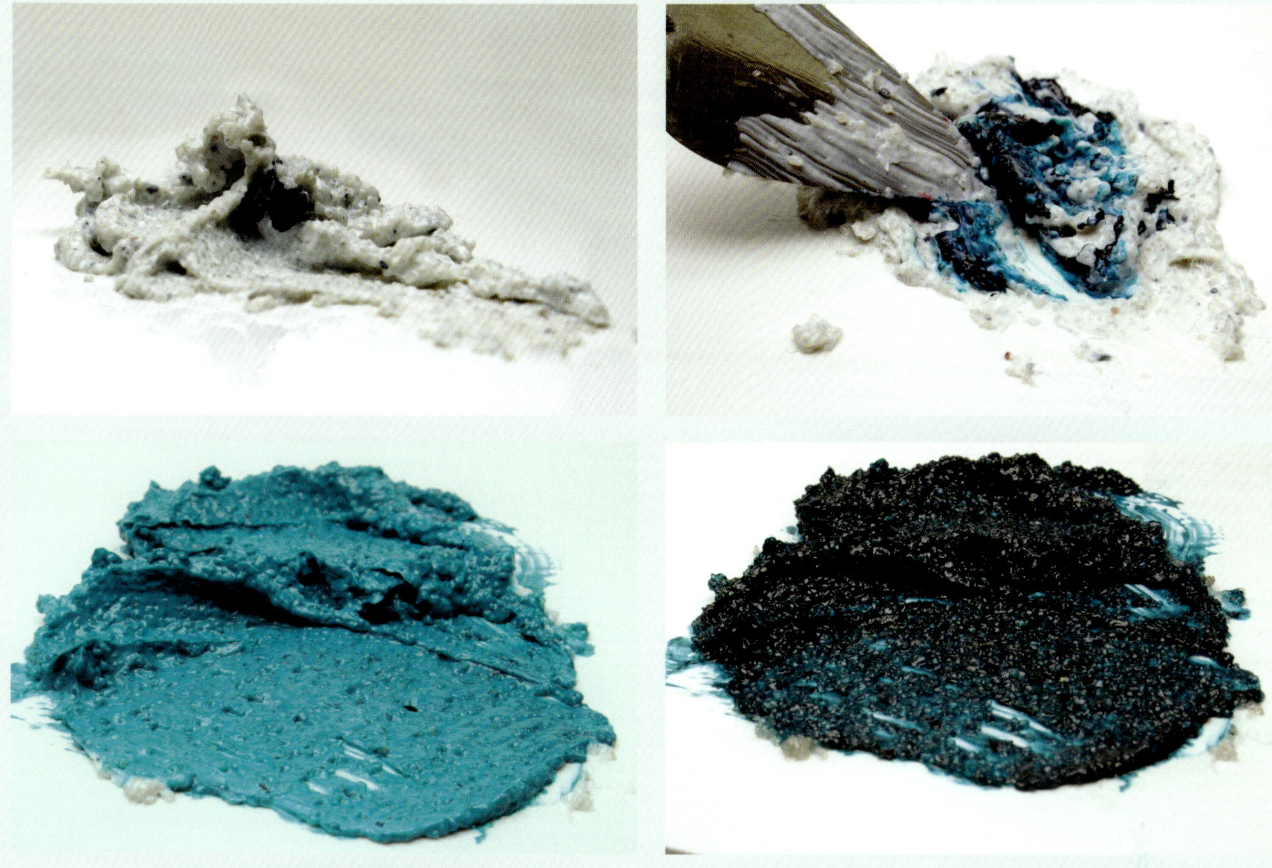

TOP, LEFT A drop of phthalo turquoise in nepheline gel extra coarse.

TOP, RIGHT Blending the turquoise into the nepheline gel.

BOTTOM, LEFT The fully blended mixture when wet.

BOTTOM, RIGHT The fully blended mixture when dry.

CHAPTER 6

USING ACRYLIC GROUNDS

GETTING YOUR SUPPORT READY FOR PAINTING is not always necessary, as most stretched canvases sold commercially are already primed and ready to go. Other surfaces, like raw canvas, wood panels, or paper may require a bit of preparation before you can paint on them.

Making the decision to prime a surface requires some understanding of the material: how it accepts paint (and water), its flexibility or rigidity, and how porous it is.

Although we tend to think of white acrylic gesso as the singular primer for acrylic painting, it is just one of many choices for an effective first layer. Gesso itself comes in several varieties, ranging from translucent, to neutral opaque gray gesso, to black gesso. There are thick grounds that can be carved, granular grounds, and super-absorbent grounds. In fact, many acrylic mediums that are not labeled as grounds can be used as such.

To prime or not to prime? It's your decision.

Lorena *Kloosterboer, Twice As Much,* 2005, acrylic on gessoed panel, 24 x 24 inches (61 x 61 cm). Collection of the artist.

Kloosterboer prepares her panel beforehand with several layers of gesso, wet-sanding each layer smooth before adding the next, to provide a completely uniform, satiny surface. The process she followed when making this work is explained on page 213.

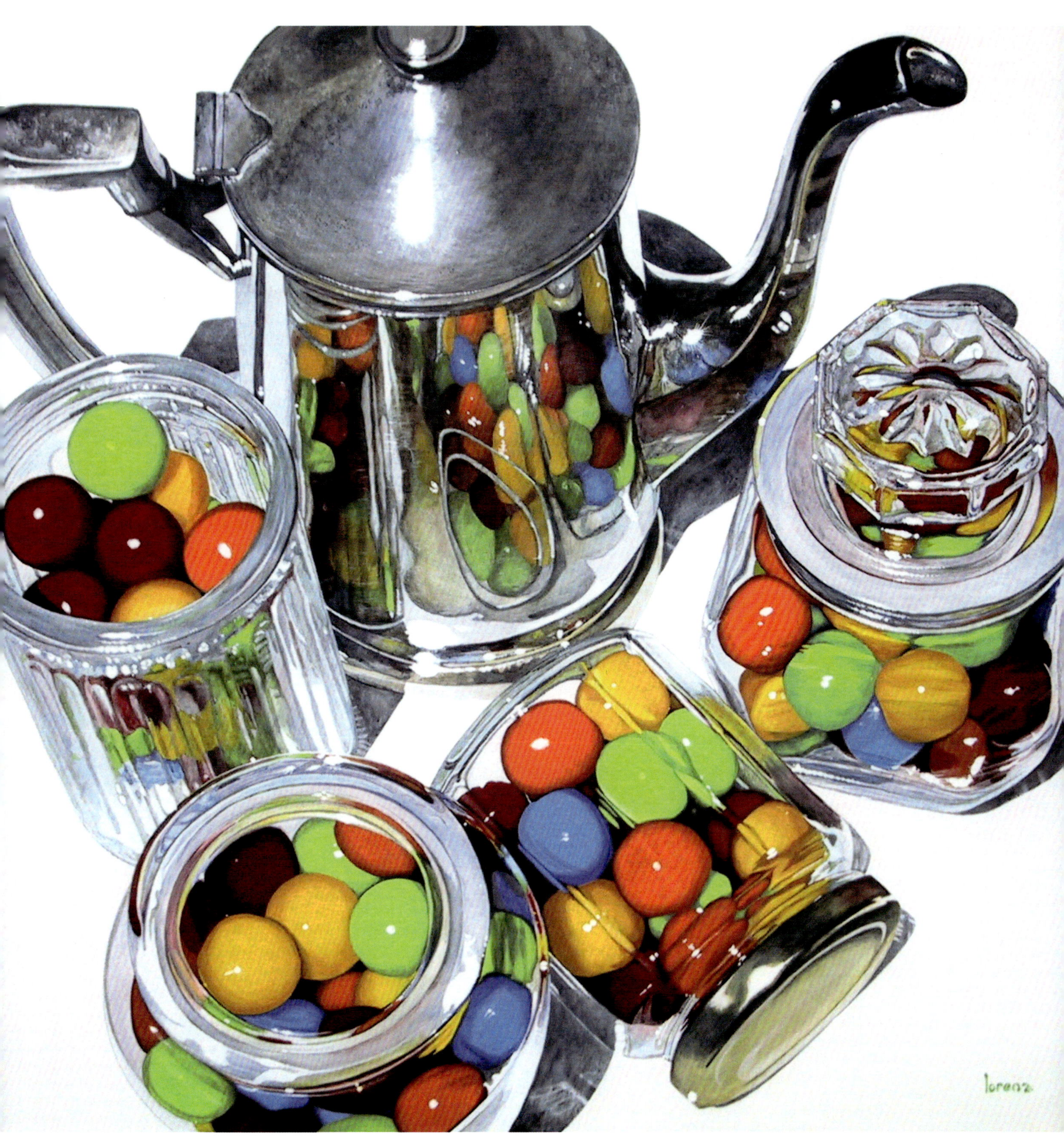
lorenz

The decision to prime is not always one of necessity, but one of choice. There are instances in which it is imperative to seal or neutralize a surface and others where the raw acrylic will bond quite comfortably with the substrate. Know your substrate, know your materials, and make your decision according to the best projected scenario for process, visual appeal, and longevity. The ground is your first layer of protection and will dictate the appearance and behavior of the subsequent paint layers. Here are some things to consider when making your choice:

- *Controlling porosity.* Priming your canvas will better prepare it to accept acrylic paint and keep the paint from soaking into the weave of the canvas. Porous substrates such as canvas, wood, and paper can absorb a lot of moisture from your paints. An acrylic dispersion ground (aka acrylic gesso) will block the flow and prevent the migration of the paint into the substrate. Also, because of their absorbency, unprimed canvas and most other very porous supports allow blending only in wet applications. A primed surface is far less absorbent, and paint can be worked for several minutes before setting.

1. *Enhancing reflectivity.* Not only does priming provide a ground surface for paint to adhere to, but depending on its color, it reflects or absorbs light. A traditional white ground provides a reflective layer that enhances color vibrancy.

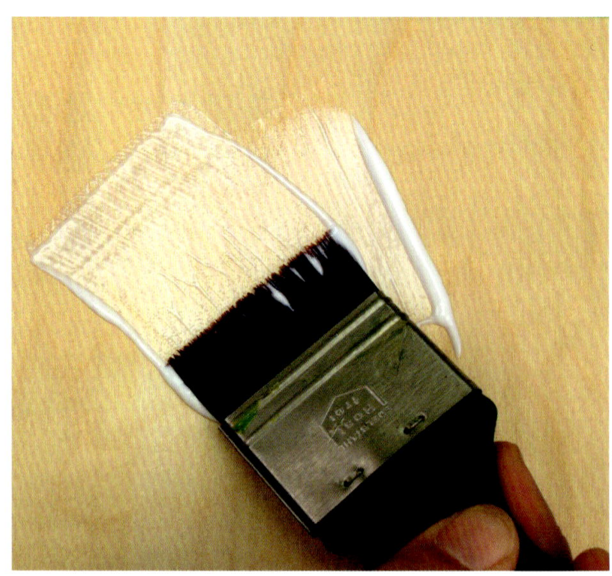

ABOVE Applying clear gesso to a raw wood panel.

- *Revealing true colors.* Black, gray, and burnt umber gessoes are neutral surfaces that allow the true hues of the colors to show themselves. Be mindful of the relative coverage of colors, as opaque colors show more clearly against these neutral grounds than do transparent ones.

If you are going to put a ground on your support, first know your support, then assess which types of grounds are best for the materials and techniques you'll be using. (If you are unsure whether a ground is necessary or desirable, err on the side of caution by applying one or two layers of gesso, because you can't go back and do it after the fact. There are no do-overs for priming once the paint is on.) Refer to the table on pages 134–135 to find the suggested ground and preparation procedure for your chosen support and process; this table gives you the broad strokes, but it's also important to use common sense.

QUALITY FIRST!

It's critically important to begin with a good ground material. Beware of a gesso that does not have the words *100 percent acrylic* on the label. If it only states that it is *for* acrylic, but does not explicitly state that it *is* acrylic, then it is not. Economy gessoes are not always 100 percent acrylic. If they are not, they can yellow, and some mixed gessoes (not 100 percent) are not flexible. Because canvas *is* flexible, the ground might flake and crack. If you are unsure regarding the flexibility of your gesso, test it beforehand. Pour a pool of it onto a flexible surface (like paper); once dry, it should flex with the paper. If it cracks and flakes away, just imagine what it might do to your painting!

RIGHT Low-quality gesso is brittle and will crack in thick layers. This is *not* the type of product you want to create the foundation of your painting!

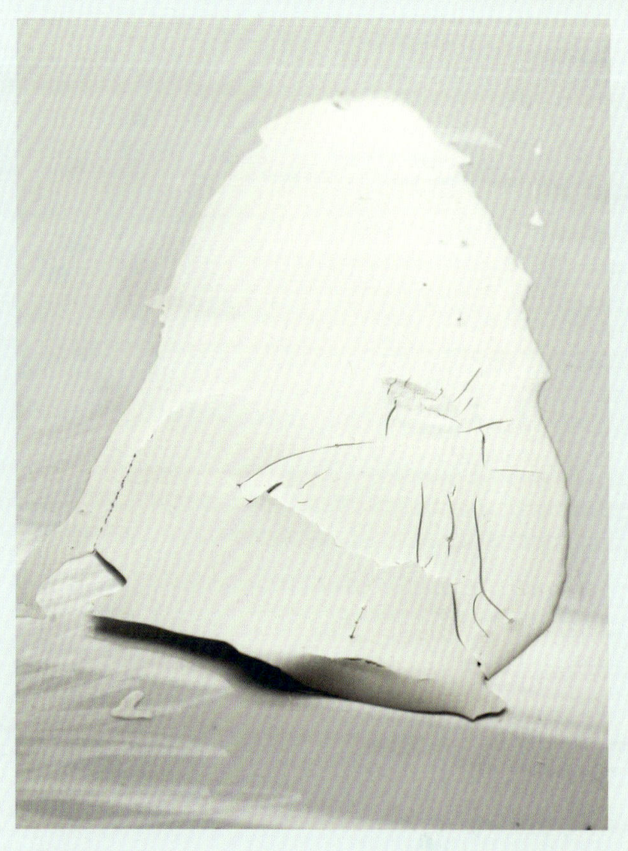

Tip: Never, ever, use wall-paint primer to prime a support. An artist's acrylic gesso (more accurately called an acrylic dispersion ground or acrylic emulsion ground) is designed to be compatible with flexible and rigid substrates, able to withstand the mechanical changes that take place during the curing process of the paint as well as shifts in the support caused by transportation or dramatic changes in humidity and temperature. Wall-paint primer, by contrast, is designed for a flat, rigid surface; it does not contain as much binder and thus will not adhere as permanently to the support, nor will it be flexible enough to weather any structural changes. This type of primer will crack or flake off and will ultimately have a negative effect on the performance, permanence, and structural integrity of the piece.

PREPARING SUPPORTS

(Adhesion: E = Excellent, M = Moderate, P = Poor)

SUPPORT	ADHESION	CLEANING	PRIMING	ADDITIONAL PREPARATION	DON'TS AND POTENTIAL PROBLEMS
CANVAS OR LINEN	E	None	Gesso: 1–5 coats depending on application; for a smoother ground, sand between layers. Acrylic polymer: A single coat of polymer will effectively inhibit water and paint absorption and increase the flow of paint on the surface while allowing the appearance of the canvas to show through.	If priming with a polymer medium, use the matte variety to provide more tooth and improve paint adhesion.	Do not use a non-flexible primer (e.g., wall primer, house paint) on canvas.
WATER-MEDIA PAPER	E	None	None required. Although not a necessary step, you may apply a single, thin layer of matte polymer medium to inhibit absorption of the first layer of paint.	Stretch paper (wet paper, tape down, allow to dry). Tape lighter-weight paper (140 lb or less) to hard surface while painting.	When framing behind glass, allow a space between the glass and the frame, as acrylic will stick to glass. Non-cured paint will continue to release volatiles and should not be framed behind glass until curing is complete.
WOOD (KILN DRIED)	E	Wipe to remove residual sawdust or dirt.	Gesso: Apply one thin coat, sand, then apply second coat. Clear coat: Apply two coats of polymer medium in thin layers; allow proper drying time between layers.	Seal any knots with shellac or similar sealant and allow 24 hours to dry before painting. Cradled panels and layered wood panels help guard against expansion and contraction of the wood.	Kiln drying protects against mildew and warping and makes the wood more absorbent.
MEDIUM-DENSITY FIBERBOARD [MDF]	E	Use alcohol-based cleanser to wash away any residual oil or grease.	Apply gesso or paint directly to board in a thin layer to seal.	Seal sides and back of panel with paint, gesso, or polymer medium to guard against water absorption.	Do not apply a very wet layer to raw board as this can cause it to swell and pucker. When sanding or cutting MDF, wear protective eyewear and a mask.
METAL	M	Use alcohol-based cleanser to wash away any residual oil or grease.	Apply metal primer. Use the appropriate primer for ferrous and nonferrous metals.	Abrading by sandblasting, etching, or wet sanding will greatly improve adhesion.	Because metal conducts heat, metal surfaces should be placed away from sources of heat, which may soften the acrylic, making it more susceptible to damage.
PLEXIGLAS	E	Use glass cleaner to remove residual oil or grease.	Clear: Apply single coat of matte polymer medium. Frosted: No priming necessary.	An alternative method for adding tooth to this surface is light abrasion.	
CONCRETE AND MASONRY	M	Use tri-sodium phosphate (TSP) to clean the surface.	Freshly applied concrete should be left to cure for at least 4 to 12 weeks prior to painting. Prime the surface with concrete/masonry sealant that has stain-blocking capabilities.	Giving the surface a ground layer of white acrylic or gesso will improve the brilliance of the colors and reduce the amount of paint absorbed into the ground. For outdoor applications, follow instructions for proper mural construction.	

SUPPORT	ADHESION	CLEANING	PRIMING	ADDITIONAL PREPARATION	DON'TS AND POTENTIAL PROBLEMS
CERAMICS (INCLUDING STONEWARE)	E–M	Remove any dust.			Painting on this type of support is purely decorative. Not food-safe or washable.
ILLUSTRATION BOARD	E	None	Although not a necessary step, you may apply a single, thin layer of matte polymer medium or gesso to inhibit absorption of the first layer of paint.	Thick applications of paint or very wet washes will cause the board to buckle. Tape the edges with paper tape.	
FIBERGLASS	E	Wipe away dust from sanding.	Seal with industrial sealant. Ask your local hardware store for a sealant that is suitable for water-based paints.	Sand or sandblast to provide tooth.	For outdoor sculptural pieces, finish with appropriate top coat.
FOAM BOARDS	E	None	None	Foam boards such as Sintra board are great lightweight surfaces. Adhesion will be enhanced by light sanding.	Depending on thickness, foam boards can be quite flexible; larger pieces should be braced before painting and framing.
MYLAR	M–P	None	It is possible to peel thick swaths of acrylic away from clear Mylar. It does, however, provide a beautiful surface for transparent painting applications.		Provided they are properly framed, works on Mylar will stand up very well. Painting on frosted Mylar provides increased adhesion.
SYNTHETIC PAPER (YUPO)	M	None	None	None	This synthetic paper is a favorite among mixed water-media artists; however, it is not an archival material. A beautiful paper to paint on, but applications of acrylic can easily peel off. Good for temporary or experimental work.
LEATHER	E–M	Clean with a mixture of bleach, fabric detergent, and water.	The paint will absorb into the leather and reduce absorption of the second coat. If painting on dark leather with transparent or semi-opaque colors, consider using titanium white as your base coat.	Pull and flex the leather as the primer layer dries, pulling the paint into the skin to reduce cracking.	Use dry-clean methods for laundering.
SUEDE	E	Clean with a mixture of bleach, fabric detergent, and water.	Suede is more absorbent than leather. Apply a thin layer of diluted paint to prime the surface.		Use dry-clean methods for laundering.
PLASTICS (ASSORTED)	E–P	Remove dust with microfiber cloth as plastic surfaces are susceptible to static buildup.	Plastics that have moderate to poor adhesion to acrylics will benefit from a light sanding.		Many plastics will bond extremely well to acrylic paint while others will repel it. Do an adhesion test to determine compatibility.

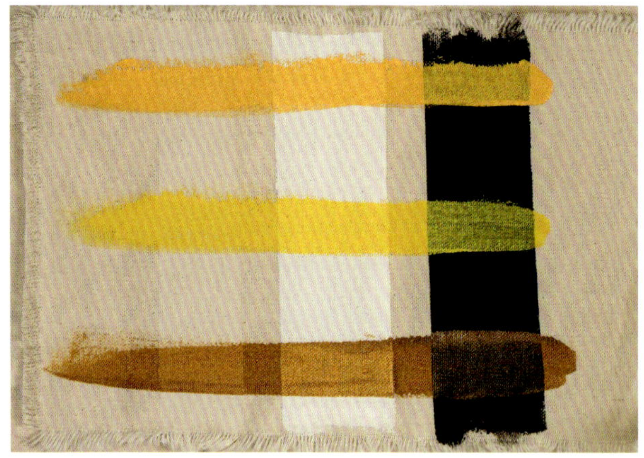

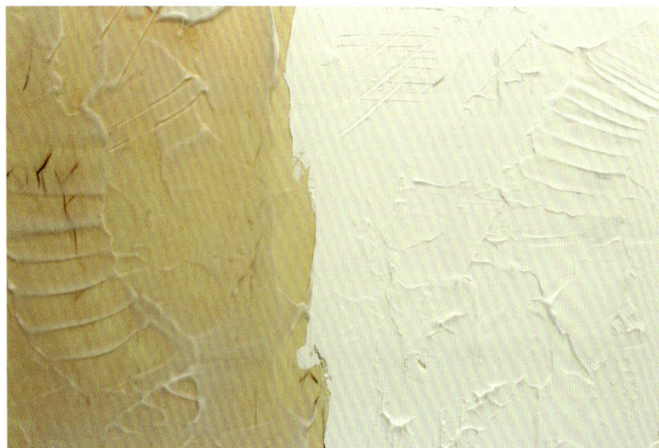

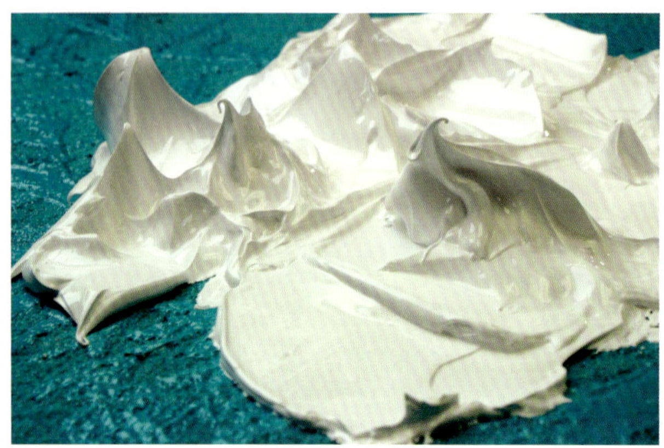

TOP Clear, white, and black gesso on canvas. Opaque (bismuth yellow), semi-opaque (arylide yellow), and transparent colors (transparent yellow oxide) will look different on each of these grounds.

CENTER Clear and white modeling pastes on a cradled wood panel.

BOTTOM Modeling paste holds expressive textures and peaks.

GESSO

A single application of gesso is sometimes all you really need when painting with acrylics. On very porous surfaces, that single layer will keep the support from absorbing too much paint. If you don't want any apparent texture on the substrate, you can apply more coats, sanding between the coats, for an ultra-smooth ground. If you will be using oil-based paints, an acrylic gesso can provide an acceptable barrier, provided that you apply at least three coats to the support. If the gesso you are using does not completely self-level, it can be thinned with water until the desired consistency is reached.

MODELING PASTES

Modeling paste works well as a ground when you want to start with a significantly textured surface. These pastes are made to be painted onto, and they can take quite a beating. The concentration of calcium carbonate in the paste allows for sanding, carving, gouging, or roughing up the painted surface in any other way.

When applying modeling paste, be aware that the sheer density of solids in the paste will cause it to dry quite slowly. The modeling paste will be touch-dry long before it has fully cured. Full curing, which can take from a few days to a few weeks, is necessary for carving and sanding. Textures produced with modeling paste will have crisp, sharp peaks and edges that are significantly less flexible than those produced with a gel medium.

To produce precision detail, extruding the paste through a piping bag allows you to guide and form the paste and gives you the control you need. Rigid stencils give you sharp edges that are extremely effective at creating bas relief designs tough enough to be used on three-dimensional surfaces.

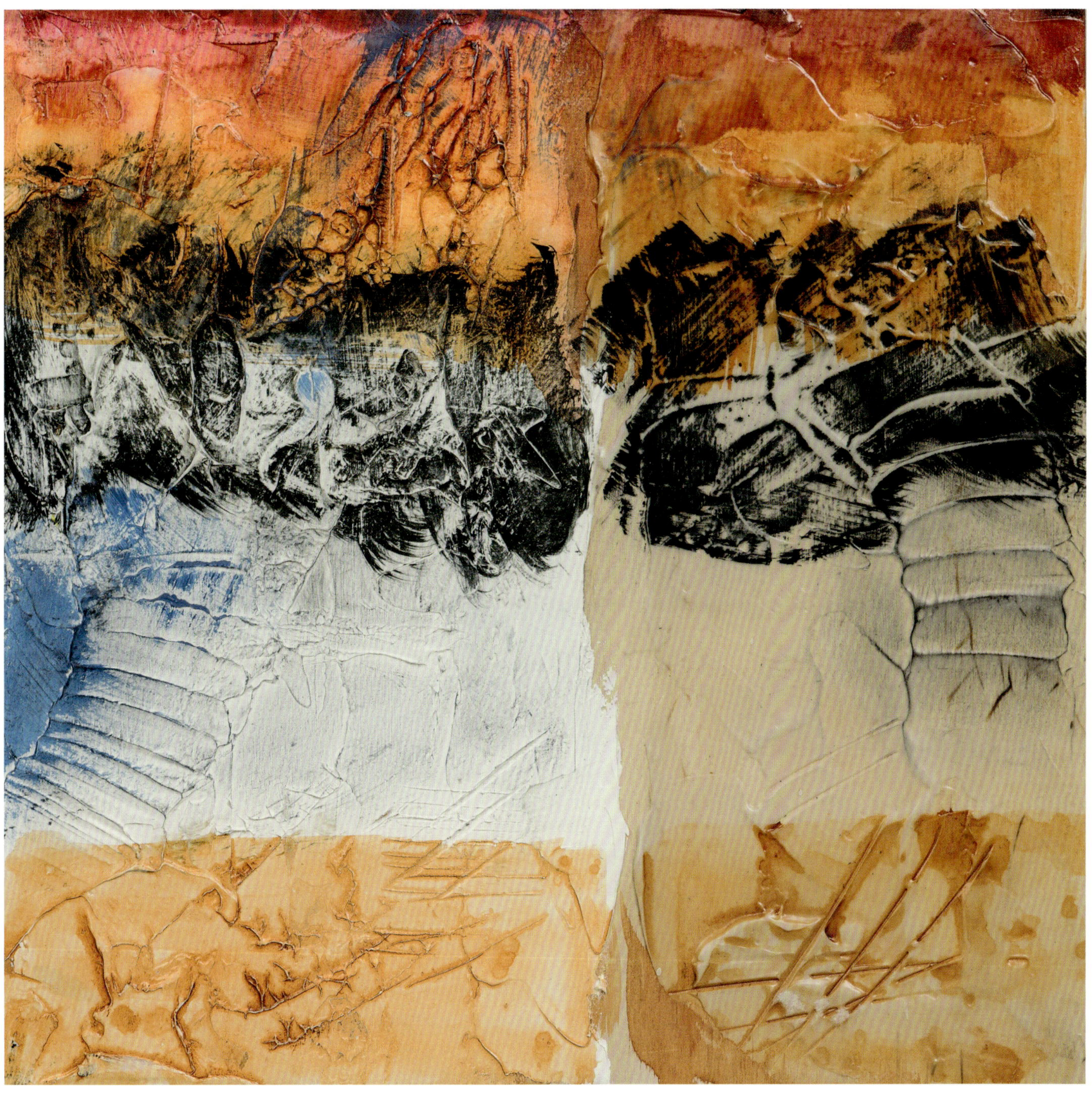

Clear modeling pastes can be tinted in any proportion with your acrylic colors. Regular modeling pastes, however, already contain titanium white pigment and thus will produce very pale results when tinted.

ABOVE White and clear modeling pastes with various surface treatments. From top to bottom: liquid-medium glaze, color added and sanded off, color rubbed on using a cotton rag, and a light color wash.

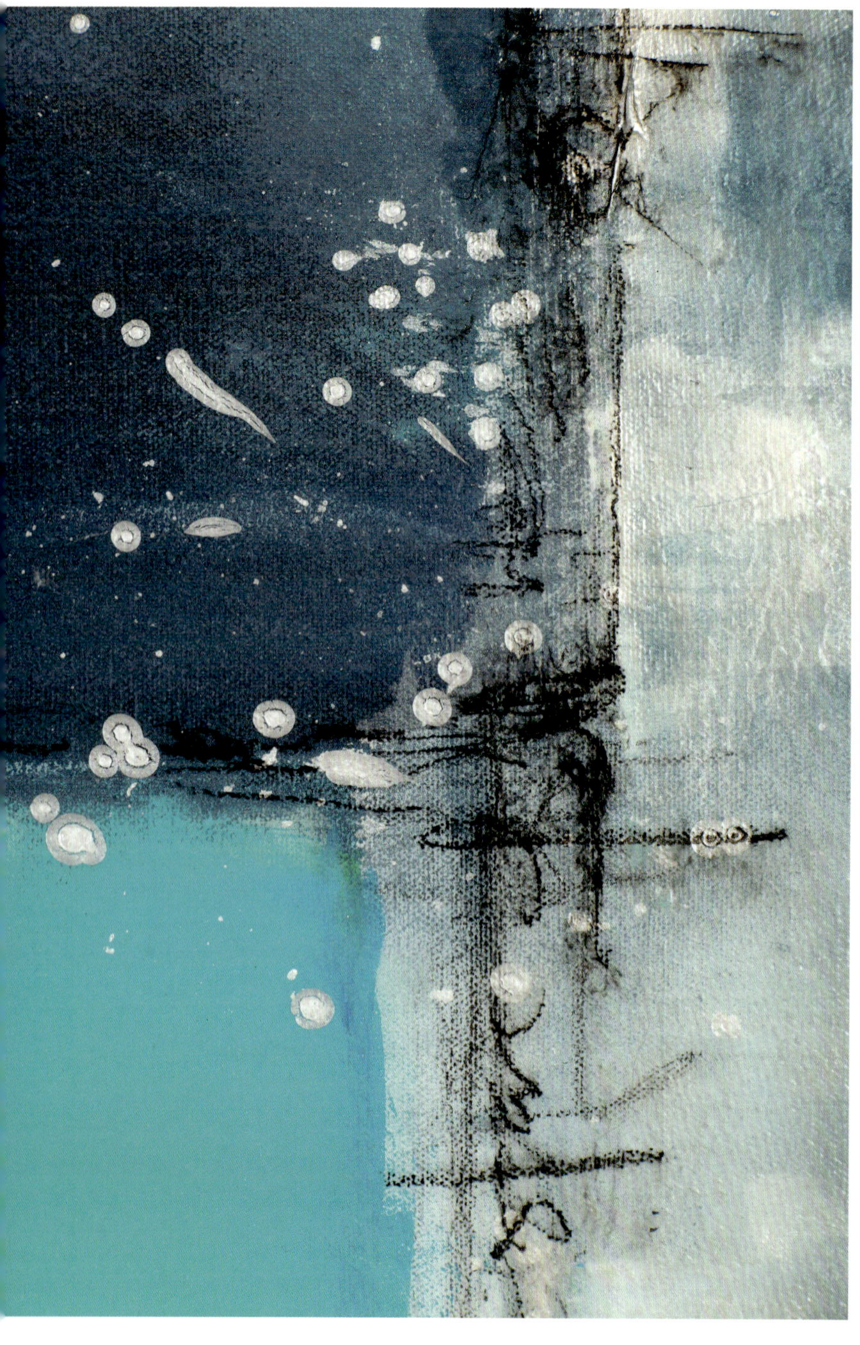

DRY MEDIA GROUND

Dry media ground is the bridge that joins acrylics to other types of art media. The rough tooth catches and holds onto loose pigments, clay, and waxy substances. How you apply it will have an impact on the behavior of the dry media you put on it.

This ground is translucent and can be tinted with acrylic colors. Overloading it with color, however, will make the medium less transparent and may cause a significant decrease in the toothiness of the surface. Wet media can also be used on the surface. Very wet applications of paint may cause the ground to become partially opaque, but this will subside as the paint dries. Overzealous wet applications are not recommended, as they may cause the ground to buckle or ripple lightly in places.

It is advisable to apply dry media ground on top of a primed surface, to provide a layer of protection between the substrate and the painting or drawing materials. Some dry media grounds or pastel grounds are quite viscous, but you can reduce their texture-holding capabilities by thinning them with water. Use a spreading tool or wide, flat brush to apply.

LEFT Across an area lightly coated in dry media ground, contrasting charcoal marks bring a sketchy linear element contrasting with the circular drops and bead shapes.

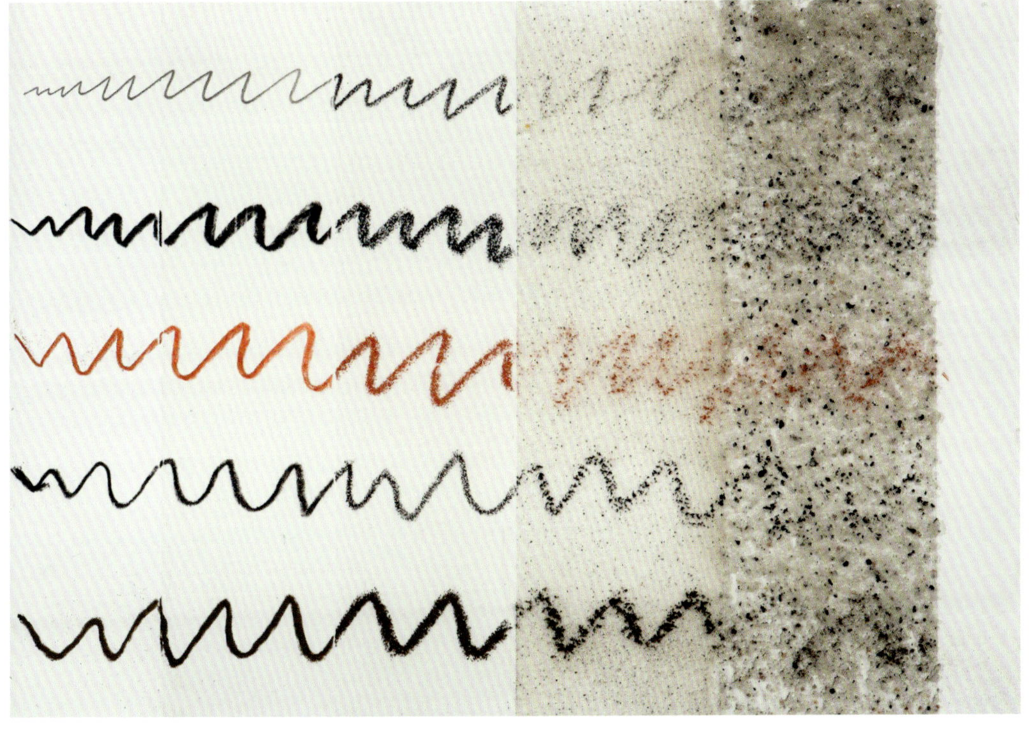

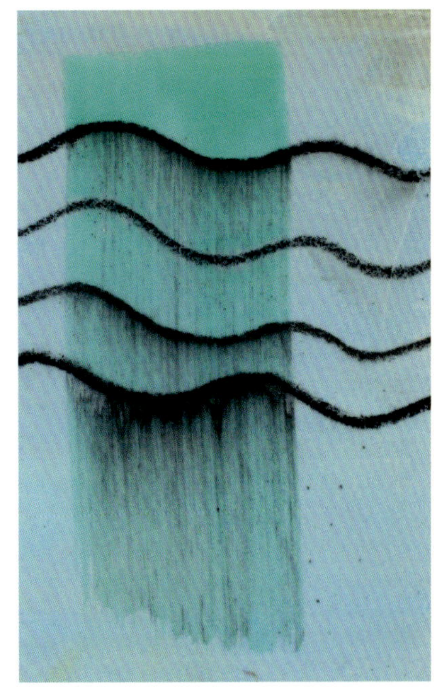

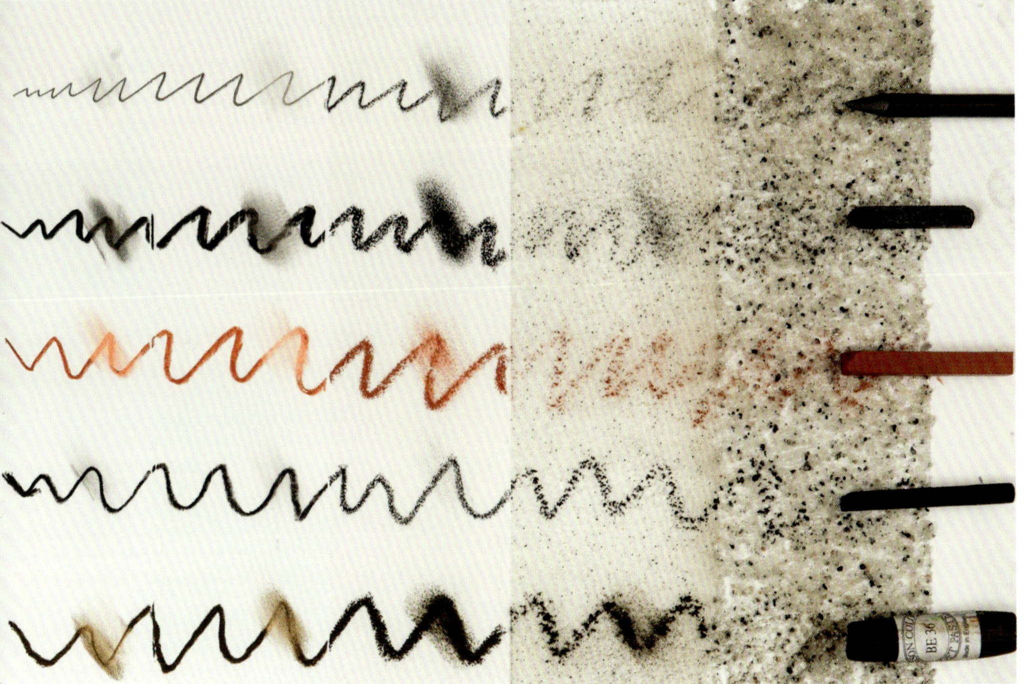

ABOVE, LEFT Graphite, charcoal, conte, pastel, and litho crayon over (from left to right) gesso, matte polymer medium, dry media ground, nepheline gel fine, and nepheline gel coarse.

LEFT Graphite, charcoal, conte, pastel, and litho crayon over (from left to right) gesso, matte polymer medium, dry media ground, nepheline gel fine, and nepheline gel coarse, lightly smudged. A light application of fixative will keep the lines crisp.

ABOVE Lines of various dry media (litho crayon, graphite, charcoal, and pastel) over dry media ground. A single brushstroke of a thin acrylic medium glaze picks up the loose medium and spreads it with ease. To prevent this from happening, apply two to three coats of spray fixative to bind the loose elements to the medium.

CREATIVE DO-OVERS

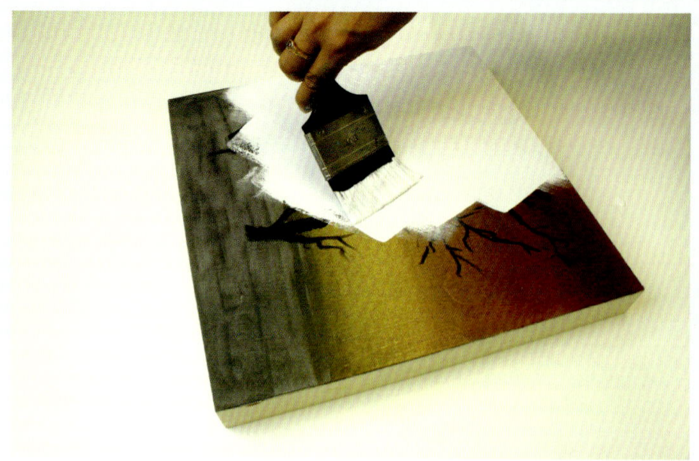

All artists have failed paintings lurking in our studios—those pieces we keep vowing to "fix." Give them a good, hard look, and rethink that impulse. There comes a point in the life of a failed painting when the only logical path forward is obliteration.

Acrylic gesso is the great facilitator of the creative do-over. Some paintings are beyond salvation and must simply be sacrificed so that you can reuse that canvas or panel. Say goodbye, take a picture if you must, then slap on a couple coats of gesso and say "hello!" to a fresh painting surface. Repurposing a canvas or panel is the most sensible way of helping you move forward from a bad painting experience. Unless the surface was previously painted with an oil- or wax-based product, applying an acrylic ground is all it takes to give you a fresh start.

When painted over with a thin layer of gesso, leftover textures are transformed into a new monochromatic platform for your ideas to play on. But if these impasto ghosts are too distracting for a new composition, spreading modeling paste on your support in a thick, smooth application can make it "new" again. Remember to allow sufficient time for the modeling paste to dry before painting. If it is very thick, this can take up to a week. You may want to sand it down to a velvety but still-toothy surface before beginning to paint.

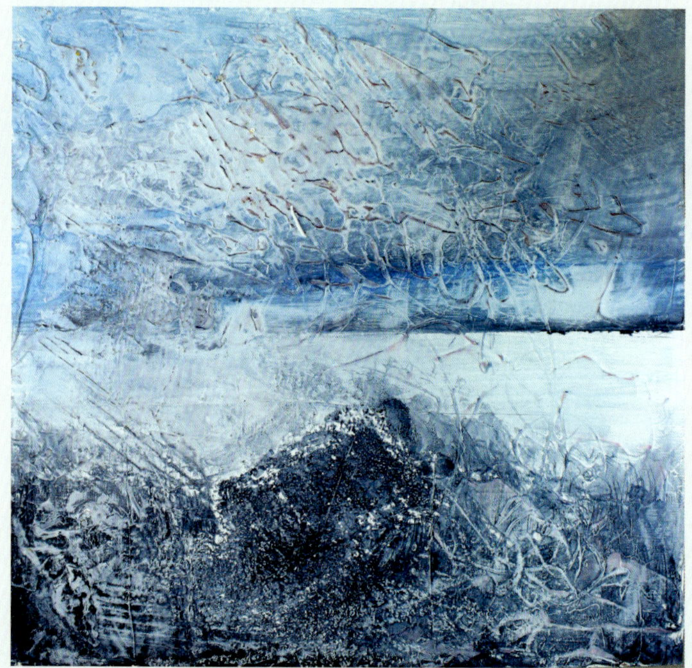

TOP Obscuring a failed painting attempt with a liberal coat of white gesso.

CENTER Gesso has been liberally applied to this failed painting with a textured surface. Now it's waiting for its next life.

BOTTOM A layer of wash, diluted and saturated, forms the new painting's first layer.

Tip: If your creative reject has been hanging around exposed for a while, it's best to give it a good wipe with a microfiber cloth to remove as much dust as you can.

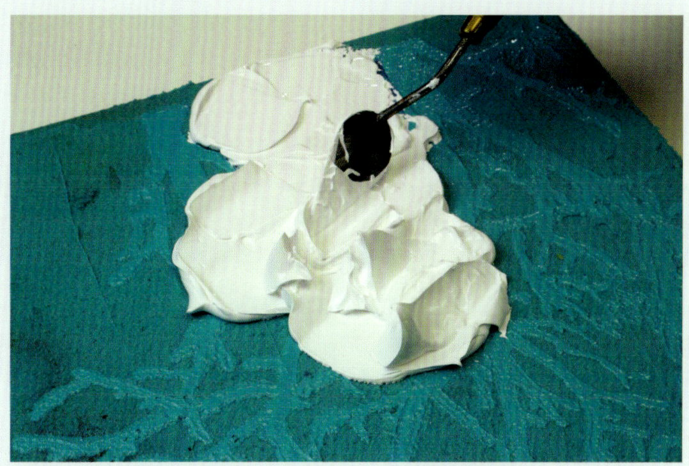

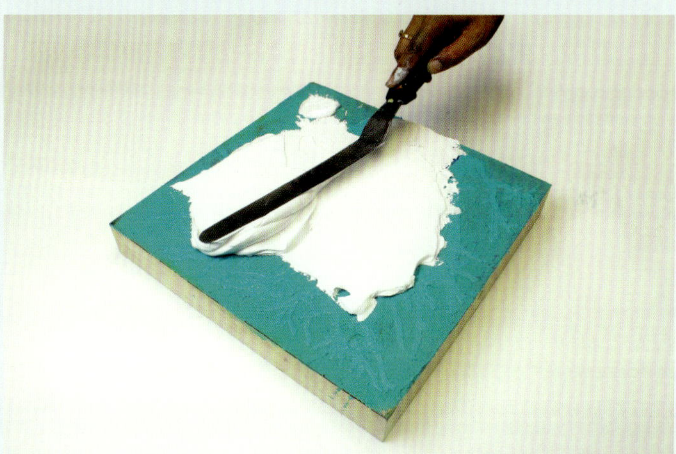

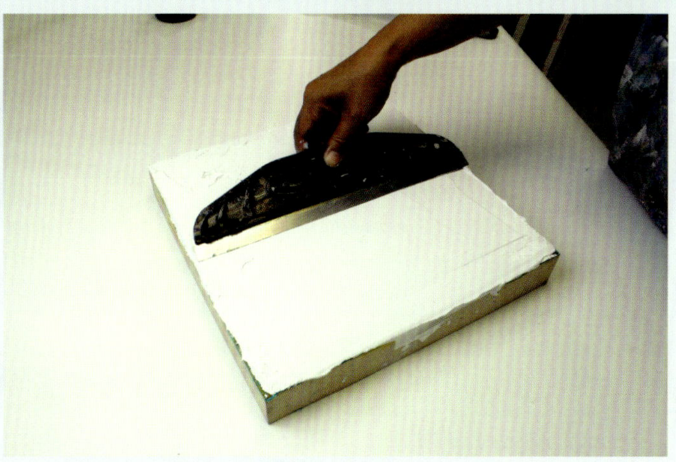

There's an alternative to reusing the texture or filling it in with modeling paste—sanding it down. For this, you will need to wear a dust mask and protective eyewear. If the paint is not fully cured, it will be too soft to sand, resulting in a tarry mess, so sanding is recommended only for surfaces that have had at least two to six months to dry, depending on the content and thickness of the paint. Textures that have been built up with modeling paste will sand more readily than those created with regular gels, which are much softer and contain fewer solids. Sanding with a fairly fine grit will leave a smoother surface and provide a good amount of tooth for the new paint layers. Because the friction from abrading the acrylic will heat up the surface and soften it, it is best to perform this process in stages, allowing the paint to cool and harden between sandings. Be careful when using an electric sander, since you might damage the support if you apply too much pressure, especially to a flexible support. Canvas stretchers should be braced from behind with a rigid board to keep the surface flat while sanding. After sanding, wipe away paint dust with a lightly dampened microfiber cloth.

TOP Unwanted texture can be obliterated with a thick application of modeling paste.

CENTER To effectively cover up the texture, apply the paste in a thick, even coating.

BOTTOM Smooth the paste with a straightedge for uniform thickness. For an even smoother surface, sand after allowing the paste to cure for at least three to five days (possibly as long as a week).

CHAPTER 7

CREATING SURFACE TEXTURES

PERHAPS THE MOST OBVIOUS USE FOR ACRYLIC MEDIUMS, especially the thick ones, is to create and control textures. Mediums collectively offer painters almost unlimited potential for dictating the appearance and character of the painted surface, from ultra-smooth to the spikiest of peaks.

Touching paintings is discouraged, which I find incredibly frustrating because being able to physically feel the surface of an exquisitely textured painting can greatly increase my appreciation and understanding of it. Sadly, gallery and museum etiquette, not to mention conservation concerns, dictate that we keep our hands at a respectful distance from the artwork, so we usually have to rely on letting light play on those surfaces instead of our fingertips.

Mediums offer almost unlimited textural potential.

OPPOSITE Ksenia Sizaya, *Autumn Peaks,* 2017, acrylic paint, soil, glass pebbles, acrylic polymer emulsion, modeling paste, and crackle medium on treated cotton canvas, 30 x 40 inches (76 x 102 cm). Private collection.

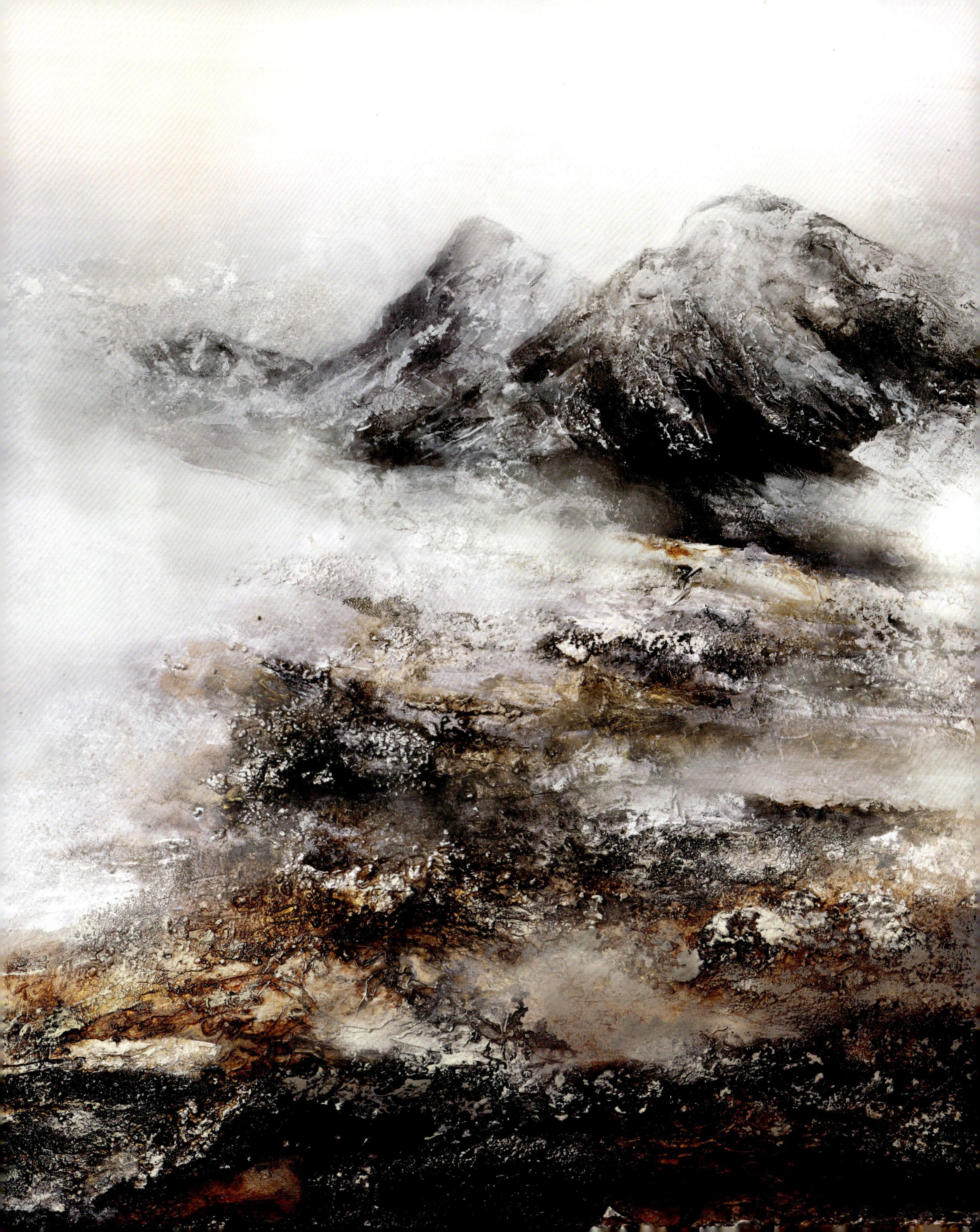

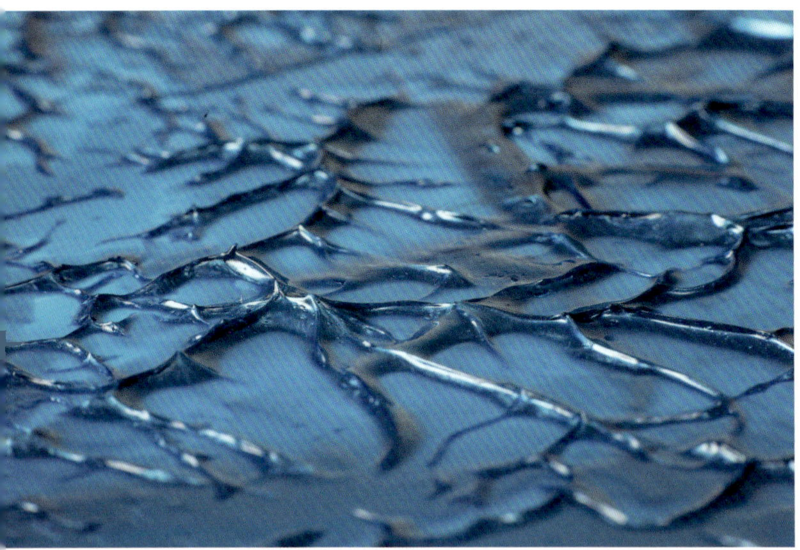

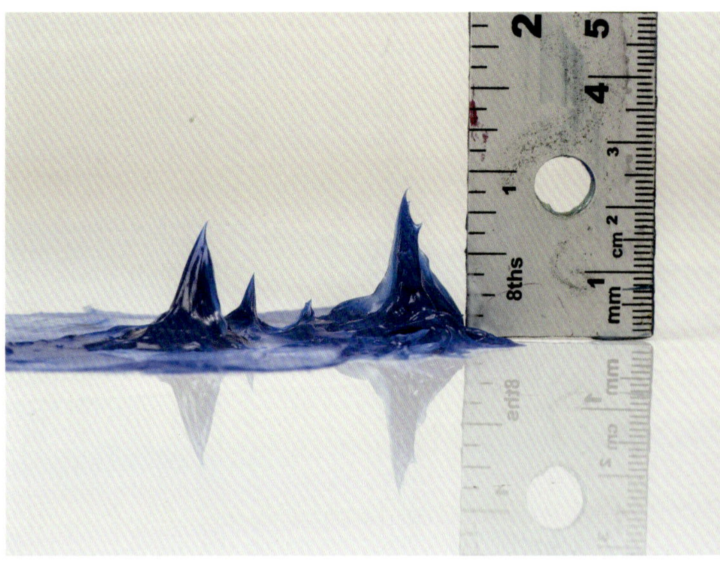

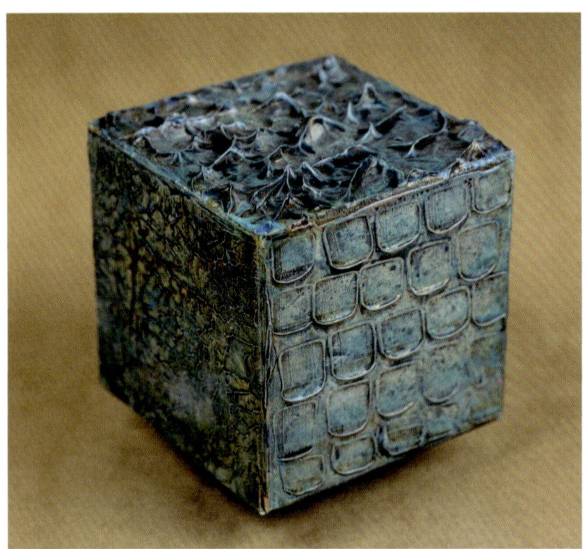

TOP, LEFT When tinted gel is sandwiched between two sheets of smooth paper and the sheets are then peeled away from each other, the suction creates sharp veins of texture.

TOP, RIGHT Measuring up: a one-inch tinted gel peak.

ABOVE A three-dimensional representation of possible gel textures. Inspiration for the fingers and the eye.

TYPES OF TEXTURE

Acrylic mediums are the masters of diversity when it comes to disrupting and dictating the every possible nuance of an acrylic painting. They transform color into a three-dimensional, interpretive tool. The shallowest of valleys between paint peaks provides enough shadow to amplify contrasting surface textures in a painting. A rough, granular surface scatters light, while a slick swath of acrylic flings it, beaming, back toward the viewer. Velvety matte surfaces appear to absorb light, rendering the surface soft and nebulous.

There are myriad tools and methods for producing textures, ranging from the traditional to the more crafty. The sections that follow offer some examples of how to produce and amplify textures in your acrylic paintings and the tools you'll need to create them.

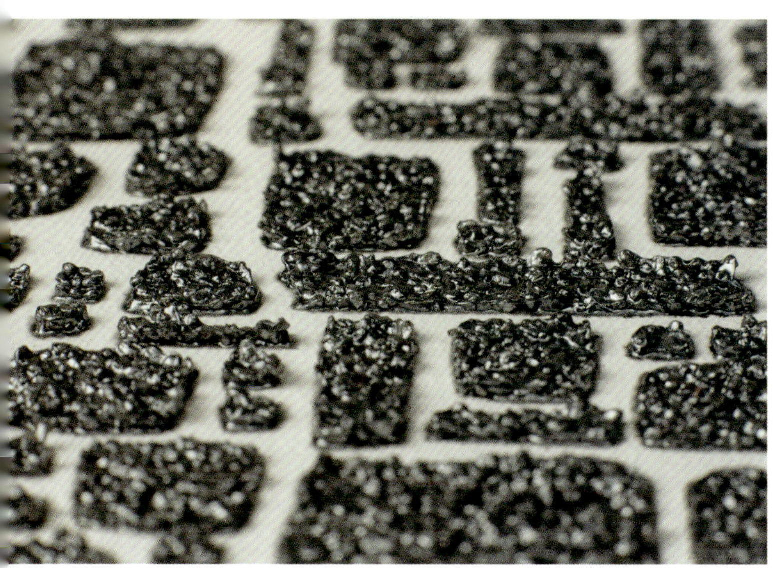

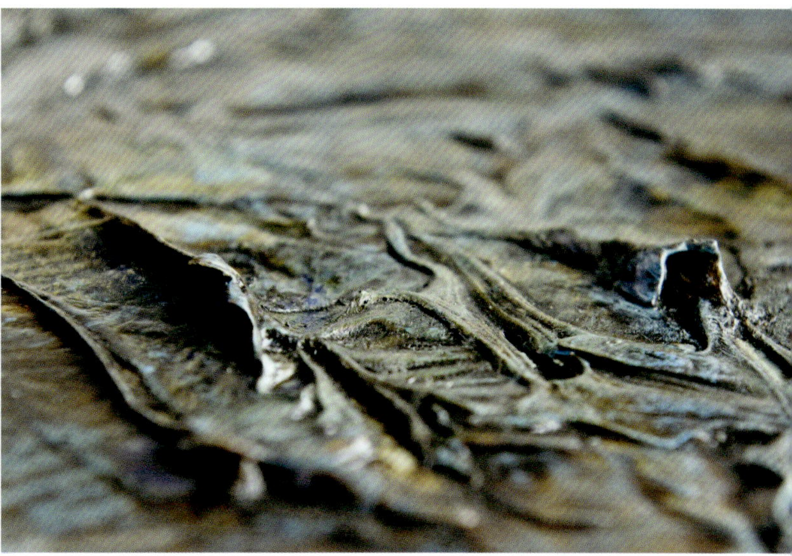

IMPASTO (*ALLA PRIMA* APPLICATION)

Impasto is the term generally associated with thickened paint that's slathered onto canvases in rippling, bold waves, showing the bold strokes of the artist's hand. It is not the only method of producing texture, although it's certainly the one most often discussed.

Alla prima application literally means undiluted, unmixed, direct-from-the-jar painting. This term is usually used to describe the application of paint colors, but can be used for mediums, as well. Whether they are thrown liberally onto a surface or systematically applied with deliberate accuracy, unaltered mediums can produce some incredibly complex and expressive textures. The signature of the artist's hand is evident in the directional strokes of paint or medium.

Thick impasto surfaces are typically built up in successive layers. Gel and modeling paste are recommended mediums. Tools for impasto include brushes, palette knives, and paint shapers such as the Catalyst tools made by Princeton Artist Brush Co.

ABOVE, LEFT Nepheline gel extra coarse + liquid mirror and carbon black produces a granular, iridescent texture on this stenciled surface.

ABOVE, RIGHT Gel textures.

BAS RELIEF

BELOW, LEFT Made using a filbert brush and some gel medium, these impasto brushstrokes create a bouquet.

BELOW, RIGHT Each petal of these flowers started as a thick brushstroke of untinted gel. Once the gel was dry, the color was built up in thin dry-brushed layers. Interference and iridescent colors give the petals extra dimensionality and contrast against the matte black background.

OPPOSITE, TOP LEFT Multiple dry-brush and wash applications, finished with a tinted glaze, enhance this bas relief surface's visual depth. Repeated dry-brush applications of slightly opaque metallic colors accentuate the edges of the stenciled shapes.

OPPOSITE, TOP RIGHT One classic example of bas relief is the face of a coin. Here, the image on the coin has been quickly reproduced with a smooth swatch of gel; areas have been etched away with the smooth edge of a color shaper and detail incised with the tip of a bamboo skewer.

OPPOSITE, BOTTOM Etching script texture into wet gel medium.

In contrast to the drama and spontaneity of impasto textures, bas relief textures are shallow and subtle. The term *bas relief* (after the Italian *basso rilievo*) refers to very low textures on a completely flat surface. I extend the idea by using the term to refer to any relief that has only a very slight height.

You can add a third dimension onto a two-dimensional surface by either building up or carving down into the material. If you're building up, you can create surface texture and detail using a variety of methods: by layering modeling pastes or gels, by using masking or stencils to produce sharply defined relief, or by adding collage elements. Relief creates areas of light and shadow, adding visual dimension. Producing a more minimal texture in a controlled manner can be accomplished with masking, dripping, incising, and some deft tool use.

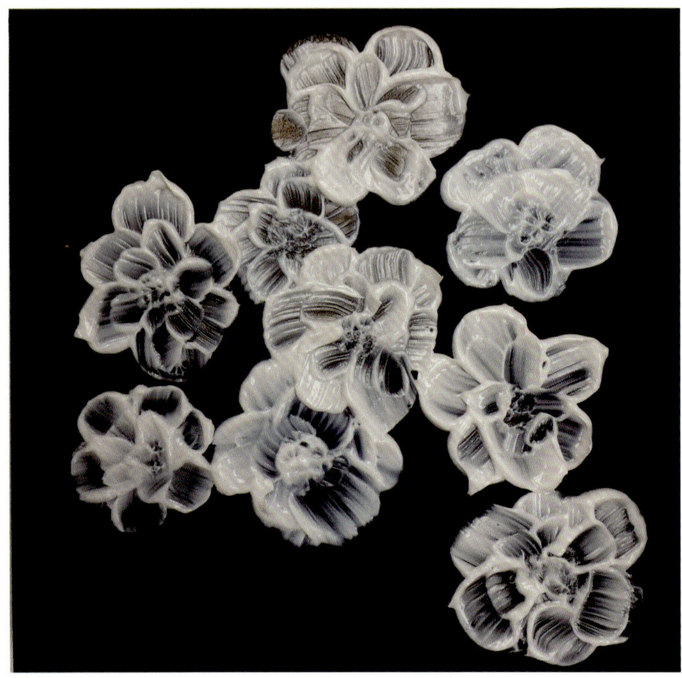

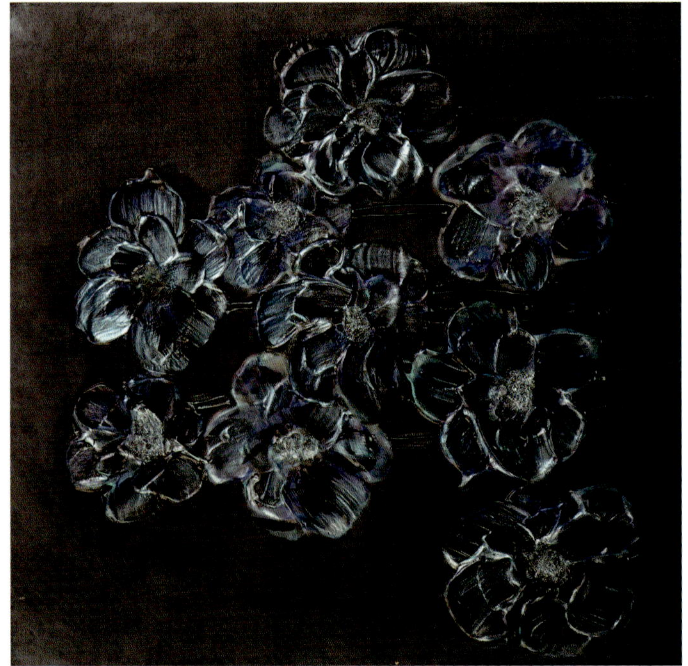

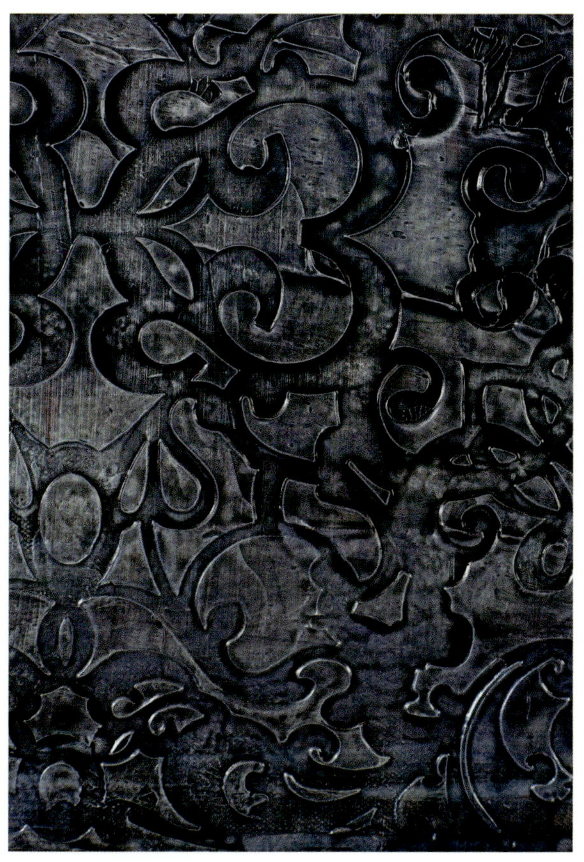

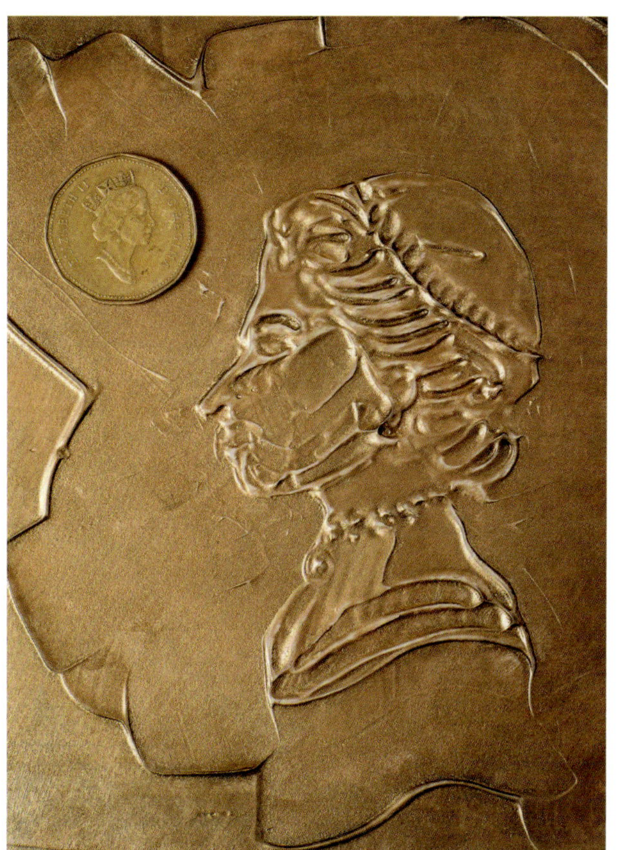

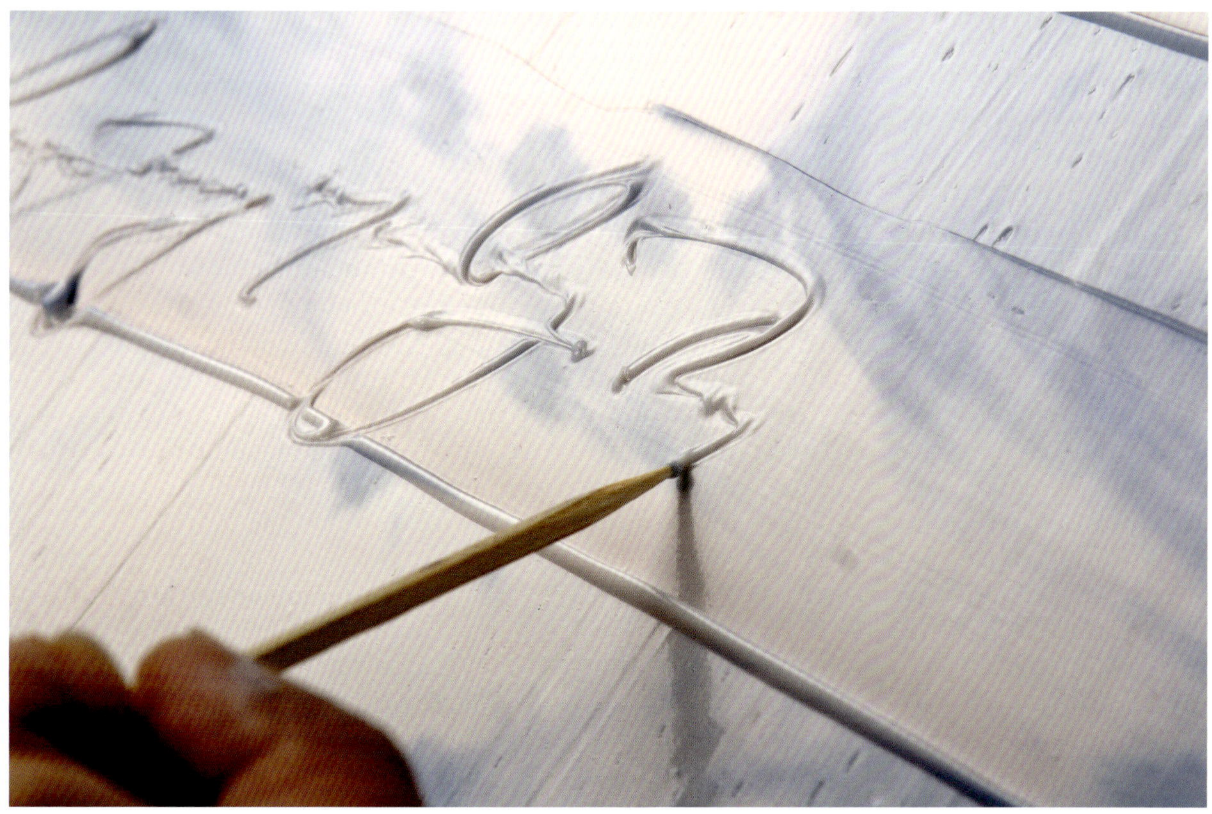

ENHANCING BAS RELIEF WITH COLOR

Augmenting the appearance of texture can be easily accomplished with a minimal use of color. A little goes a long way in amplifying subtle textures. These images show just a few methods for enhancing bas relief with color.

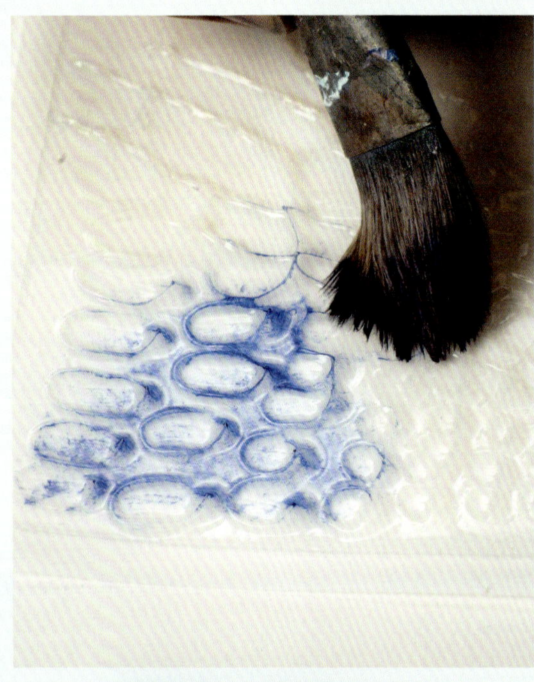

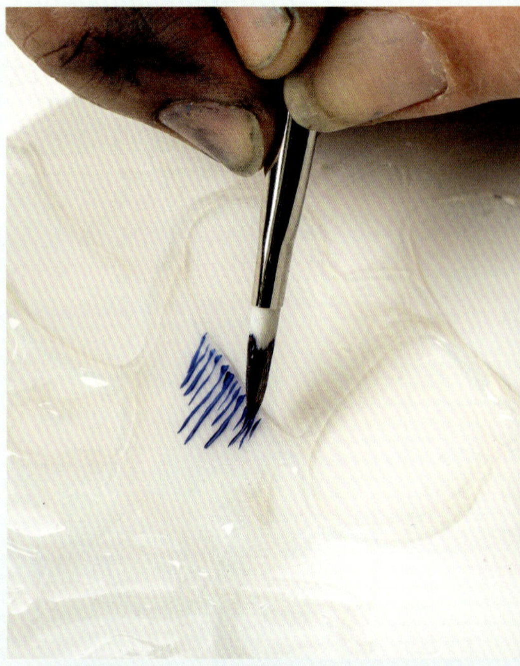

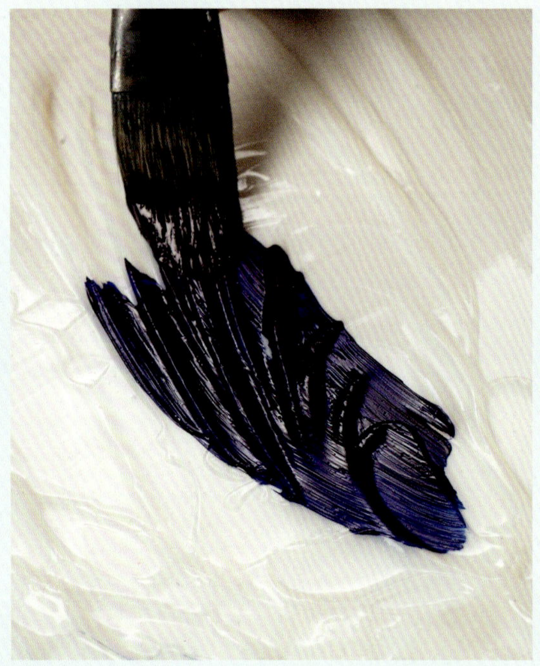

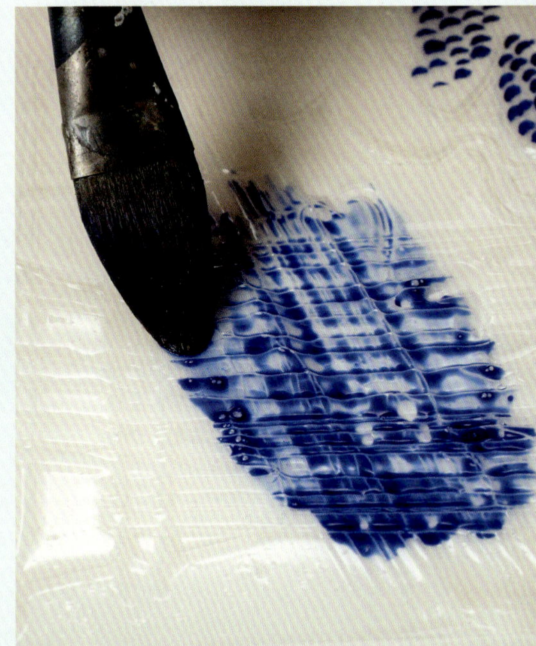

TOP, LEFT TO RIGHT Application of color with a very dry brush reveals the valleys and ridges of dried gloss gel medium. • Detailing around clear gloss gel ridges.

BOTTOM, LEFT TO RIGHT A smooth *alla prima* application of color reveals the underlying texture. • Water with just a single drop of color pools in the valleys between elevated gel cross-hatchings.

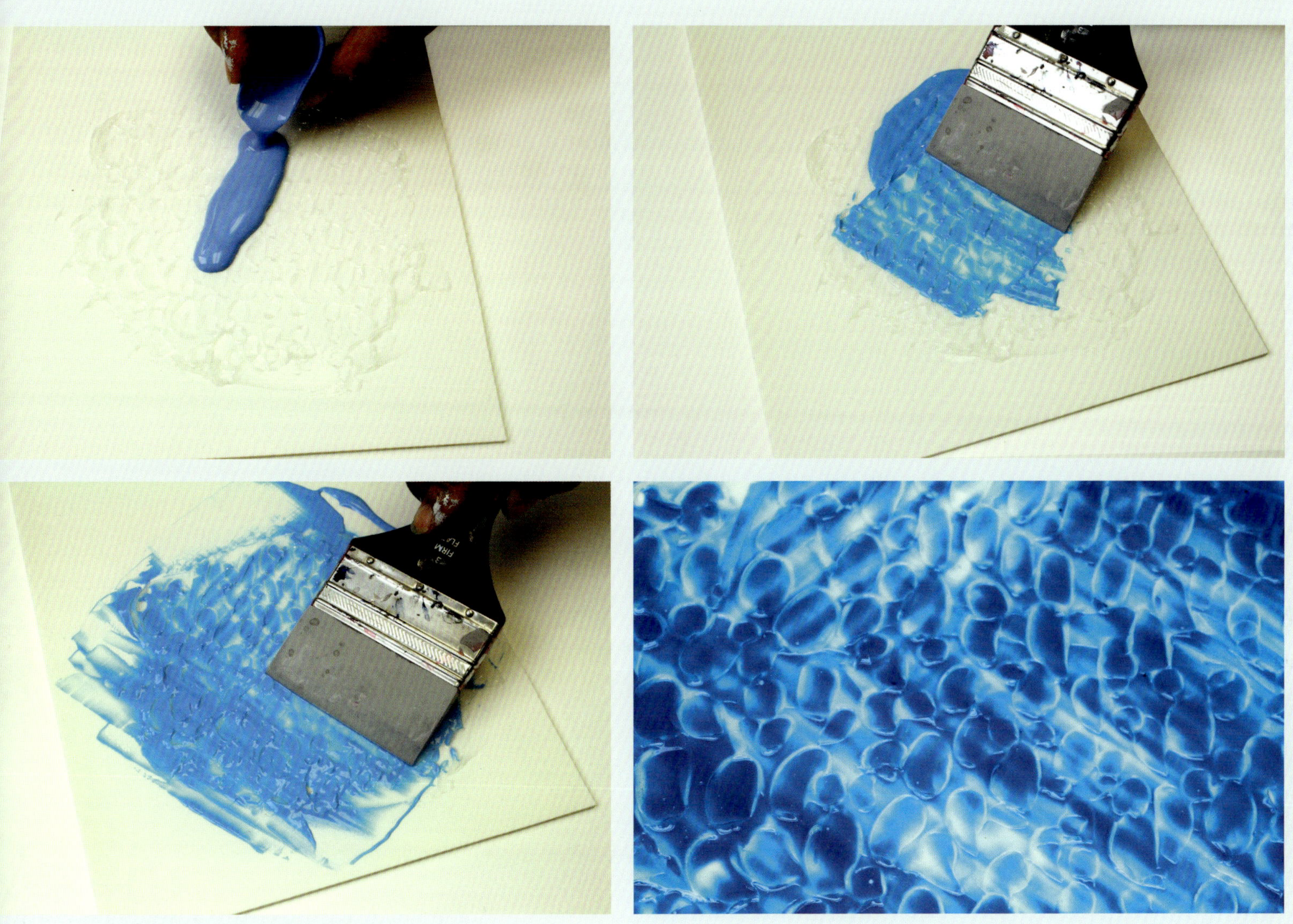

TOP, LEFT TO RIGHT A tinted glaze of liquid medium brings out the nuances of a surface textured with clear gel. • Pushing the glaze into the crevices and valleys of the texture.

BOTTOM, LEFT TO RIGHT Using a flat tool like this color shaper, you can push the glaze into the texture without adding brush marks. • The now-dry glaze rests in varying thicknesses on the textured ground.

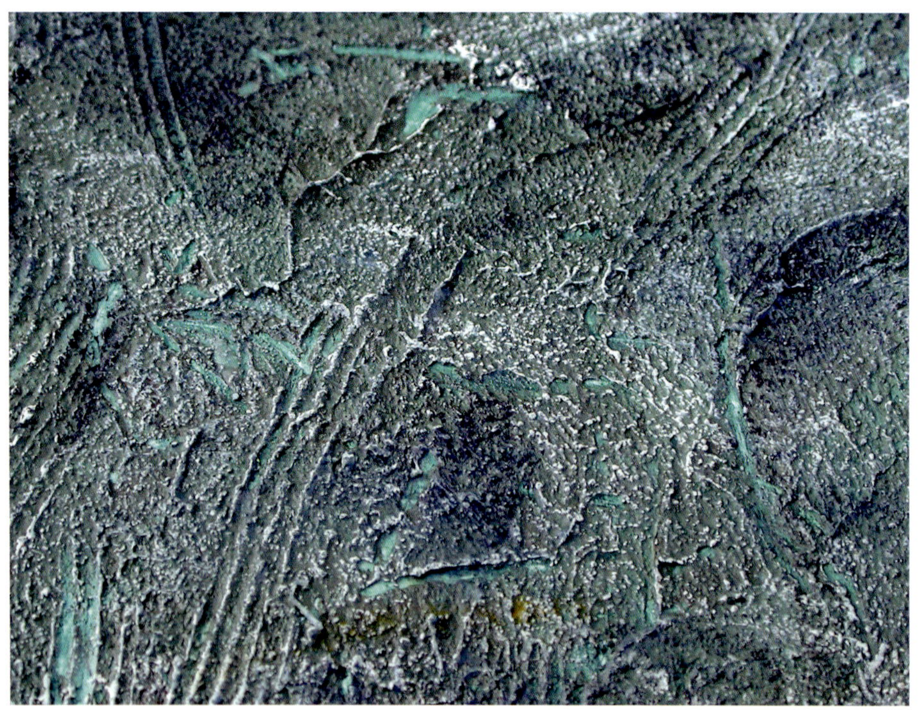

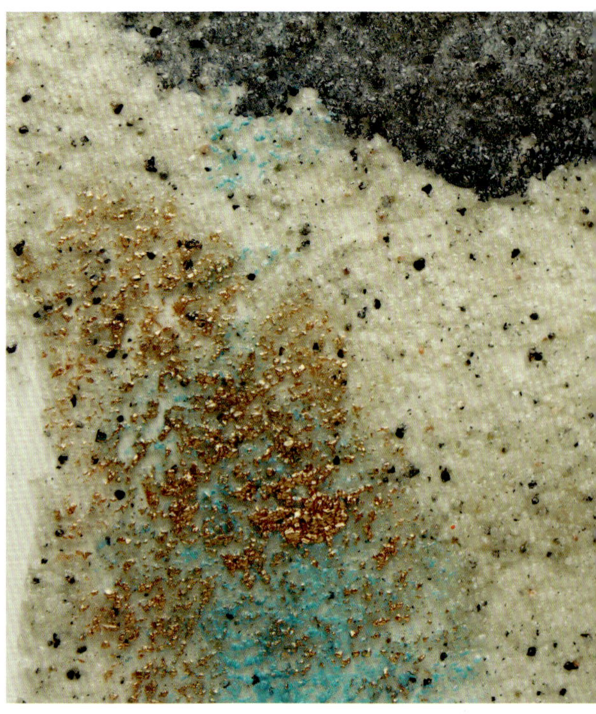

GRANULAR TEXTURES

A surprising variety of granular mediums are available on the acrylics market, and they seem, so far as I can tell, to be quite underused. There's a particulate for every taste, ranging from the subtle and uniform to the wildly erratic and coarse. Like the spice in a recipe, they add new dimension to plain gel medium. Opaque pumice, translucent nepheline syenite, glass, metal, plastic, and other materials are used to give grit to acrylics.

Granular mediums break up the surface of a painting, scattering light and adding tooth. Finer particulate granular mediums provide lovely surfaces on which to blend colors; their velvety tooth minimizes streaks and breaks up the color while causing barely perceptible visual color mixing.

Granular mediums with larger grit particles can be a little loose and messy. To keep those errant particles in

ABOVE, LEFT A mixture of tinted fine and coarse nepheline gels was scraped with a palette knife, then dry-brushed with light and dark colors to bring out this granular surface texture.

ABOVE, RIGHT Nepheline extra coarse gel. Part of the process of getting to know a medium is to experiment with paint application methods as well as various types of color. Dry-brushing, color wash, and *alla prima* painting were applied, leaving the clear medium in the center untouched.

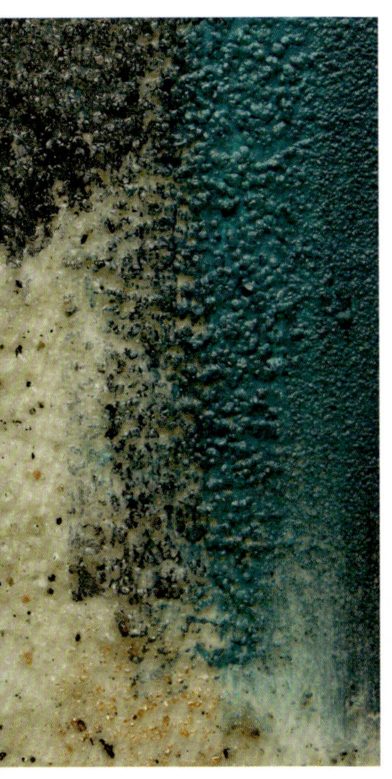

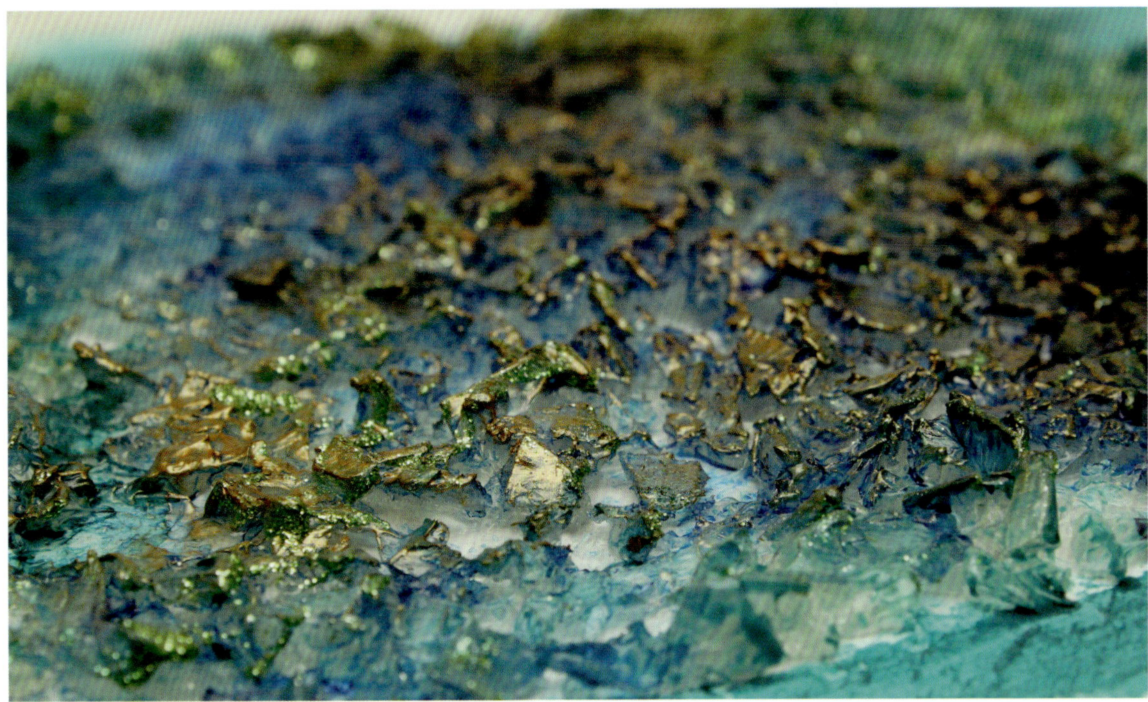

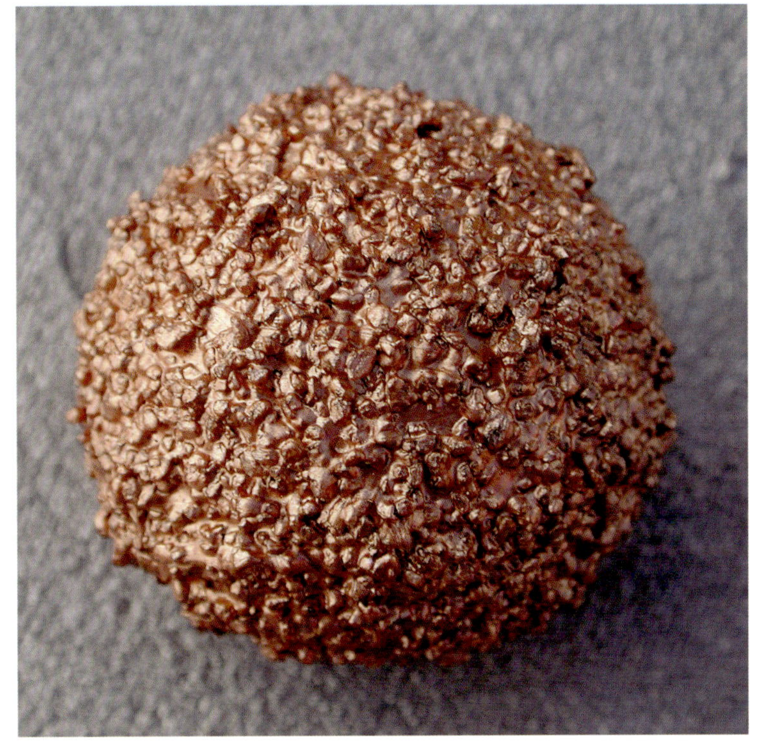

check, you can add more gel medium to bulk up the adhesion. Dry brushing and applying thin washes onto coarser mediums highlights the rough edges and three-dimensional appearance of the particles, adding a real sense of depth to the paint surface.

ABOVE, RIGHT PETE plastic gel texture, with dry-brushed and wash applications of color.

RIGHT Nepheline gel extra coarse with iridescent copper on the ball and nepheline gel fine on the surface.

CHAPTER 8

LAYERING AND COMBINING MEDIUMS

COMPARED TO OIL PAINTS, ACRYLICS DRY QUICKLY, and this aspect of their character makes them ideal for layering. Every member of the acrylic family—colors, liquid mediums, gels, textured pastes—has a good relationship with all the others, a common bond that allows them to be combined with each other in almost any proportion, and to great visual effect.

Adhesion is the key to combining and layering acrylic colors and mediums. A little patience is essential, as well, as time strengthens and maintains the bonds between layers. Besides being mindful of curing times, you also need to consider transparency, texture, and relative viscosities when building a complex fusion of acrylics.

Concocting a multilayered, multi-material composition can do a lot to augment your art. Marion Fischer's *Once Removed,* opposite, incorporates clear gesso, liquid acrylics, glazing mediums, semigloss polymer medium, self-leveling gel, gloss modeling gel, and gloss gel medium retarder. (The image within the blue frame is a photo transfer.)

Acrylic mediums are truly at their best in this arena, and exploring their interrelationships will absolutely amplify your understanding and mastery of acrylic painting as a whole.

All acrylics share a common bond.

OPPOSITE Marion Fischer, *Once Removed,* 2014, acrylic on wood, 24 x 36 inches (61 x 91 cm). Photo: Marion Fischer. Private Collection, Ottawa, Ontario, Canada.

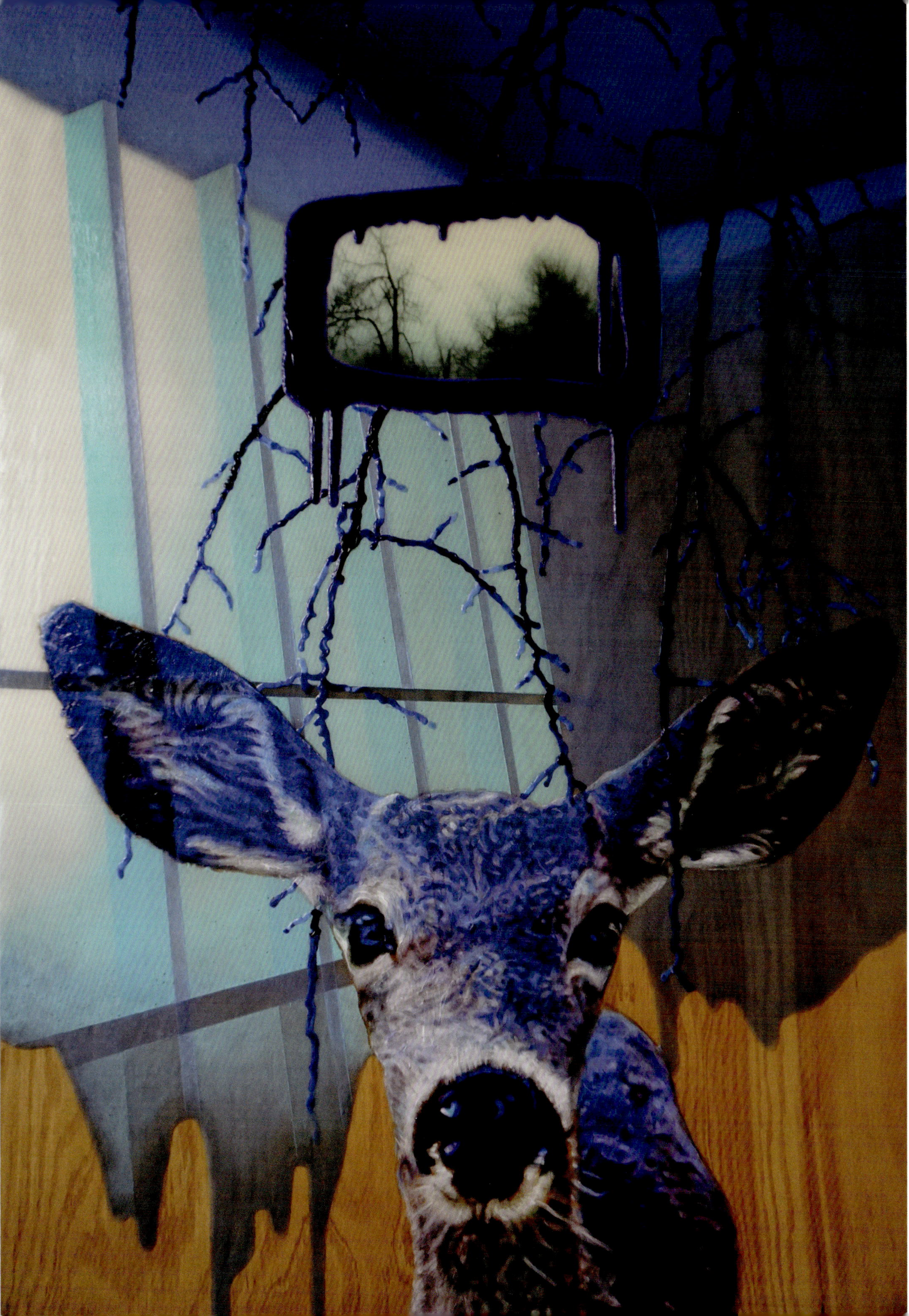

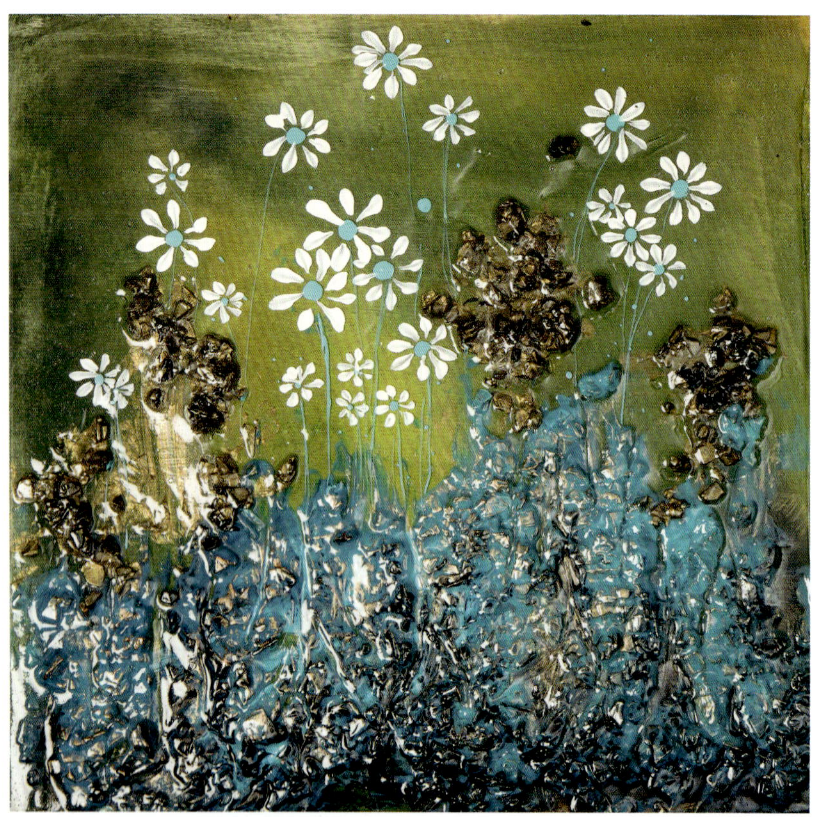

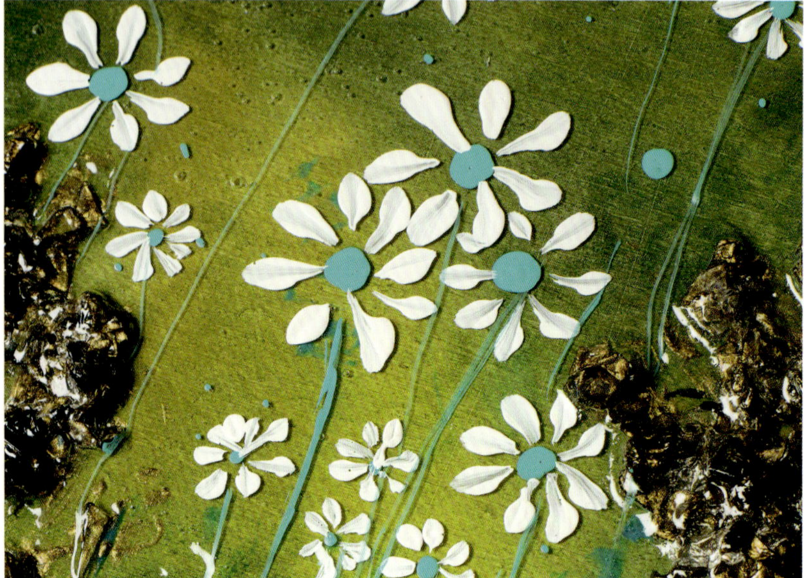

LAYERING

Layering is an exercise in patience, some guesswork, and a little reverse thinking. It's great fun to build up a texture, cresting its surface with dry-brushed color and then surrendering it to the mercy of a smothering layer of thick, gooey gel. Like mountaintops poking up through dense cloud, those impasto peaks insinuate themselves up beyond the gel surface, leaving the braille shadow of textures beneath.

There are infinite possibilities for layering mediums and colors, but here are some tips to help guide your process:

- To maintain transparency between layers, increasing luminosity and depth without obscuring the underpainting, use gloss mediums (liquid or gel, pure or tinted) instead of matte mediums.
- Transparent color in a glaze layer will reveal the underpainting, while more opaque colors can obscure it.
- Semigloss or matte mediums will produce translucent (rather than

TOP This painting was produced as a demonstration for layering radically different mediums. The sharp edges of the shards of recycled PETE plastic gel throughout the painting—and especially evident in the lower, blue portion—are softened where enveloped by self-leveling gel. The daisies, painted in liquid acrylics on the surface of the gel, appear to float over the background.

BOTTOM Opaque flowers painted on a very thick layer of self-leveling gel cast shadows on the underpainted surface, appearing to float above it.

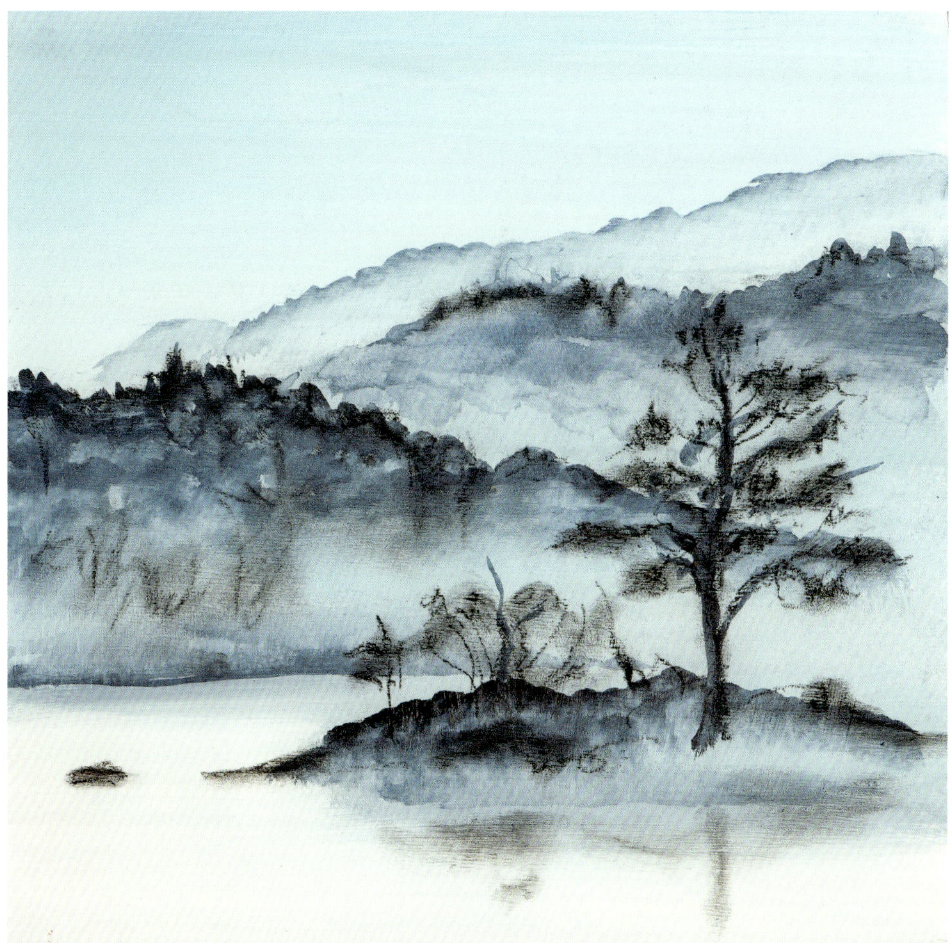

transparent) layers. Keep in mind that thick or repeated applications will gradually hide the underpainting.

- To maintain transparency in glazes, use very diluted colors.
- Using white modeling paste between layers will act as a mask, completely blocking out the underpainting.
- Use granular mediums sparingly if you intend to maintain translucency between layers.
- Allow sufficient time for a thicker layer to cure and clarify before adding another layer on top. Thin layers can be painted over when they are touch-dry, but any application thicker than a couple of millimeters should be allowed to dry for several hours before being painted over.

You can touch up or obscure an area in a glaze layer by painting in a thin layer of opaque color. This method is not recommended for large areas of uniform color but can be effective in busier compositions and detailed areas. You can use a similar method, adding touches of opaque color or medium within a layered glaze, to make the color or medium appear to float within the acrylic.

Using a few layers of a matte medium, you can "fuzz out" an area while not completely obscuring it. This can soften a harsh transition.

ABOVE A quick sketch with a single color and some charcoal accents was the starting point for this misty composition. Several layers of liquid matte medium were strategically applied to enhance the misty effect.

PROCESS

ACCIDENT AND IMPROVISATION

My painting *Solstice Anchorage* evolved through experimentation from the thumbnail sketch onward. It's a combination of planned layers with accident and improvisation. Not including water, I used five types of acrylic mediums in this painting: gloss gel, self-leveling gel, gel with recycled glass, dry media ground, and gloss polymer medium. Some of the mediums were combined, others layered. Each stage was an experiment.

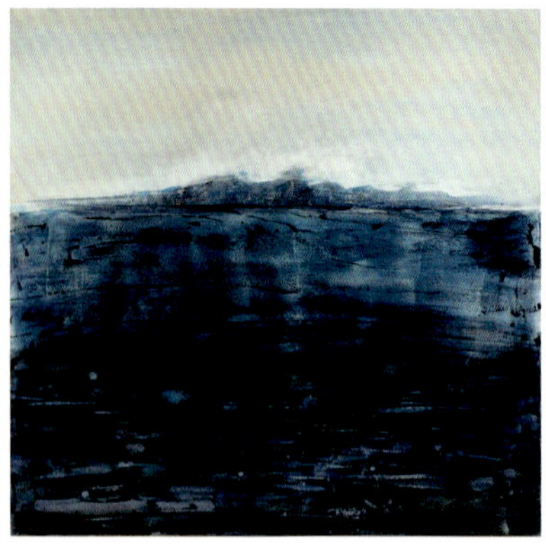

1
Here is *Solstice Anchorage* halfway through the painting process.

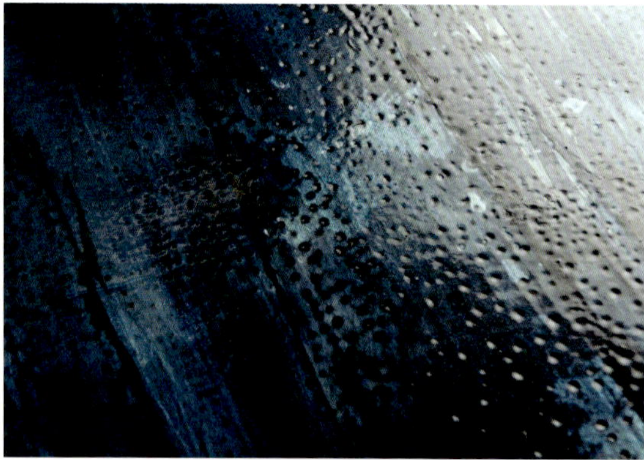

2
Sometimes, accidents occur when mediums are combined. In this case, I spritzed water onto a layer of wet self-leveling gel to smooth it out. Instead, the fine mist of water created a pockmarked field when it dried.

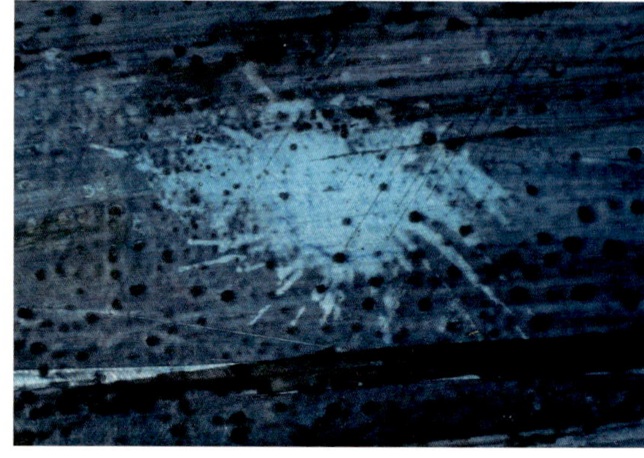

3
If after applying a loose wash layer you wait a few minutes and splash or drop some water onto it and then wait a few minutes more, the water will reactivate the wash layer to reveal the layer that lies beneath. That's what happened here—accidentally! But I left it as a new compositional element.

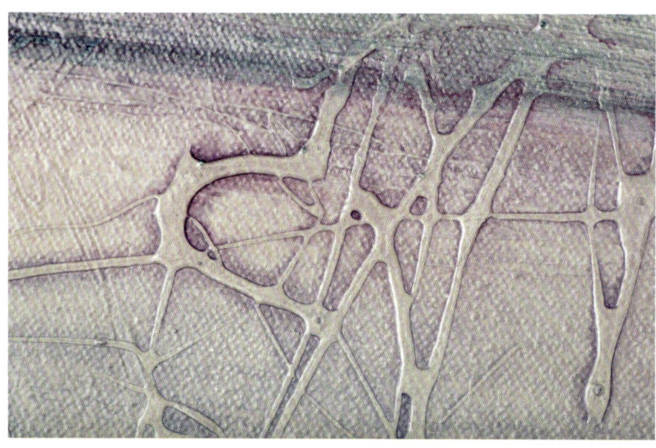

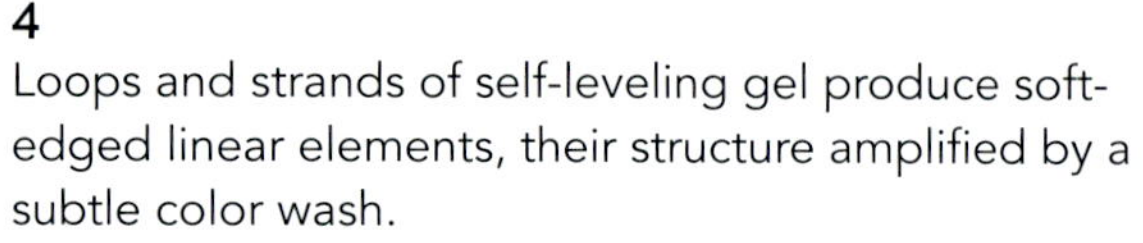

4
Loops and strands of self-leveling gel produce soft-edged linear elements, their structure amplified by a subtle color wash.

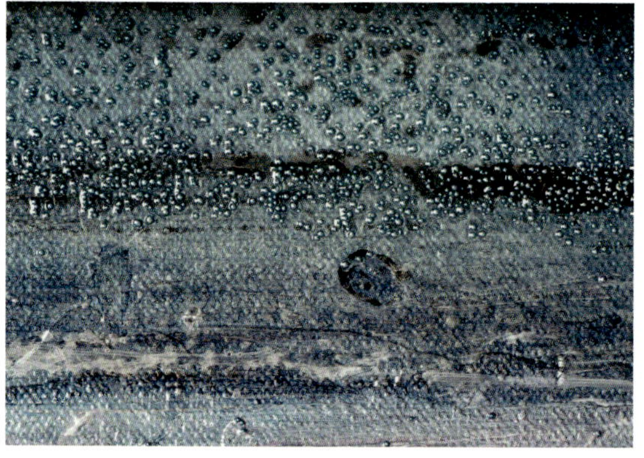

5
A cascade of glass and metal beads, like a foamy surf, lies embedded in a layer of gel.

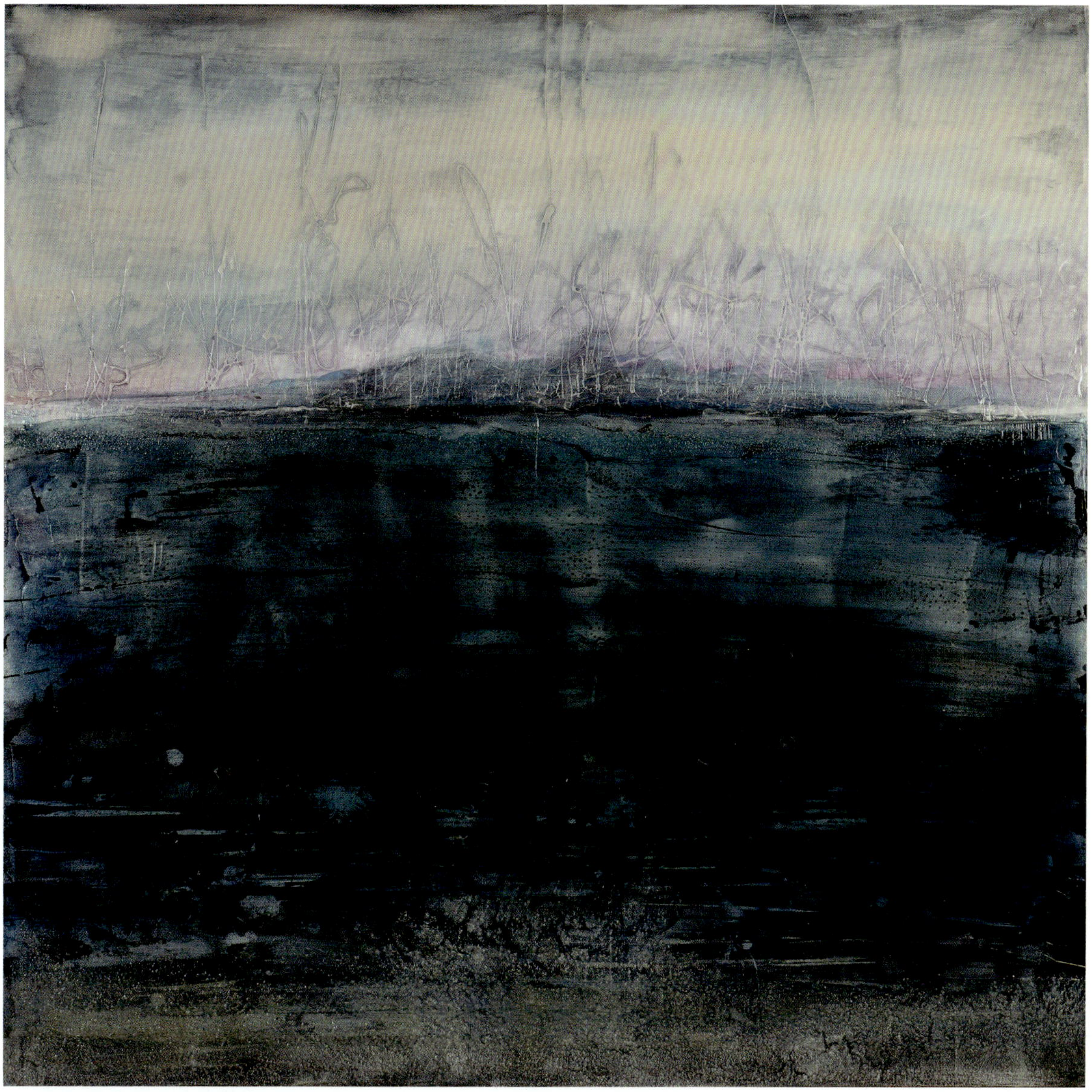

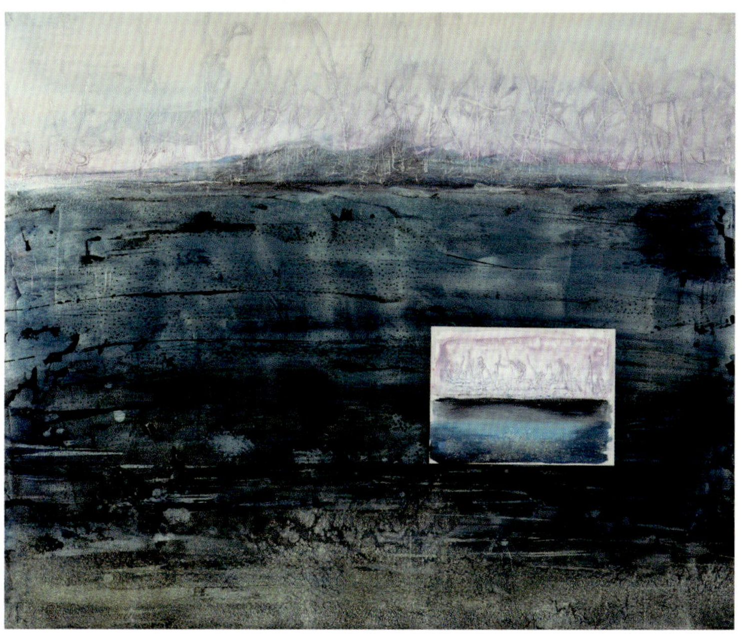

ABOVE **Rhéni Tauchid, *Solstice Anchorage,* 2017, acrylic and beads on canvas, 48 x 48 inches (123 x 123 cm). Collection of the artist.**

LEFT Here's the completed painting with the thumbnail. (A closeup of the thumbnail appears on page 29.)

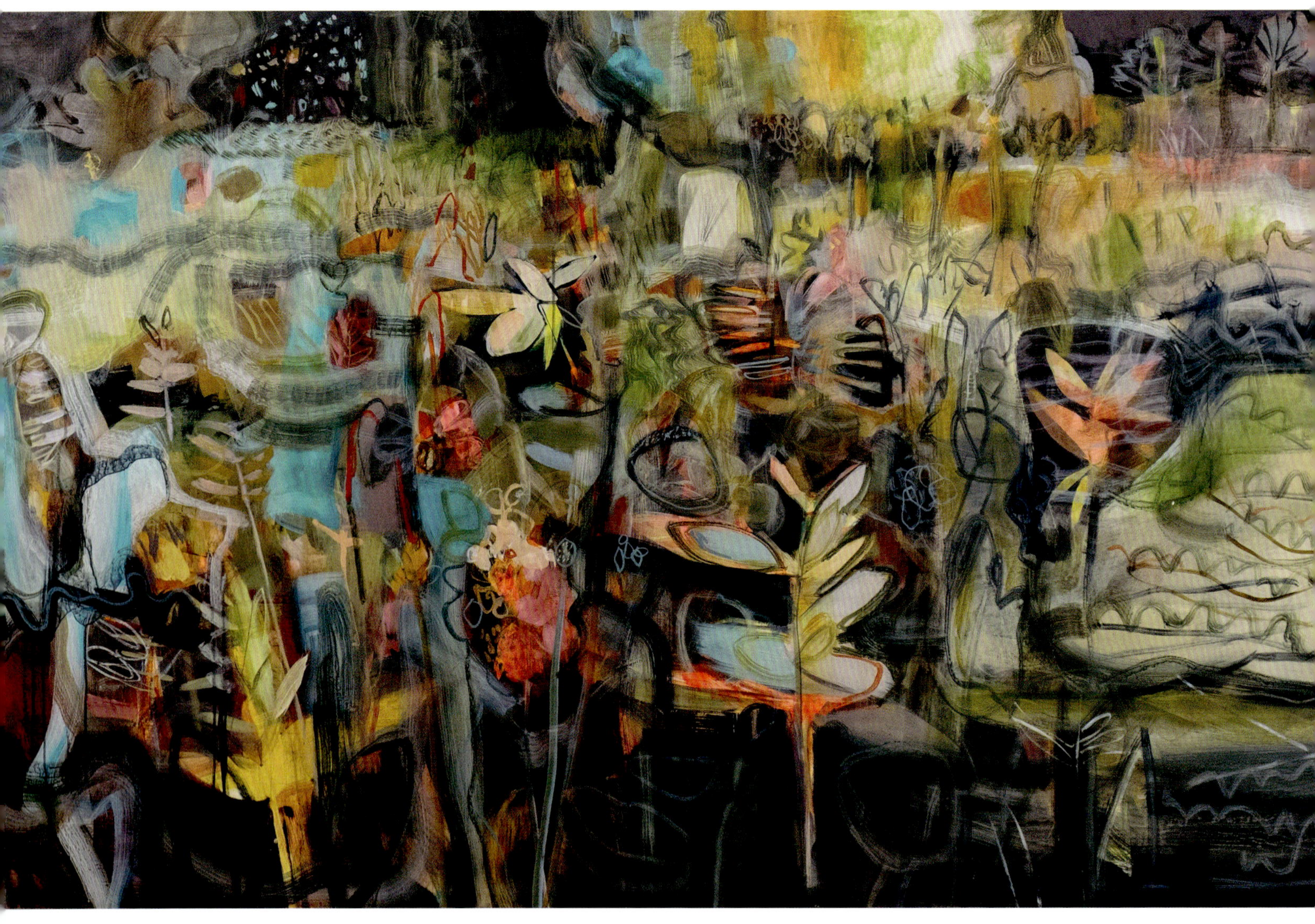

Lori Richards, *Wild Pursuits,* 2016, acrylic on wood panel, 48 x 72 inches (122 x 183 cm). Private Collection, Ottawa, Ontario, Canada.

About her painting Wild Pursuits, *artist Lori Richards says, "This painting is all about layering. It is large enough that you feel you can walk right into the buzzing summer field. Abstracted flowers, grasses, and seed pods vibrate with energy, and wavy horizontal patterns give shape to the air currents. To achieve this three-dimensional effect I used paint and mediums in various combinations from slippery transparent 'watercolors' (polymer medium with liquid acrylics) to thicker opaque impastos (clear modeling paste with high-viscosity acrylic paint). Gel medium (matte and semigloss) is a constant on my palette and gloss self-leveling gel makes a lovely base for smooth, effortless brushstrokes."*

COMBINING

Acrylic mediums are very similar to each another, and you can combine two or more mediums to produce unique mixtures. For example, say you want a granular medium that self-levels. You can achieve that by mixing a nepheline gel with a self-leveling gel. To ensure good bonding you must pay attention to how materials can be combined. Here are some things to keep in mind:

- *Blending mediums of differing viscosities.* First, take a look at the way in which each of the mediums holds—or doesn't hold—texture. Let's say you're combining a polymer medium (low viscosity, short rheology, holds minimal texture) with a regular gel medium (high viscosity, short rheology, holds significant texture). In this case, the key characteristic to focus on is the rheology. Depending on the proportions, combining these two mediums will result in either a thinner gel or a thicker polymer, holding different degrees of texture while maintaining a short rheology.
- *Blending mediums of differing rheologies.* The rheology and, by extension, the potential viscosity of the combined mediums will depend on proportion as well as viscosity. If, for example, you mix a regular polymer medium with a self-leveling gel, the thick self-leveling gel will lose some of its viscosity. And because the polymer medium contains a quantity of texture-holding thickener, the self-leveling gel will also lose some of its self-leveling capability. The resulting mixture will have a medium (shorter) rheology, a lower viscosity, and a slightly less smooth finished surface. If self-leveling gel is mixed with water or with a self-leveling type of liquid medium (e.g., low-viscosity polymer or another medium that does not contain a thickening agent that holds texture and that also has a short rheology), the result will be a lower-viscosity medium with a long rheology and a higher flow rate. The more water (or thin medium) added, the shorter the mixture's rheology will become.
- *Blending mediums of differing lusters.* As has been discussed, pure acrylic resin is glossy and transparent. Matting agents scatter and block the light, creating films that are more translucent. By mixing gloss and matte mediums together you can control both the luster and transparency of your medium.

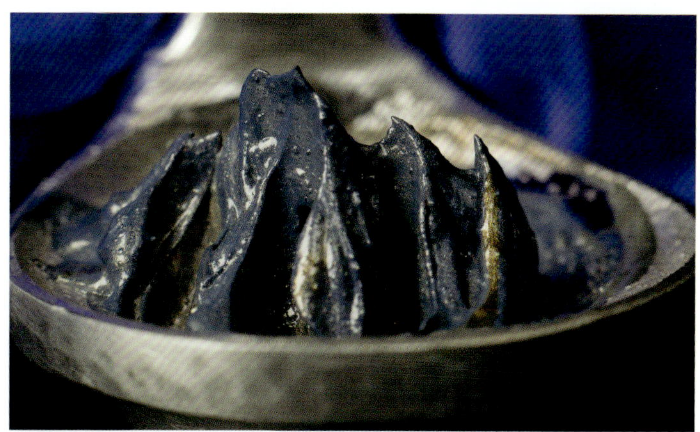

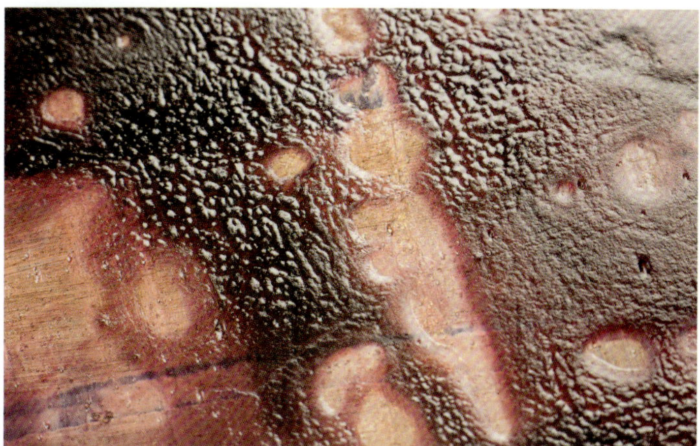

TOP A combination of light granular medium and gloss gel mixed with Payne's grey and iridescent bronze has a distinctive pewter-like appearance.

ABOVE Accidents like this orange peel-like surface can occur when different mediums—in this case, self-leveling gel and low-viscosity polymer medium—are mixed and dried unevenly. Results like this are impossible to predict, but they're not necessarily a bad thing.

PROCESS

COMBINING MEDIUMS

Four mediums are at work in this whimsical piece. The background is clear gesso tinted with teal and then scraped to look like painted wood. The tangerines are a mixture of self-leveling gel and nepheline gel fine tinted with a naphthol orange and transparent permanent orange mixture, capped off with extruded leaves of gloss gel and green acrylic colors. Here is the process I followed to create the tangerines:

RIGHT **Rhéni Tauchid, *Tangerine Teal*, 2017, acrylic on birch panel, 12 x 12 inches (30.5 x 30.5 cm).**

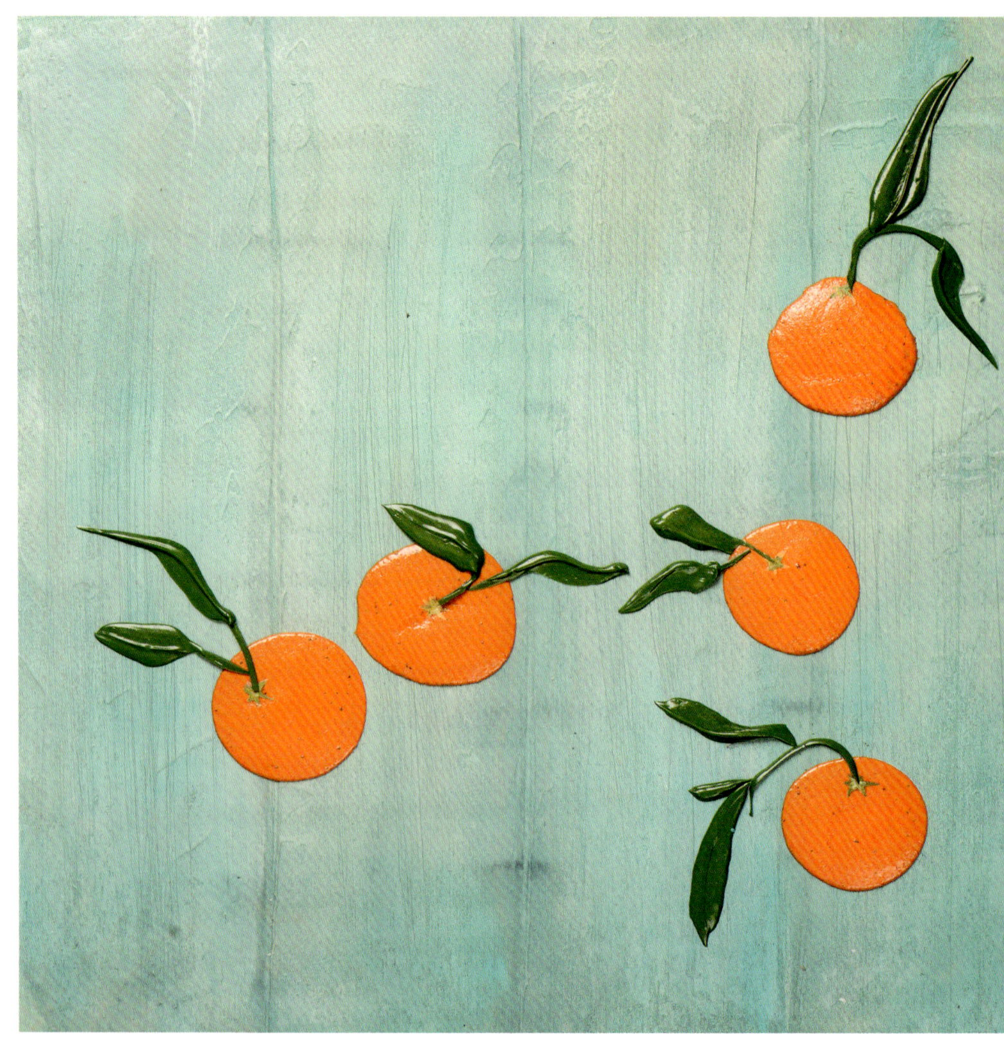

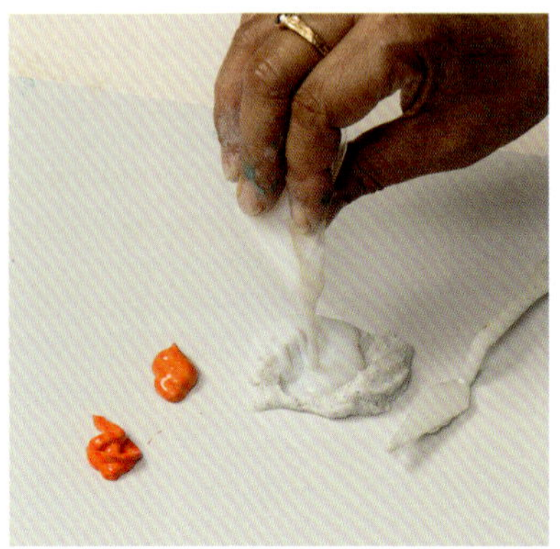

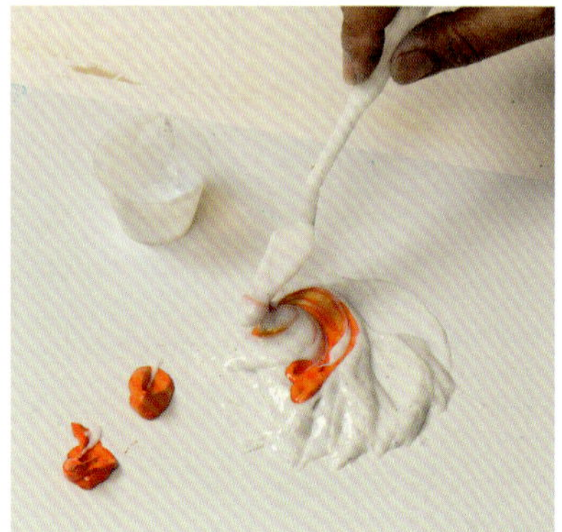

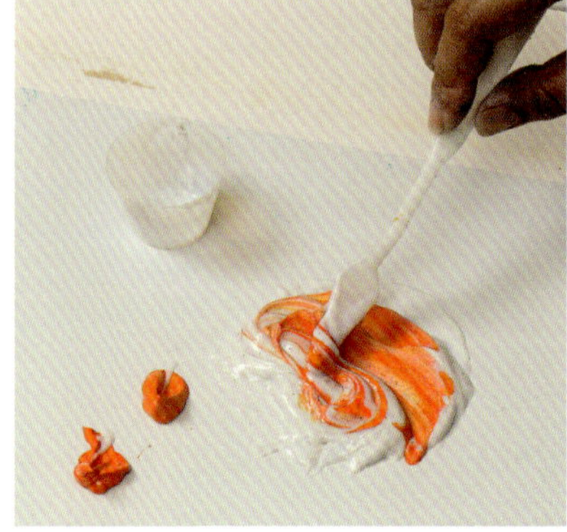

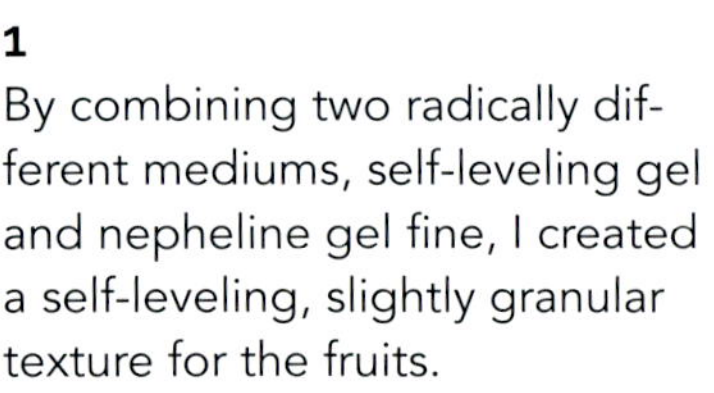

1
By combining two radically different mediums, self-leveling gel and nepheline gel fine, I created a self-leveling, slightly granular texture for the fruits.

2
Next, I needed to find just the right shade of orange for the tangerines' hue. A mixture of naphthol orange and pyrrole orange captured the tangerines' brightness and sunny warmth.

3
As I mixed, I tested for the right consistency. The mixture had to be fluid enough to pour.

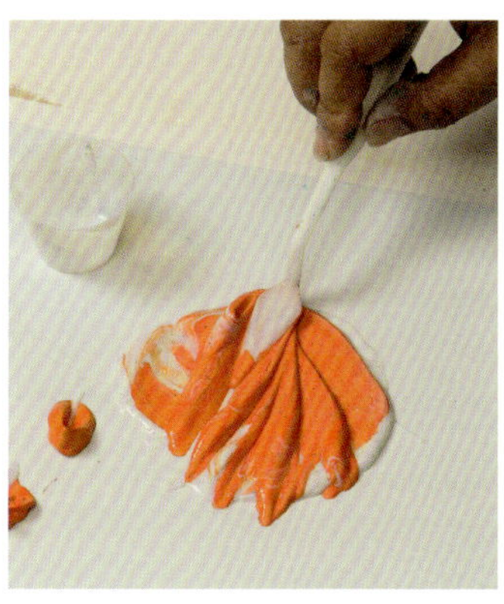

4
A palette knife is an ideal tool for mixing the color into the combined mediums without creating too many air bubbles.

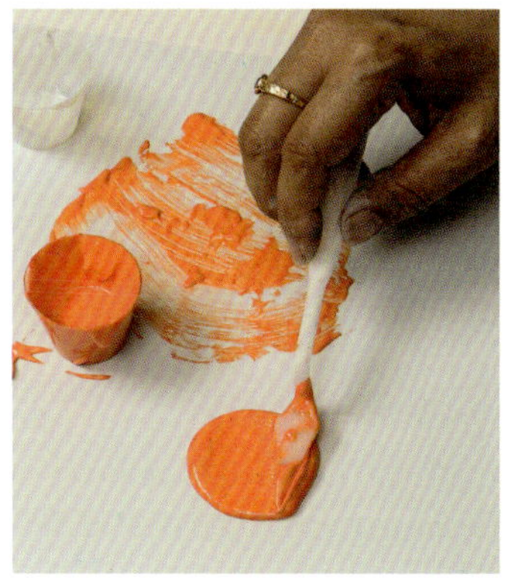

5
Next, I formed the fruits. Making these on a nonstick palette allowed me to create a selection of shapes to work with—and to move them around on the substrate before gluing them down with gel.

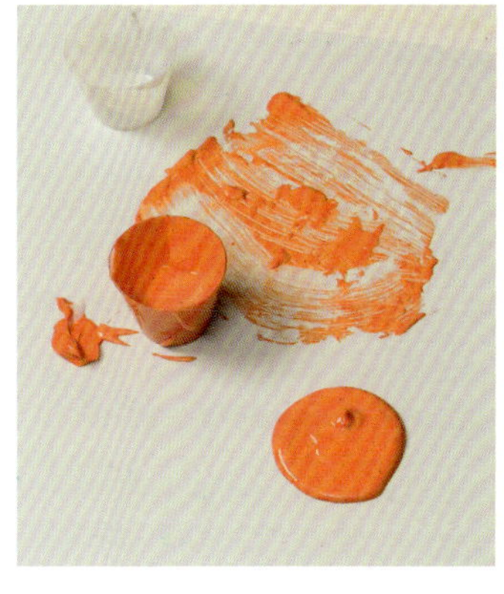

6
Shapes like these take 24 to 48 hours to dry. They can then safely be peeled from the palette and placed on the substrate.

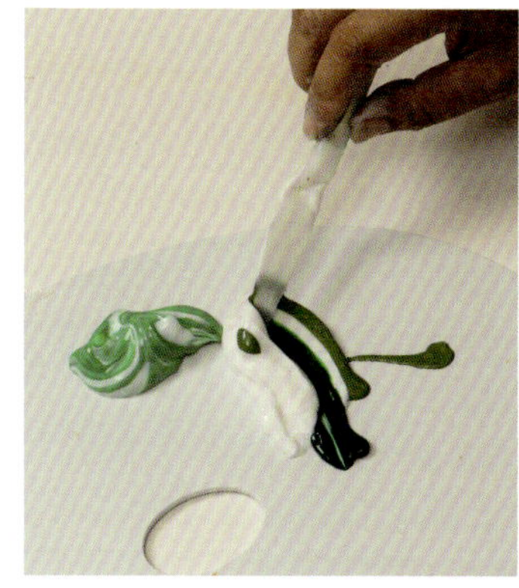

7
For the leaves, I mixed a little gloss gel with chrome oxide green and phthalo green yellow shade, producing a dark, rich green with a healthy sheen.

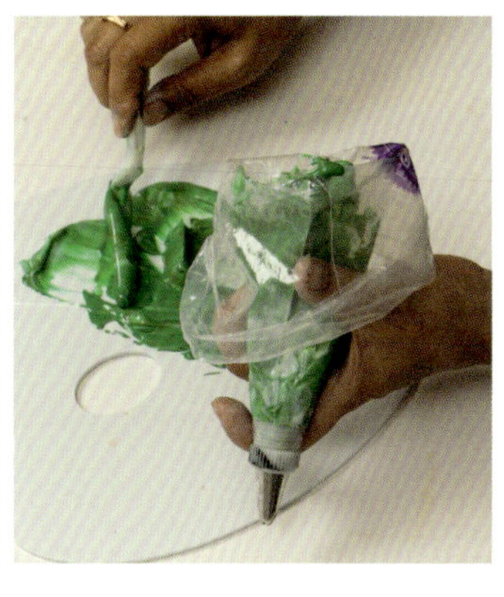
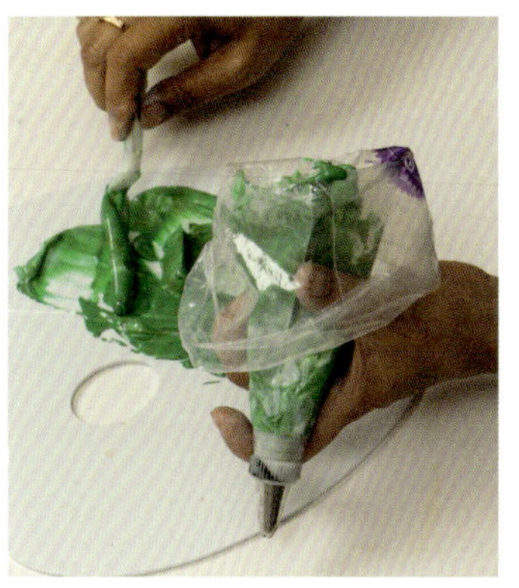

8
To make the leaves, I used a pastry tube filled with a fistful of paint.

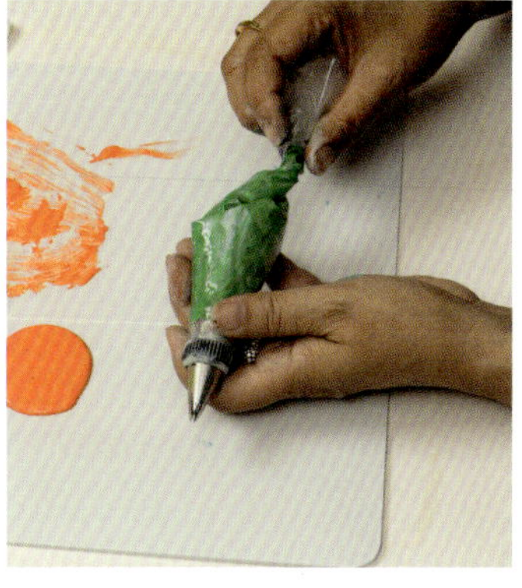

9
Twisting the open end of the pastry bag and tucking it into the palm of my hand prevented any paint from leaking out.

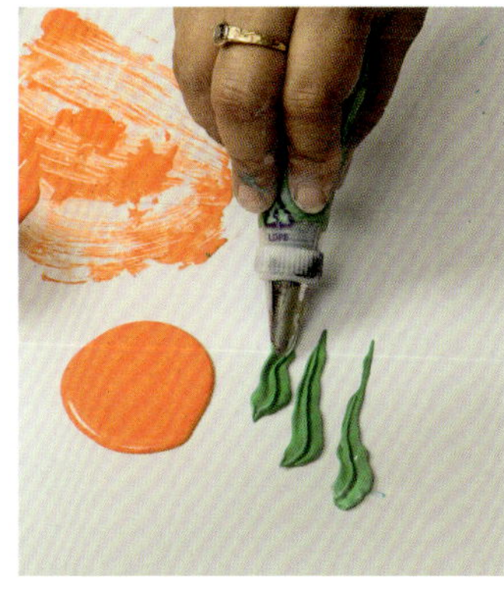

10
The specialized nozzle of the pastry tube is specifically designed for making leaf shapes.

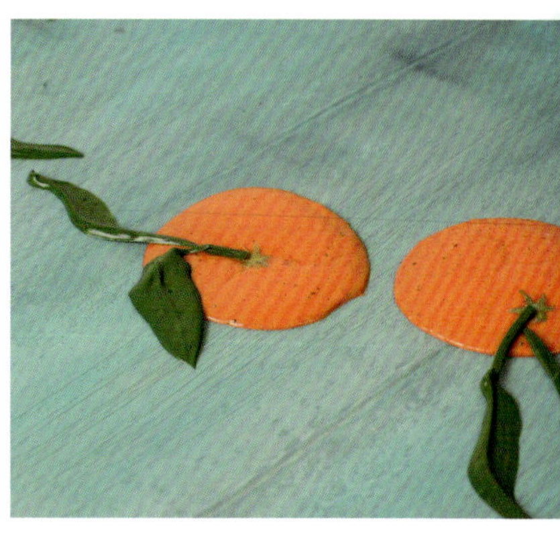

11
This detail of the finished work shows the tangerines on the teal background.

CHAPTER 9

GOING BEYOND THE BASICS

THERE IS SO MUCH THAT CAN BE ACCOMPLISHED using modern acrylic mediums that goes well beyond the mediums' traditional functions—and even more that painters, crafters, and makers of all sorts have yet to conceive of. This chapter looks at a number of unconventional ways in which mediums are being used. Some of these practices have been around for a long time; others are more cutting-edge.

Product labels often indicate the uses for which particular mediums are intended. It takes a sense of adventure and curiosity to travel beyond those recommendations. In some cases it's about taking a known acrylic medium and using it in an unexpected way; sometimes it's about using alternative tools to produce new effects. Such tools can expand the ways in which acrylic paints express your creativity.

Mediums are being used in unconventional, cutting-edge ways.

OPPOSITE **Diane Black, Untitled, 2017, acrylic pouring medium on panel, 10 x 14 inches (25 x 36 cm). Collection of Tri-Art Mfg.**

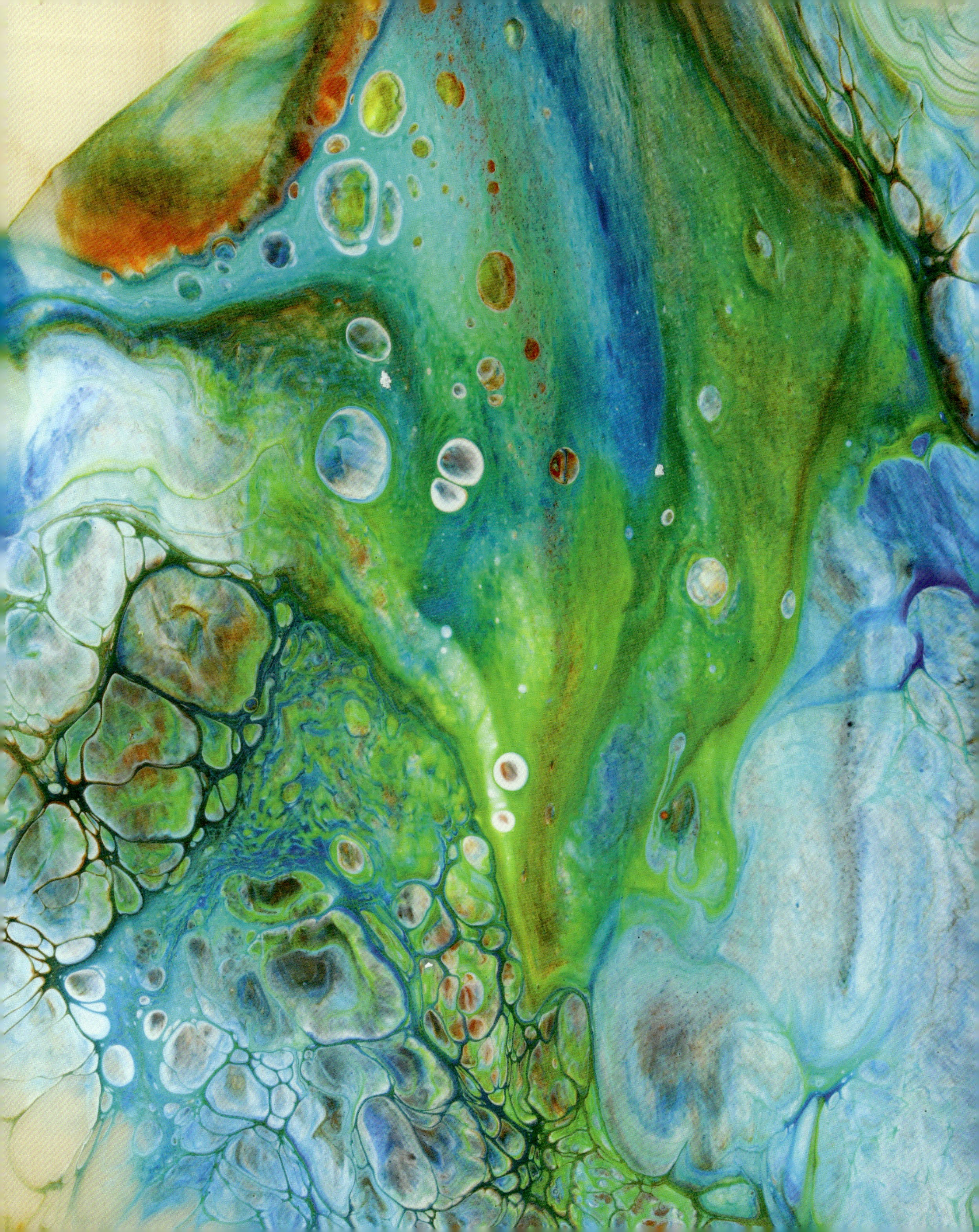

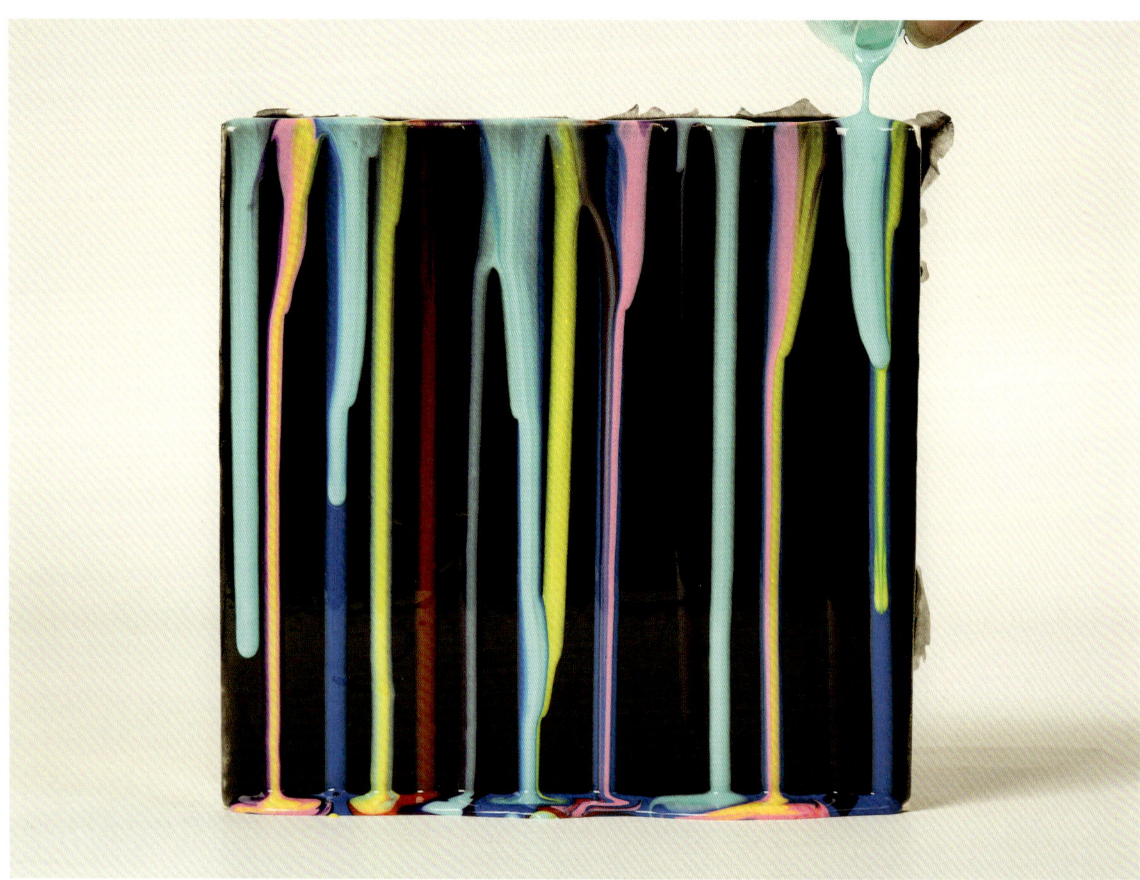

POURING MEDIUMS

Viscous, oozing pools of color—there is something viscerally enticing about these types of applications, and they're so satisfying to create. The results—which can range from the very subtle to the intensely dynamic—are always unexpected.

Pouring acrylics is fun, but it can be stressful if you are precious about your materials. Don't fret, though: mediums are (relatively) inexpensive, and if you use the right tools, the excess that falls from the pouring can easily be salvaged for later use. Mediums suited to pouring include pouring or self-leveling gels, low-viscosity polymer mediums, gesso, glazing mediums, and final finishes—basically, any mediums that are thin or have a long rheology. The medium's viscosity will determine how quickly it will pour, an important consideration if you want to maintain control.

ABOVE Allowing gravity to form the lines of pouring medium.

Don't be precious about your materials.

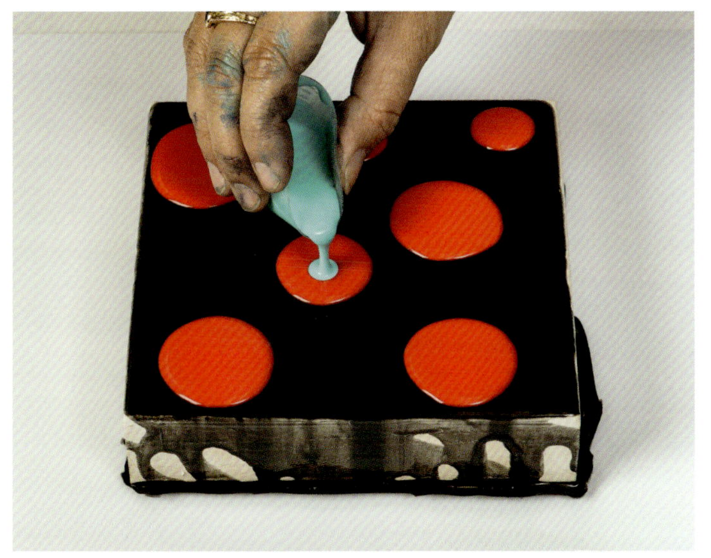

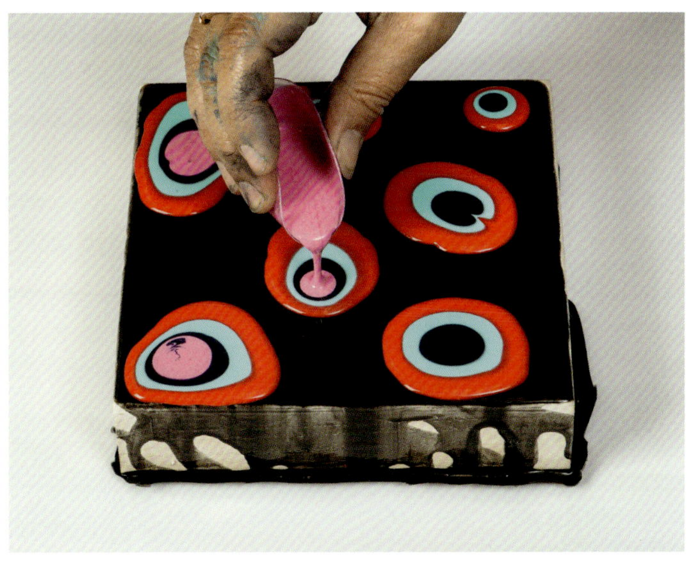

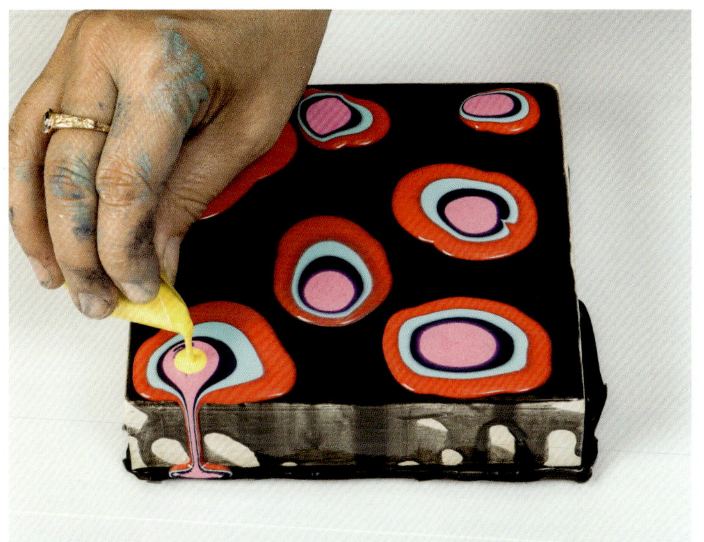

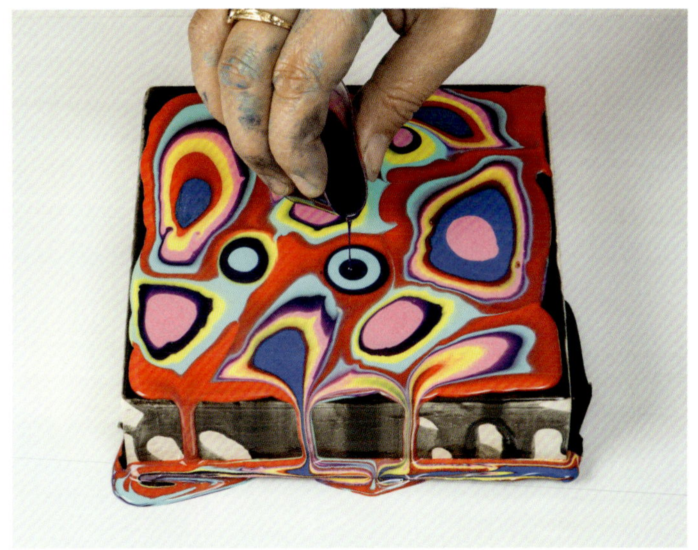

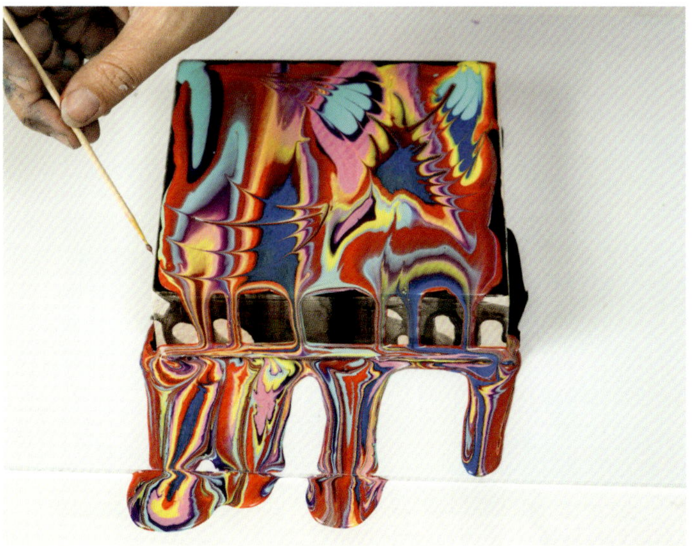

TOP ROW Pouring medium is hypnotic in its slow, oozing movement.

CENTER ROW Colors mixed into the pouring medium flow and float, remaining distinct from each other.

LEFT Tipping the panel gives gravity free rein. A bamboo skewer pulls the colors through each other to produce a marbled effect.

The more fluid your paint mixture, the more susceptible to gravity it becomes. You will want to work on a level surface to avoid pooling or having most of your paint drip off the support! Depending on the thickness of the pool of acrylic, some crazing may occur as it dries. This can be difficult to control if you are combining various mediums or adding water to your mixture. Stick to thinner layers if you intend to manipulate the poured medium by tilting the support. Add more medium (or water) as you work only if you need it.

Creating an absolutely smooth pool of paint is not as easy as you might imagine. In fact, it's next to impossible. And because so many factors contribute to how a paint film dries, it's almost impossible to reproduce any single application exactly. These factors include, but are not limited to, the following:

- Ambient temperature
- Ambient humidity
- The porosity of the support
- How recently the paint was mixed
- The age of the paint and medium
- Application tool used
- The particular combination of medium and color
- The addition of water to medium
- The degree to which the underpainted surface has cured
- Painting on a surface that is not completely level

POURING DO'S AND DON'TS

Controlling mediums with a long rheology, as most pouring mediums have, requires that you pay close attention to their application.

WHAT TO DO:

- Mix color into the pouring medium several minutes to hours prior to application to avoid trapping air bubbles.
- Ensure that your painting surface is level prior to pouring.
- Pour close to the surface to avoid splatters
- Spread with flat tool.

WHAT *NOT* TO DO:

- Don't mix your paint/medium too vigorously. (You'll create lots of bubbles.)
- Don't add too much water.
- Don't pour from significant height (big bubbles!).
- Don't combine mediums/paints of different viscosities.
- Don't spread with brush (too much tug).
- Don't touch the surface until dry—unless you are fond of fingerprints!
- Don't continue working surface as it dries (disaster!).

PROCESS

FLOATING COLORS IN POURING MEDIUM

Fluid painting is a simple but mesmerizing process that for many is the entryway into working with acrylic mediums. The random swirling patterns and color pools are achieved through pouring and dripping colors and then adding a new dynamic, using movement to make them travel across the support.

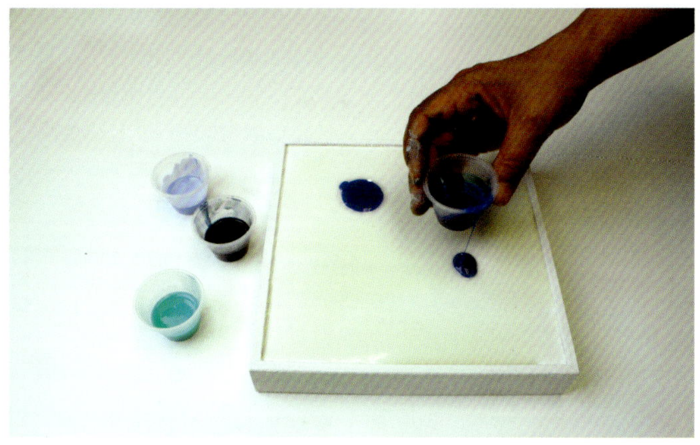

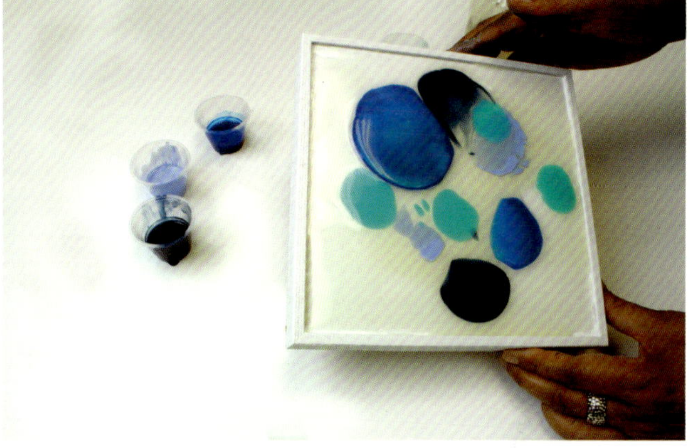

1
First, apply the initial layer of pouring medium onto your surface. (This cradled wood panel has a small lip, making it perfect for very fluid applications like this one.) Tint small quantities of medium with the desired colors, mixing them in thoroughly before beginning to pour.

2
Take turns applying drops of colored medium and clear medium, then begin to tilt and move the support, allowing the colors to flow over each other.

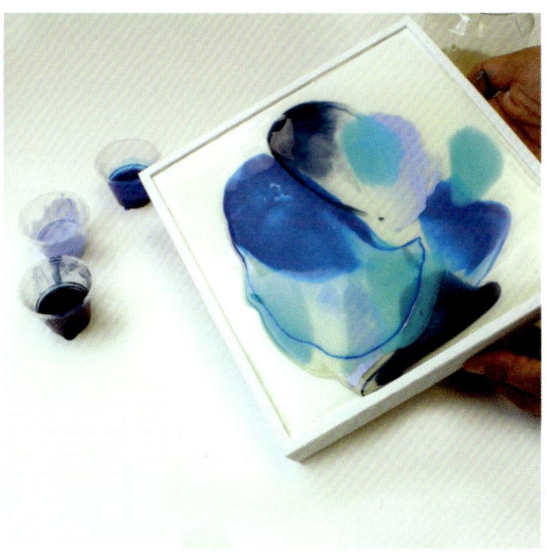

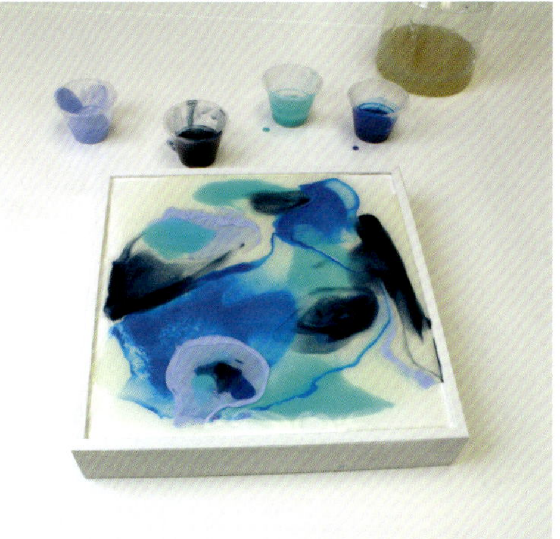

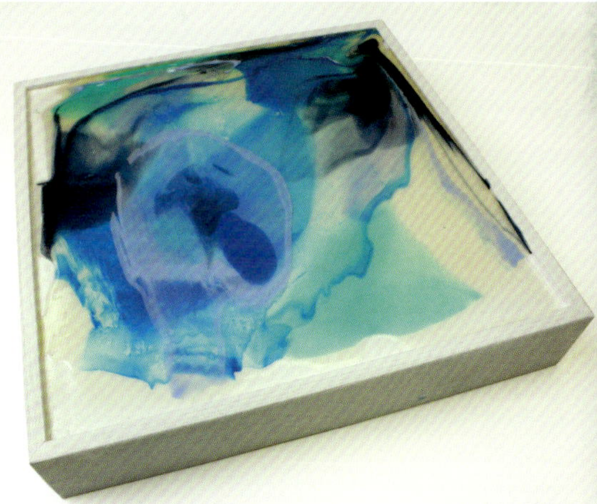

3
The more clear medium is added between layers, the more the colors will appear to float over each other.

4
Once you achieve the effect you are looking for, allow the paint to rest for at least twelve hours.

5
The now-dry pour looks very similar to the wet version because of the transparency of the pouring medium when wet.

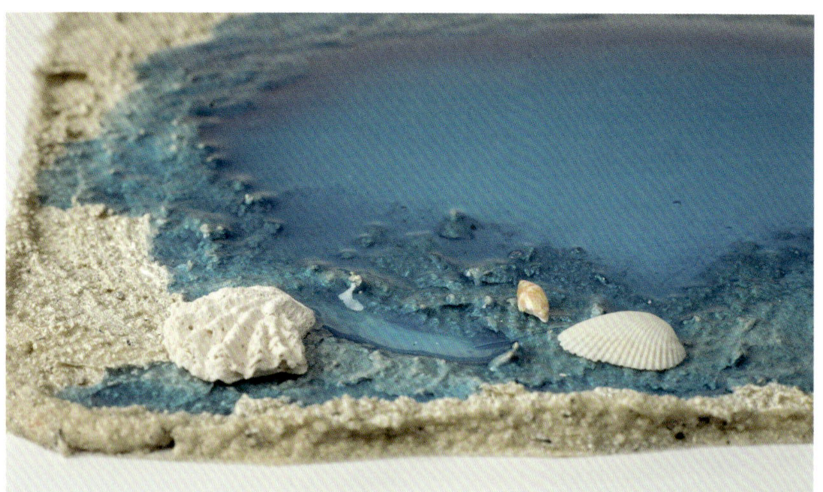

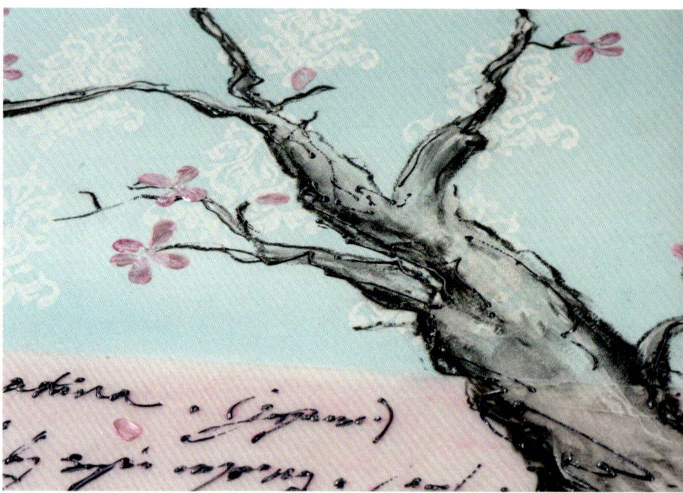

ABOVE, LEFT AND RIGHT A lagoon and beach created with pouring medium (tinted with phthalo blue green shade) and nepheline gels (tinted with unbleached titanium). • Many mediums are at work here: a light liquid medium resist on watercolor paper with a dilute acrylic wash background, a layer of dry media ground, extruded gel detail over a charcoal drawing, and acrylic-skin petals glued to the surface with matte liquid medium.

"Pouring mediums" are any mediums that are pourable, although there are some that are specifically designed for this purpose and are best for this procedure. Such mediums are different from self-leveling gels in that they tend to dry to a harder, less flexible finish, are often glossier, and will not tolerate heavy tinting. For a smooth finish when pouring colors into pouring medium, mix each color separately with the medium prior to application. The colors will then have the correct consistency and viscosity to be seamlessly integrated into the medium.

USING MEDIUMS AS RESISTS

Acrylic paint is water based, and dried acrylic paint film is water resistant. You can use these two realities to create interesting effects. While it's also a fact that acrylic sticks to acrylic extremely well, this does not actually happen until the paint is fully dried (cured). Until that point, the fluid nature of the paint means that the wet additions are movable on a painted surface and will absorb at different

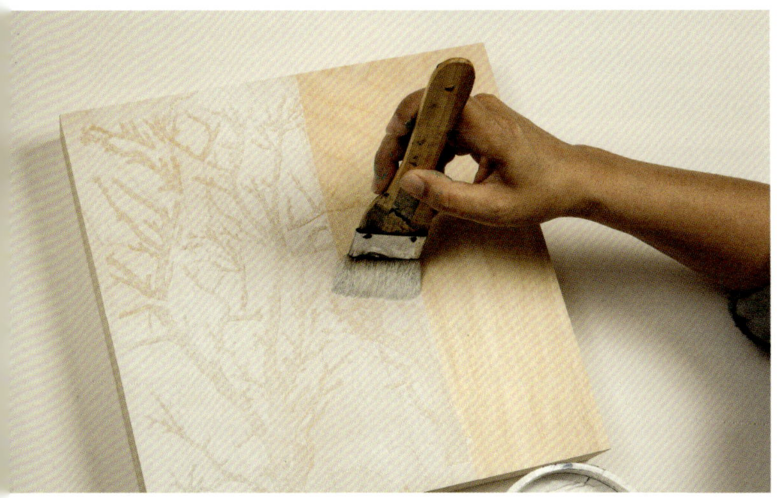

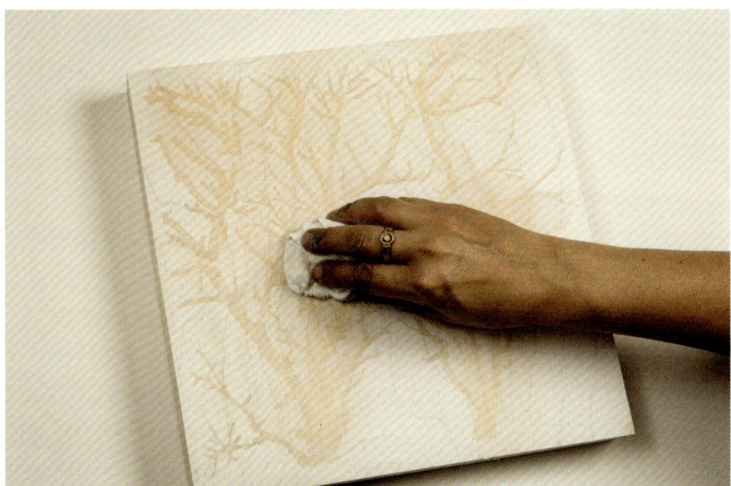

ABOVE, LEFT AND RIGHT Gloss polymer stencil on raw wood panel. Once dried, the medium acts as a resist element. The light white wash soaks into the unprimed, and thus absorbent, wood but not into the areas blocked by the stencil. • Here, the stencil has been removed. Gently dabbing and rubbing away excess paint helps keep the resist area clear.

levels on surfaces of varying porosity. For example, paint will absorb more readily into paper or unprimed canvas than it will on a glossy painted surface. There is time to "play" upon the surface of a raw, primed, or even painted support.

Playing with absorption—the push and pull of gloss versus matte—can produce serendipitous accidents that give dimension to an otherwise flat surface. Paint adheres to glossy and matte surfaces differently. A wash-thin application of color over an absorbent matte area will saturate the surface more readily, resulting in deeper color than will result when the same wash is painted over a slick, glossy surface, because the smoothness of the gloss finish acts as a resist.

A fun exercise, and an effective method for blocking out color in an underpainting, is to use a gloss polymer medium to block out areas on an uncolored ground. Once dry, these areas will resist the subsequent application of color, which will soak into the unblocked areas. This works particularly well on watercolor paper or unprimed wood panels.

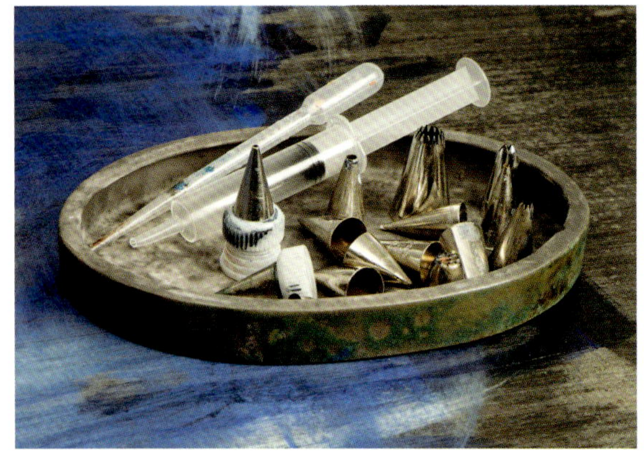

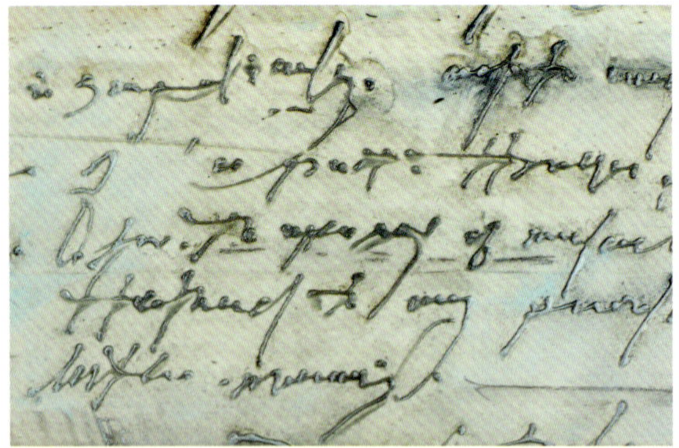

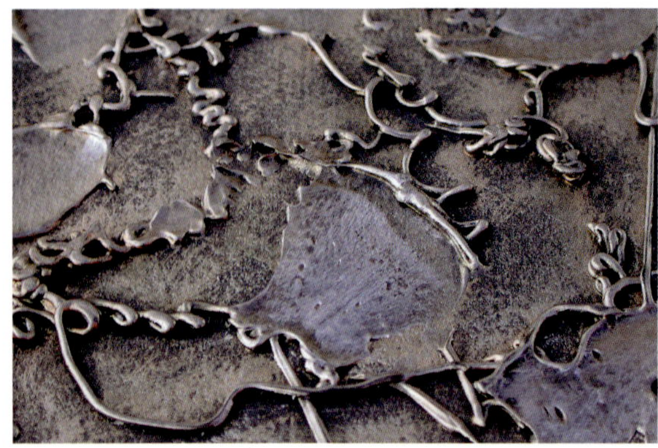

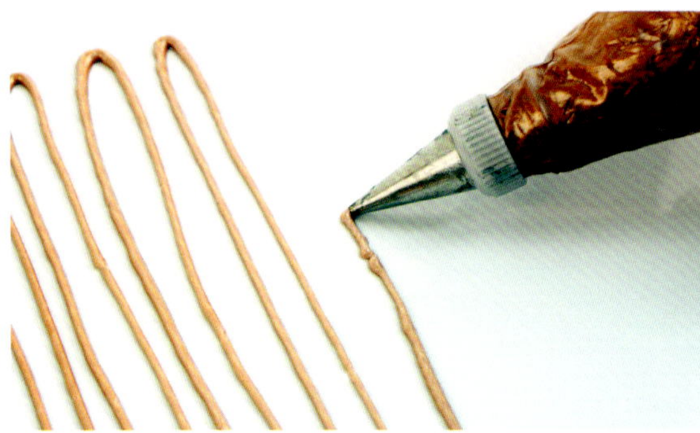

TOP TO BOTTOM Tools for extrusion. • Matte gel extruded text amplified with dry-brushing and washes. • Gel extrusions, partially smoothed with a palette knife and painted to look like metal. • Extruding gloss gel and iridescent copper "thread" onto a nonstick palette.

EXTRUDING MEDIUMS

Many years ago, my need for precise control of a line of texture led me into a kitchen supply store, and my love affair with extruding tools began. Pushing paint through a piping tip creates a uniform line whose girth and form are dictated by the shape and size of the tip's opening. Plastic or plastic-lined pastry icing bags make perfect watertight reservoirs for acrylic paint, keeping it wet and workable for a long time. Adding raised text was the starting point toward perfecting my use of a pastry tube. (Handwriting is entrenched in our muscle memory, so it is a simple way to practice this technique.)

Here are some tips for using a pastry bag for extrusion:

- Do not overfill the pastry bag. Fill it one-third or one-half full, twisting the open end and tucking it into your palm.
- Holding the bag downward, tap the paint in the bag a few times to remove any air bubbles.
- Touch the tip to your surface while applying the paint. This will help you control the flow of your line.
- To keep your line a consistent width, apply even, continuous pressure to the paint reservoir while extruding.
- If you need to stop for a short while and do not want the paint to dry in the tip of the tube, rest it tip down in a shallow cup of water. Make sure the open end of the bag is twisted closed to keep the paint inside the reservoir wet. When working with a small tip, stuff the end of a round toothpick or nail into the nozzle to keep the paint from forming a dried plug.
- For a medium to perform well in extrusion, it must be thick and be able to hold texture. If you are using a granular medium, use a tip with a larger opening to avoid blockage.
- Other tools can be used for extrusion, such as large plastic syringes, squeeze bottles, or, in a pinch, Ziploc bags (just cut off a corner!).

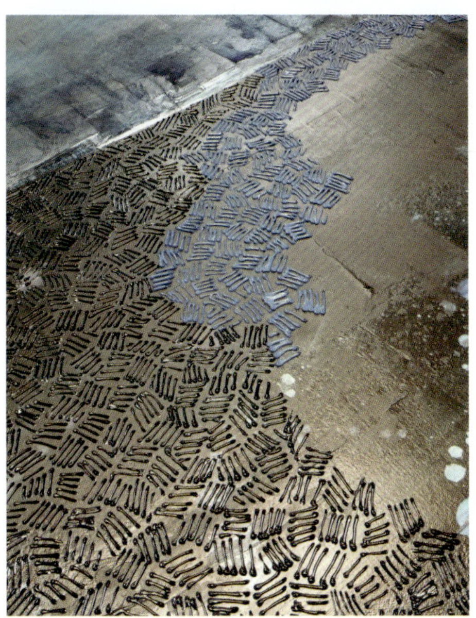

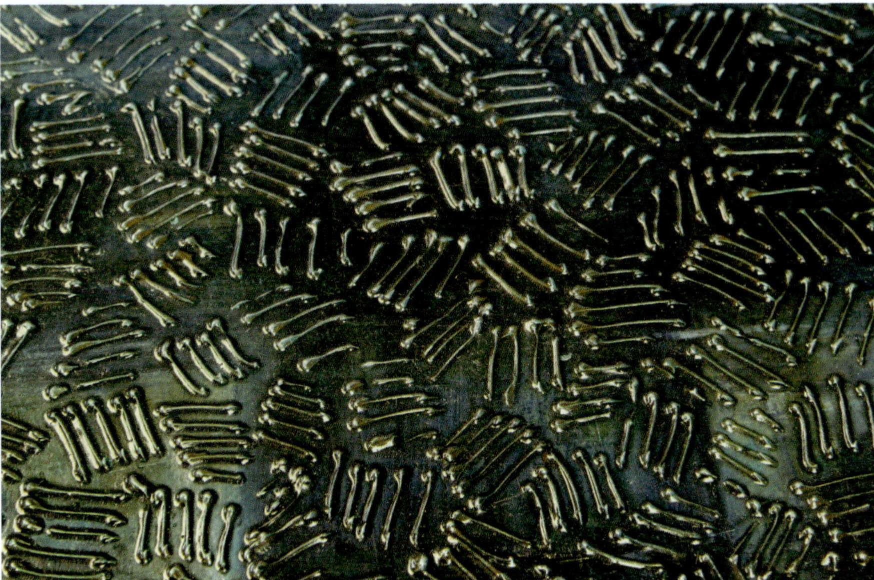

TOP **Rhéni Tauchid, *Deep North Revisited*, 2016, acrylic on canvas, 40 x 60 inches (102 x 152 cm). Private collection, Kingston, Ontario, Canada.**

A profusion of crosshatches (gloss gel plus Payne's grey), extruded one line at a time, provides a visual and textural contrast to the flat gloss-gel layers below. A sweep of bright metallic color amplifies the dimensionality of the crosshatches.

BOTTOM, LEFT Adding crosshatch details to *Deep North Revisited*, one extruded line at a time. The newly added area, still wet, is much lighter than the dried area.

BOTTOM, RIGHT Extruded crosshatch lines of tinted gel on colored surface, brushed with iridescent color to bring out the texture.

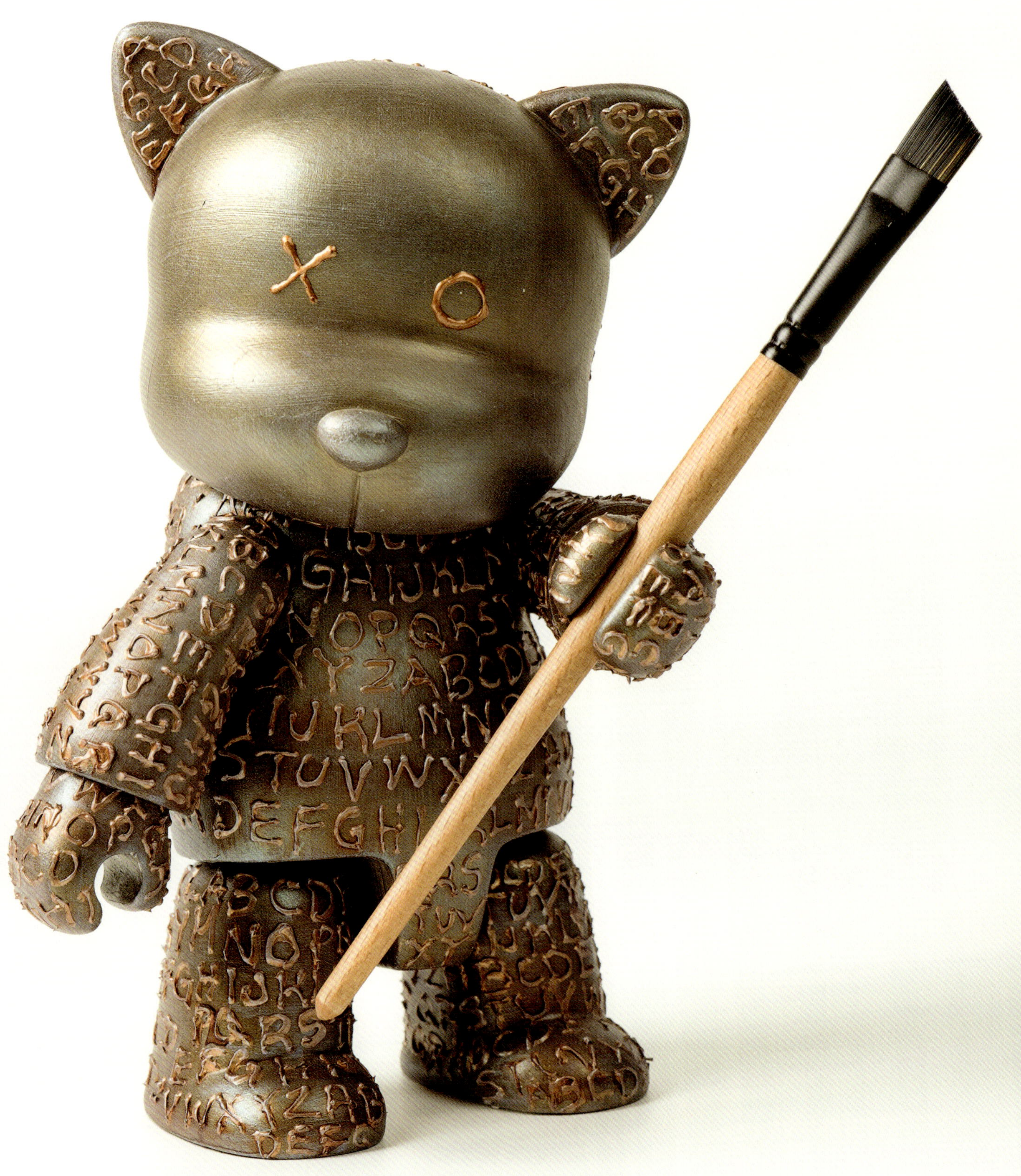

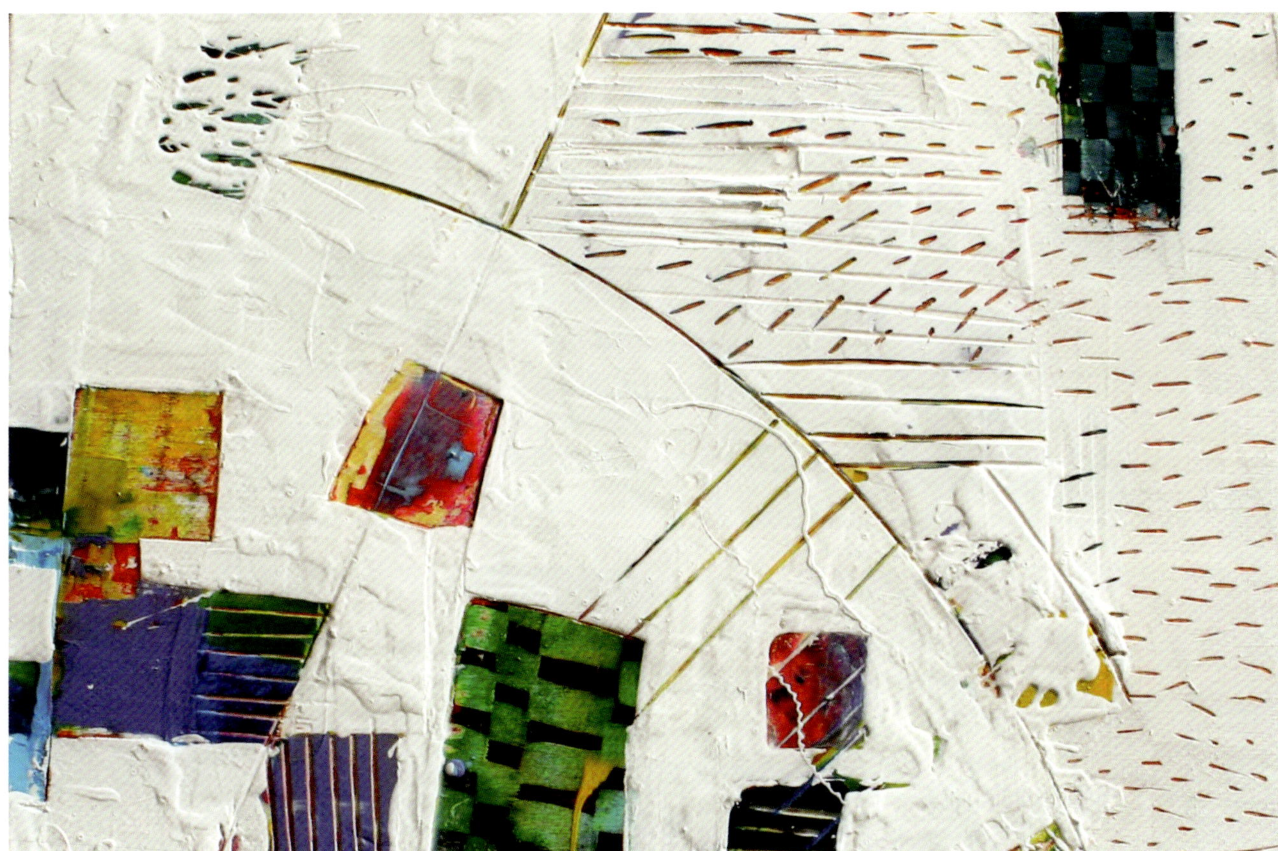

ABOVE **Pauline Conley, *Visiting Day,* 2011, acrylic and self-leveling gel on board, 24 x 24 inches (61 x 61 cm). Private collection, Ottawa, Ontario, Canada.**

The underpainting of Pauline Conley's *Visiting Day* consists of poured semitransparent layers of self-leveling gel. Suspending the layers of pigment in this way allows the ambient light to mix the colors. The top, obscuring layer is pale and opaque. The underpainting is revealed through small "windows" in the top layer and through incisions, made with wood-carving tools, that reveal some of the mysteries beneath. The carved lines also energize the calm color fields created by the layers of gel.

OPPOSITE **Rhéni Tauchid, *Alphabetakitty,* 2014, cast vinyl Artoyz toy with acrylic colors and mediums, 8 inches (20 cm) high. Collection of the artist.**

Do-it-yourself Artoyz Qee vinyl paintable toys are lots of fun to paint. On this kitty, I started with two layers of liquid iridescent bronze acrylic. Next, I added letters made of a mixture of gloss gel and iridescent copper paint extruded through my pastry tube. I finished off with a light dry-brushed layer of interference green paint.

INCISING THE ACRYLIC SURFACE

The reverse of extrusion is incision—etching or carving into a surface. Smooth, high-viscosity mediums with short rheologies can be inscribed into when wet, leaving valleys and crevices for new layers of color to sink, pool, or be pushed into. Etching into a semi-wet swath of medium is intensely satisfying, revealing underpainting and visual and textural contrast.

You can also cut into dried mediums—some more easily than others—using a sharp linoleum cutter or craft knife. Mediums with more solids, like modeling pastes, will cut more smoothly than gels. Very soft, flexible gels may be more stubborn, and, unless your blade is very sharp, the tug can produce a messy incision.

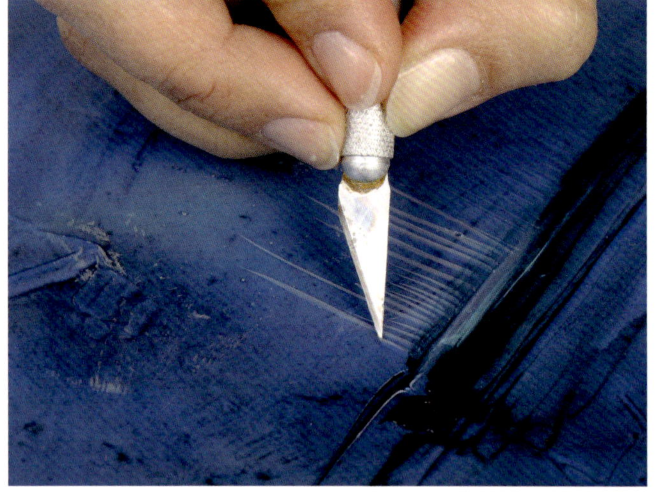

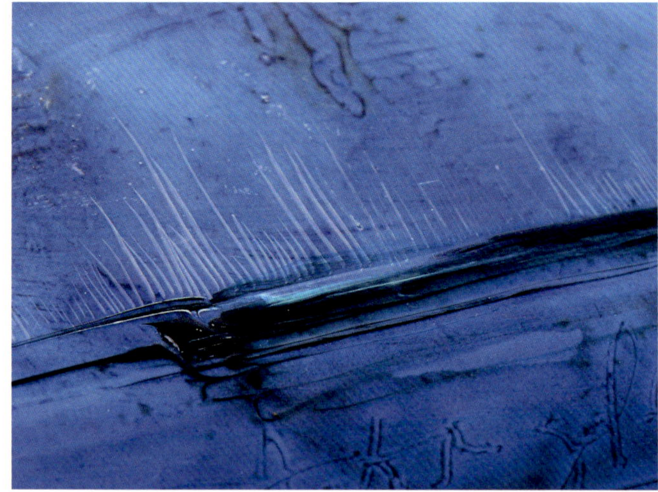

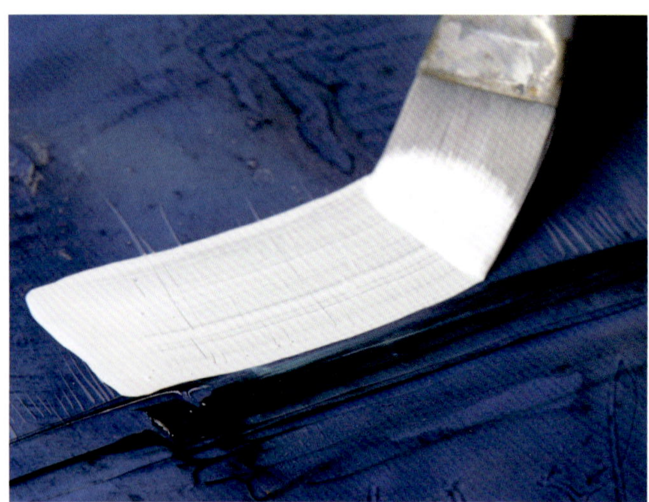

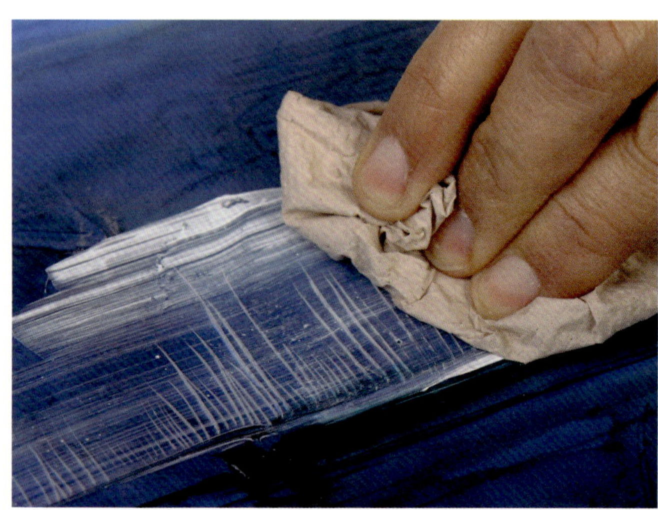

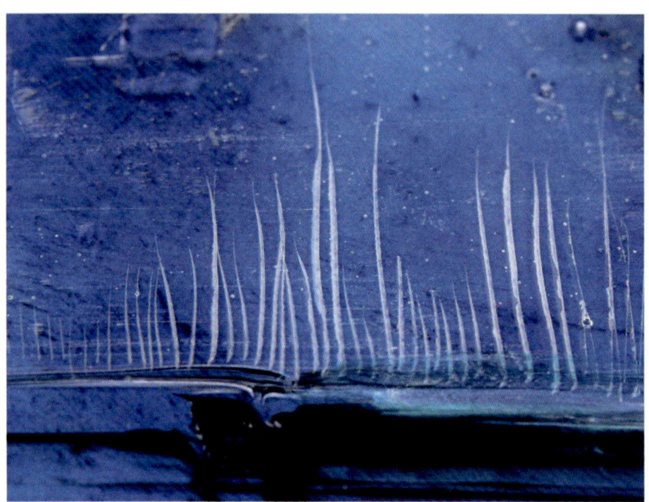

To effectively distress an acrylic surface, you have to wait until it is cured; otherwise the paint surface will be too soft and will tear instead of being cleanly incised, sanded, or scraped. Because they are brittle when dry, mediums with lots of inert fillers, like modeling pastes and gessoes, can be sanded or abraded with ease.

TOP, LEFT AND RIGHT Cutting into dried paint film at a sharp angle. • Incisions in dried paint.

CENTER, LEFT AND RIGHT Pushing paint diluted with water over incisions. • Wiping away excess paint.

BOTTOM, LEFT Paint trapped in the incisions.

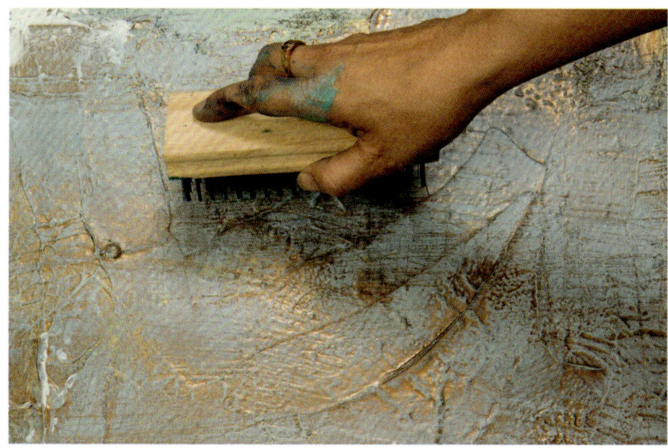

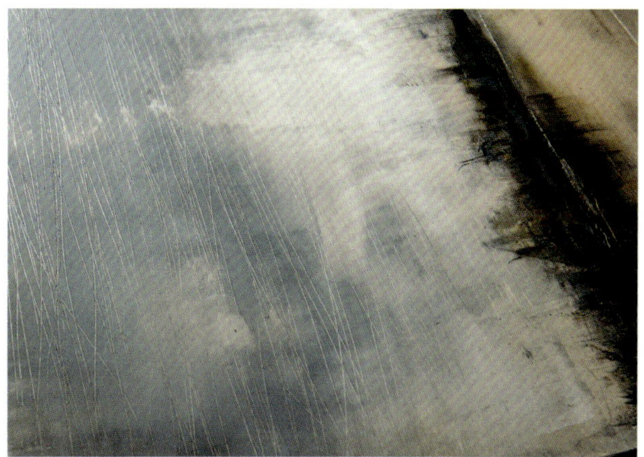

ABOVE The stiff metal bristles of this tool dig into the dried film of tinted gesso, bringing out the burnished coppery painted surface below. The result is a contrast in luster and reflectivity.

ABOVE, RIGHT A wet layer of translucent medium (dry media ground) has been cut through with the edge of a palette knife to reveal the lustrous underpainting—like the marks left by skate blades on a thin veil of snow over ice.

PRINTING WITH ACRYLIC MEDIUMS

Acrylic mediums really can do it all. Additives help acrylics become printing paints, but acrylic mediums can also become printing plates. A relief made of medium—or any medium that has been incised—can act as printing plate when "inked" with acrylic color.

For the most effective results when printing with acrylics, the acrylic paint should first be mixed with a retarder or flow release agent—especially important if you want to create more than a single print (monoprint). Retarding gels add another dimension to blending. These have the consistency of petroleum jelly, viscous and gelatinous. They impart a slippery quality to thicker paints and are particularly useful for printing applications. Acrylic colors can be adapted for screen printing when mixed with a screen-printing medium.

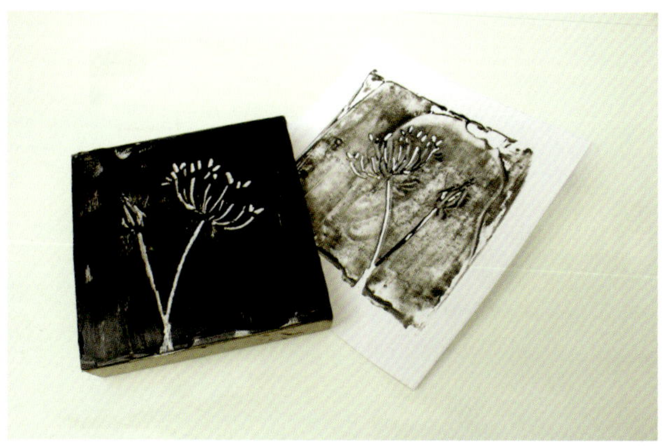

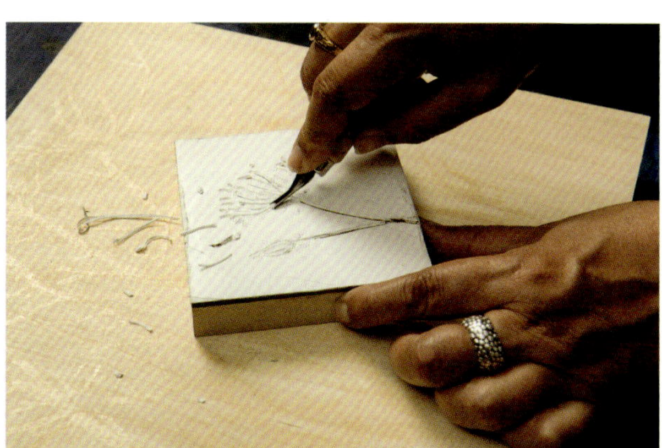

CENTER, LEFT A thick layer of modeling paste, fully cured, can be carved into with linoleum or wood cutting tools and used as a printing plate.

BOTTOM, LEFT Modeling paste used as a printing plate.

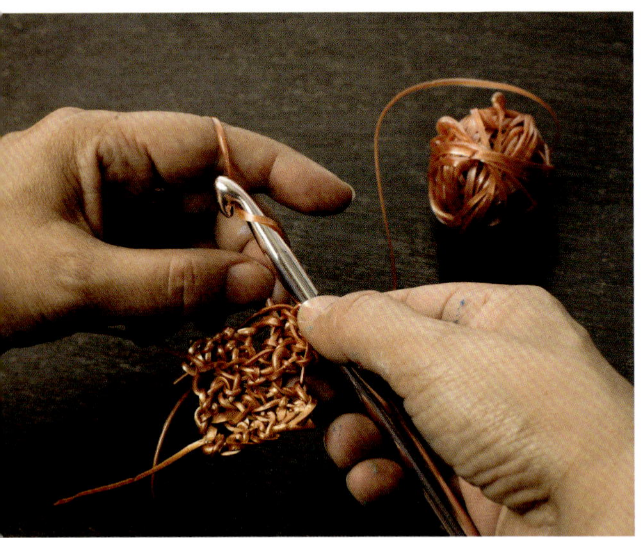

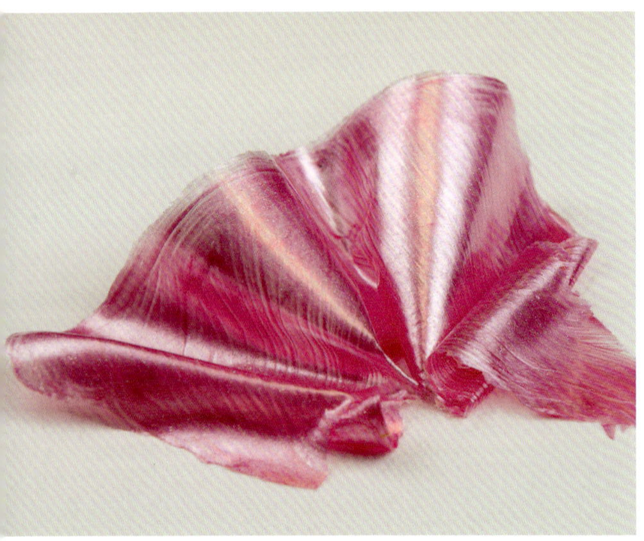

TOP Crocheting with extruded acrylic "yarn."

BOTTOM This skin of smooth, shiny gloss gel medium and metallic pink looks like fabric for a party dress.

SCULPTING WITH ACRYLICS

We don't generally think of acrylics as a solid, sculptural material, but this is where the medium truly distinguishes itself as revolutionary. Other paint media can do nothing more than lie on their supports, but acrylic paint films have a flexibility, resilience, and cohesiveness that allows them to break free of any support.

Acrylic skins are not a new material. We make them every time we paint with acrylics on any substrate, and we peel them off our palettes and tools, usually tossing them away. A paint skin that is intentionally created to be independent of the substrate, however, is radically divergent as a creative material.

We know, of course, that acrylics are plastic, but we often forget that fact when using them in their wet state. Once dried and cured, they exhibit amazing flexibility and tensile strength—capable of being stretched and wrapped and of maintaining those shapes. These traits allow acrylics to be used as solid sculpting materials, greatly expanding the scope of what can be created with them.

When you work with acrylics three-dimensionally, the ambient temperature of your environment will have an impact on the materials. Acrylic paint is easiest to work with at room temperature. Flexibility and surface tack increase in very warm environments, and in colder temperatures the acrylic becomes stiff. In extreme cold, it can crack or shatter if bumped. When transporting a flexible acrylic sculpture in cold weather, do not fold it or let it blend until it has reached room temperature and regained its flexibility. This rule also applies to paintings on unstretched canvas rolled for ease of transportation.

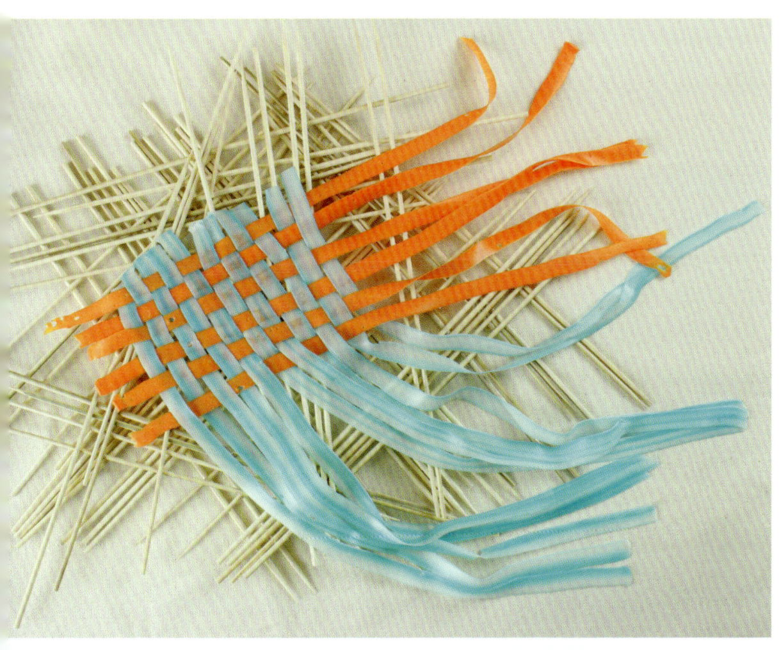

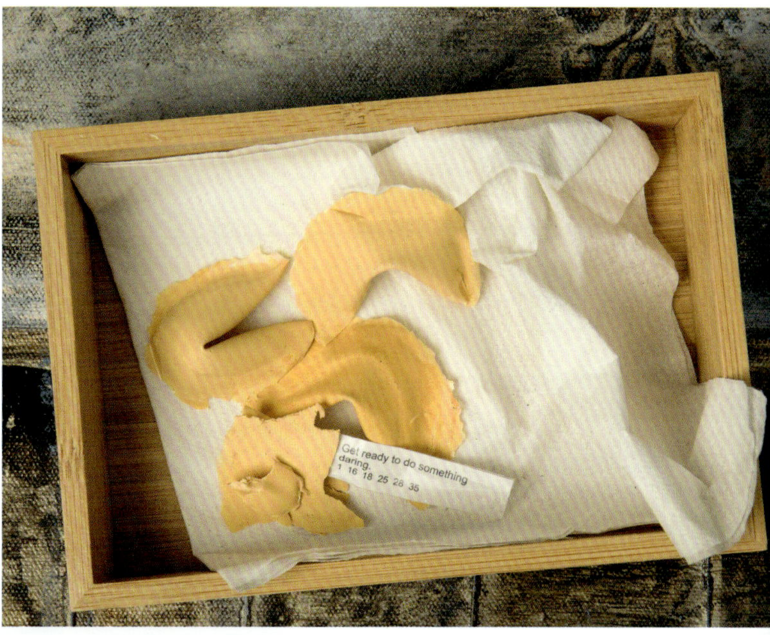

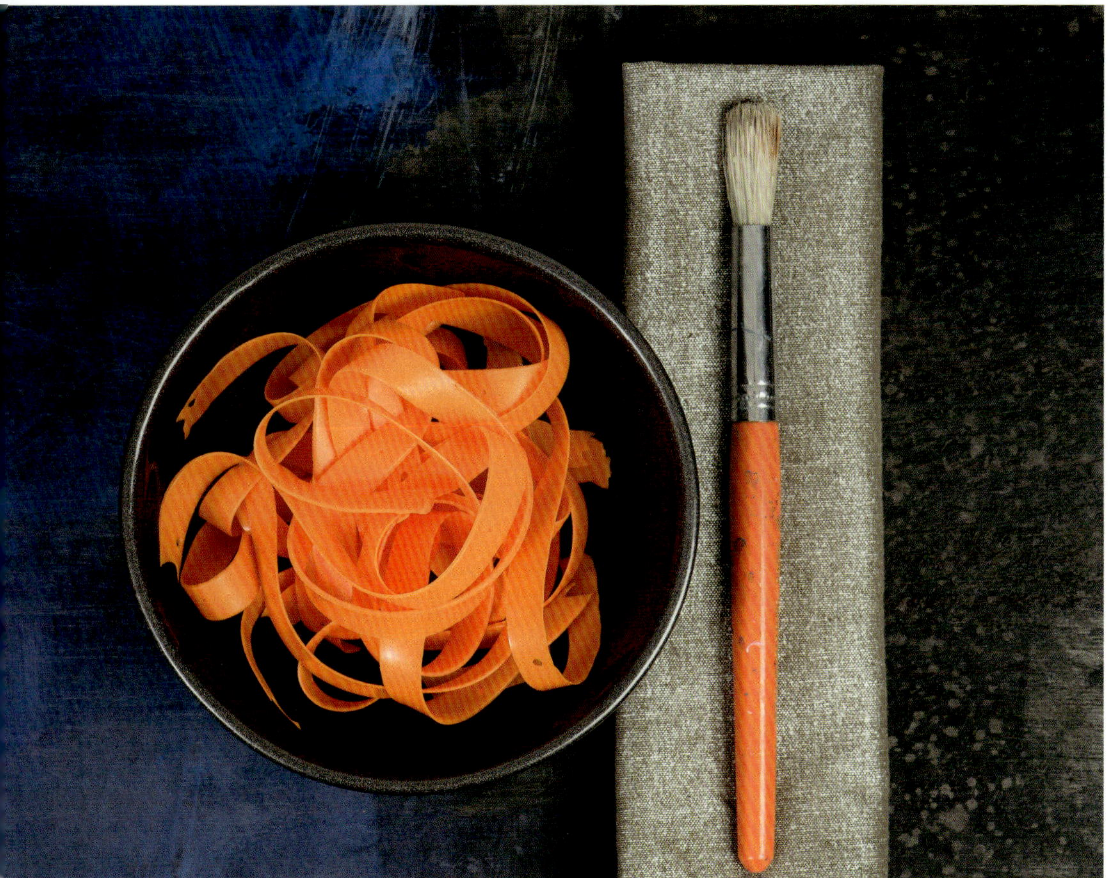

ABOVE, LEFT Woven bands of color-infused matte gel medium.

ABOVE, RIGHT Thin wafers made with clear gesso, lightly diluted with water and tinted with color, have the exact shade and texture of fortune cookies.

LEFT Extruded bands of matte gel are like acrylic "fettuccini."

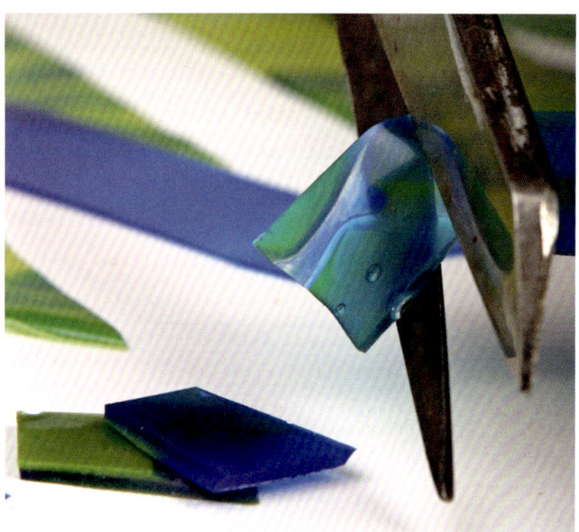

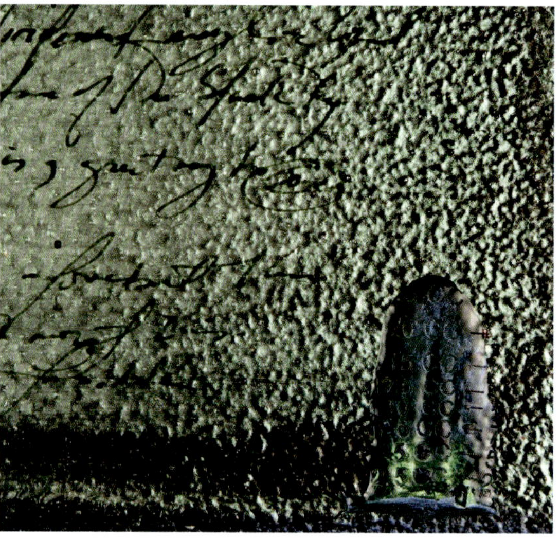

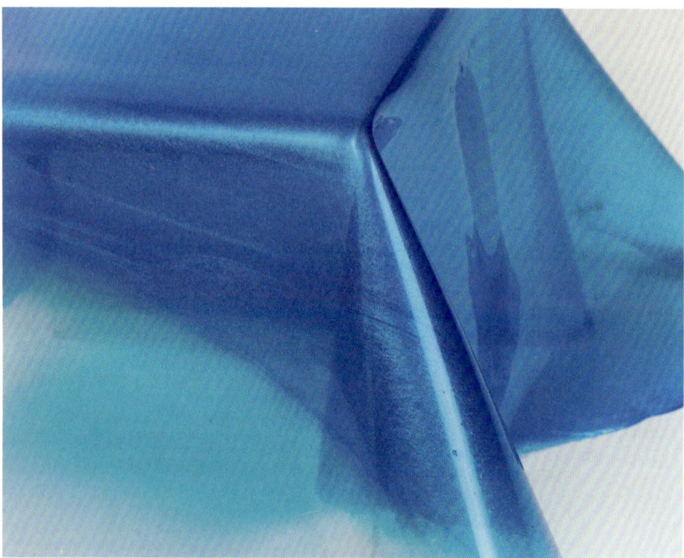

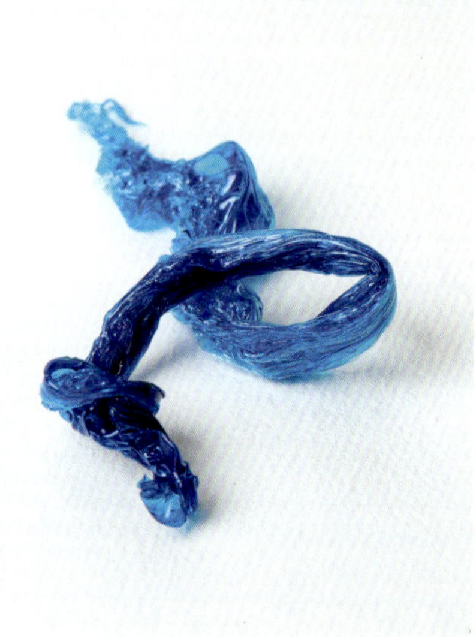

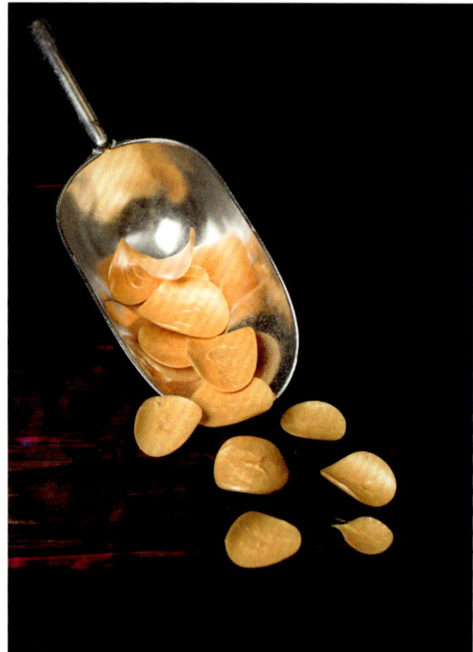

TOP, LEFT AND RIGHT These small squares of acrylic gel skin make a great alternative to glass mosaic tiles. • Incorporated into a larger work, the small pull of acrylic medium at lower right was lifted from the outside surface of a plastic measuring cup.

CENTER, LEFT AND RIGHT This skin of self-leveling gel can be smoothly folded. • Crinkled-up liquid-medium paint skin looks and feels like cling film.

LEFT Dried chips of gold-tinted pouring medium.

OPPOSITE **Rhéni Tauchid, *Lamp,* 2006, large acrylic skins on an iron framework, 48 (h) x 10 (w) x 10 (w) inches (121 x 25 x 25 cm). Private collection, Kingston, Ontario, Canada.**

The two LED bulbs inside this lamp are barely warm enough to soften the paint material, which is composed of gloss gel mixed with transparent red oxide, iridescent copper, and Payne's grey.

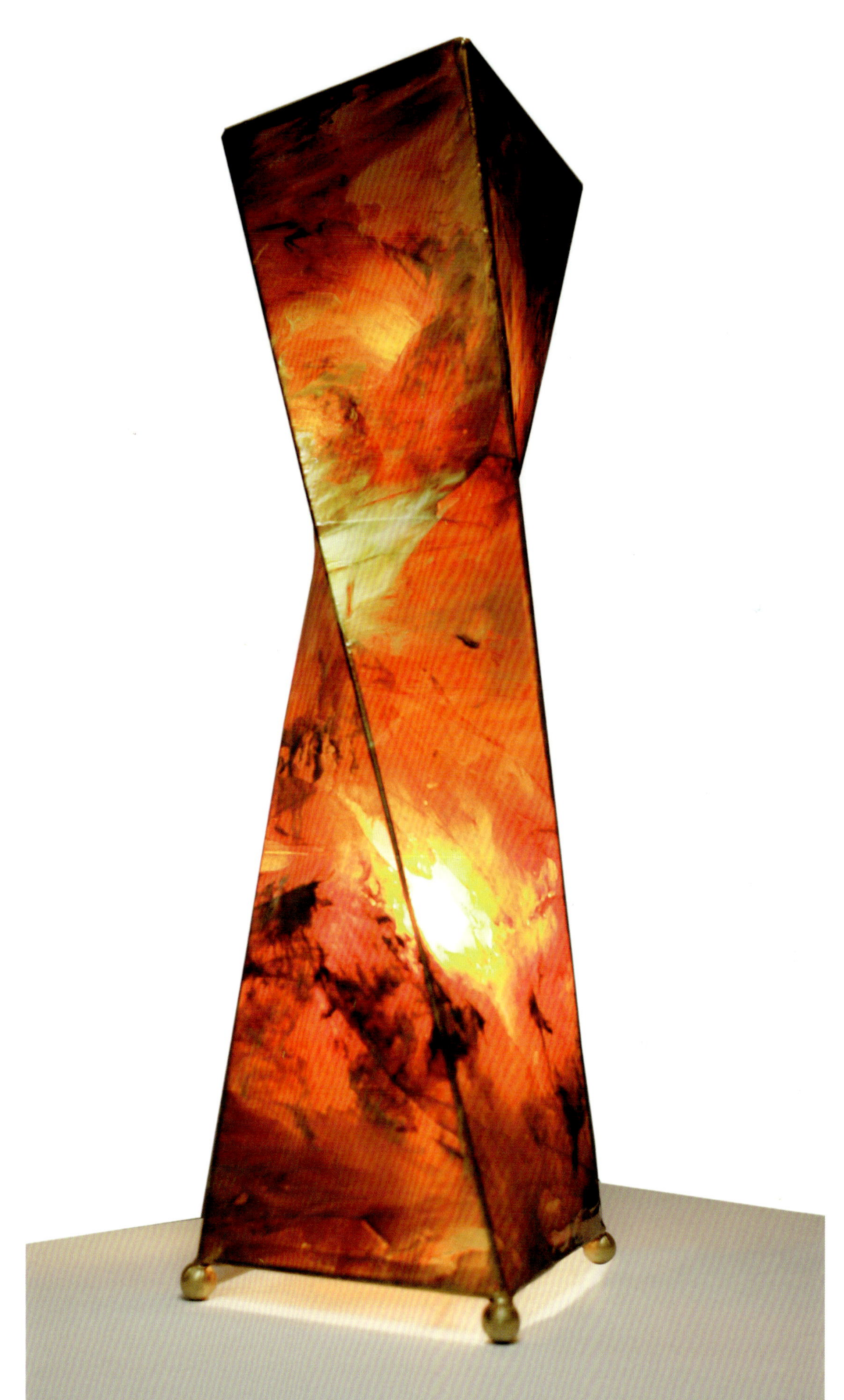

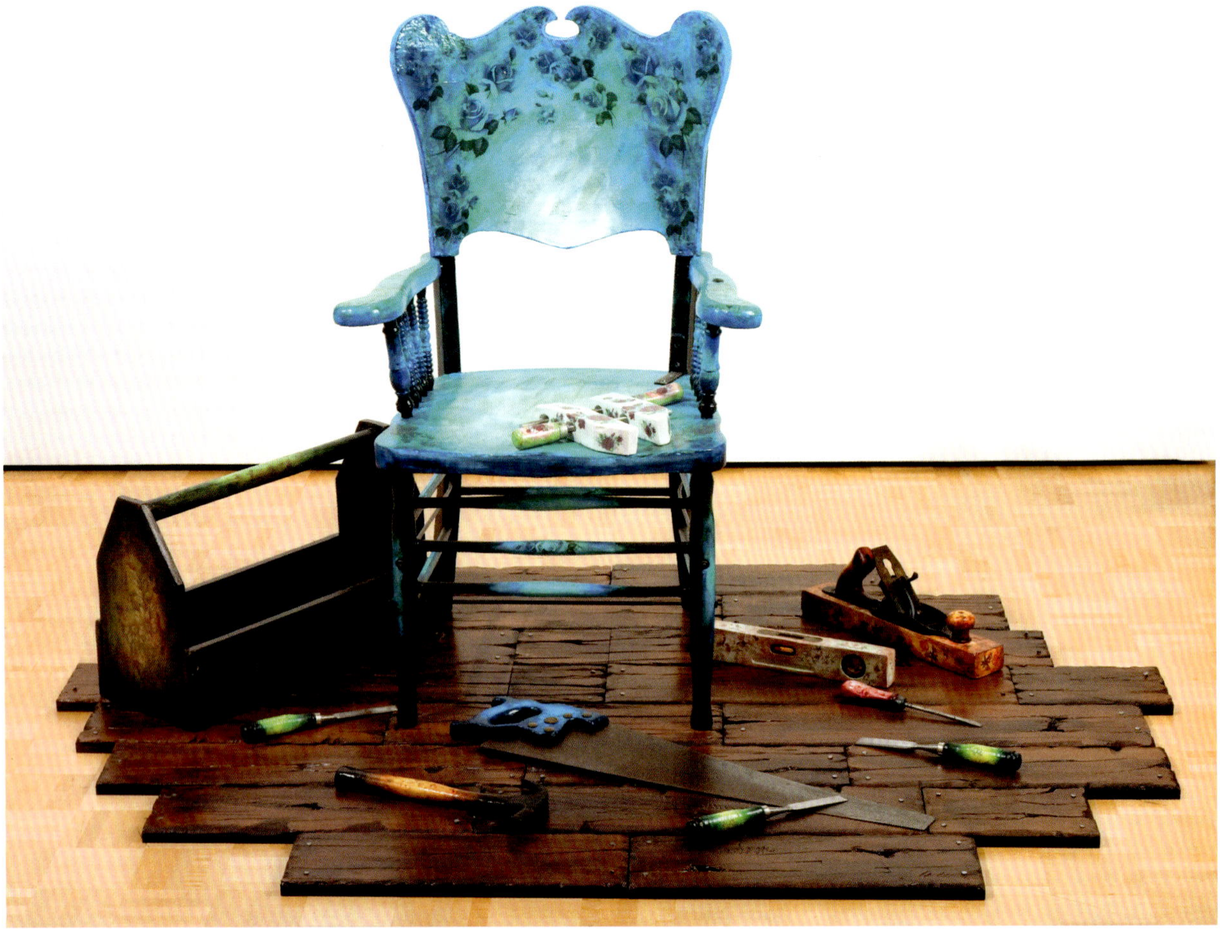

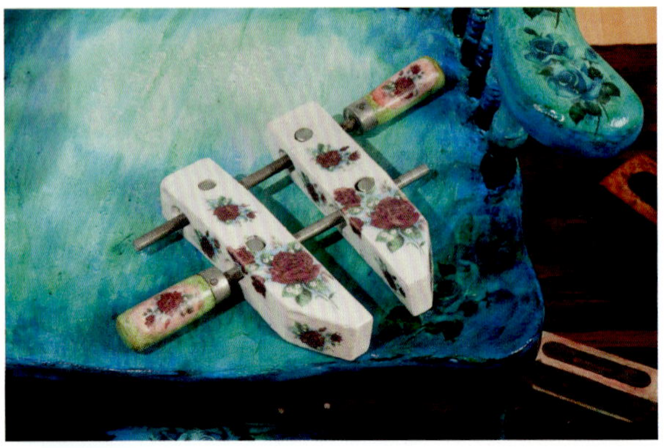

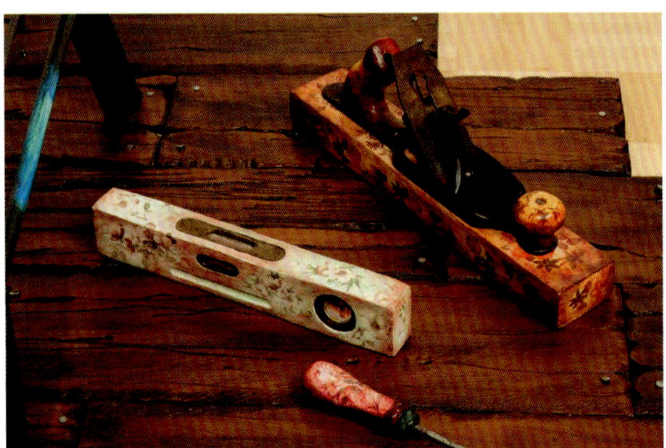

TOP **Marc Courtemanche, *Trompe l'Oeil,* 2010, acrylic sculpture installation, approx. 50 x 42 x 38 inches (127 x 107 x 96.5 cm). Collection of the Saskatchewan Arts Board.**

CENTER When making the wood clamp, Courtemanche wanted to make sure that it could be used—that it would be functional just like a "real" clamp. The decals reference ceramics and highlight its fragility.

BOTTOM All these tools are functional. When viewers look through the glass of the level, they can see all the different layers of paint used to create it. Though functional, the acrylic tools are much lighter than the originals on which they're based—which confuses the viewer's perception.

Is a chair that's made of paint a real chair or a sculpture of a chair? It took artist Marc Courtemanche a year to paint/sculpt Trompe l'Oeil. *Here's how he did it: "First, I selected a chair that I wanted to transform into acrylic paint. Then I made molds of all the parts of the chair out of plaster. Then, using modeling paste mixed with acrylic paint, I started painting into the molds, one layer at a time. This chair took one full year of painting to complete all of the chair's parts. Each part was then sanded, painted, and prepped for assembly. I used modeling paste as the glue, as well as some screws to help hold things together. The decals were added later, with an overlay of color for toning. This chair is made* out of *acrylic. It is not a 'painted chair' but a 'paint chair.'"*

PROCESS

MAKING AND STORING ACRYLIC SKINS

To make an acrylic skin, simply apply a layer or layers of acrylic paint or medium onto any surface from which it can easily be peeled off. Allow the acrylic to dry for several days before removing it. Peel from the outer edges in. Once the skin is off the support, it can stick to itself quite easily, so take care to prevent this. Storing acrylic skins can be tricky (and sticky). Once cured, they can be stored between sheets of baking parchment or freezer paper.

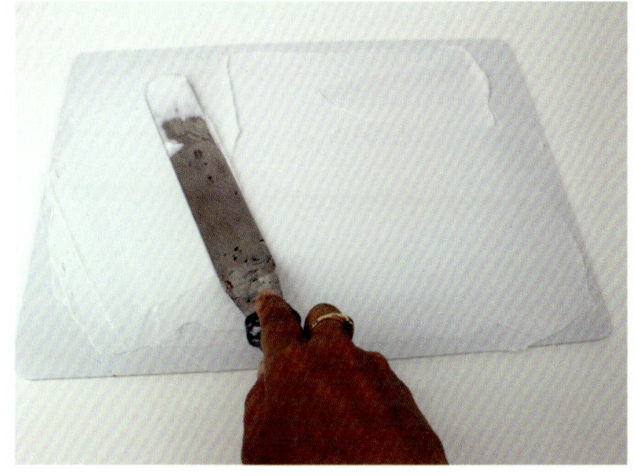

1
Here, gloss gel medium is spread in a uniform thickish layer onto a nonstick palette.

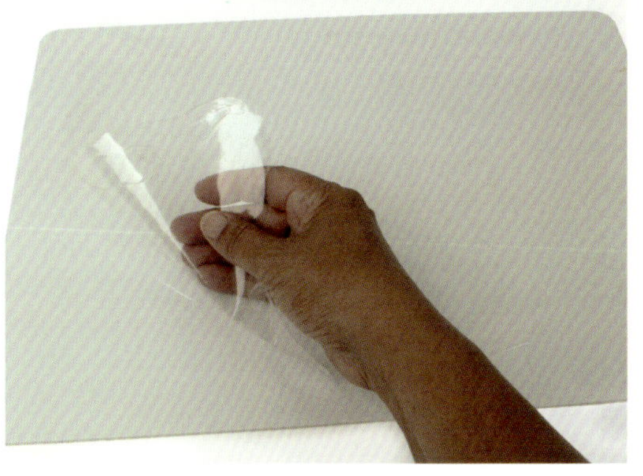

2
After twenty-four hours, the gel is dry and completely clear and peels away cleanly from the palette's surface.

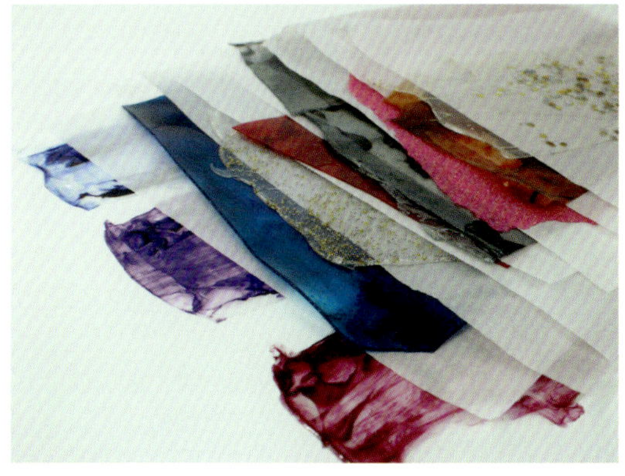

3
Store acrylic skins between sheets of baking parchment so that they will not stick to one another.

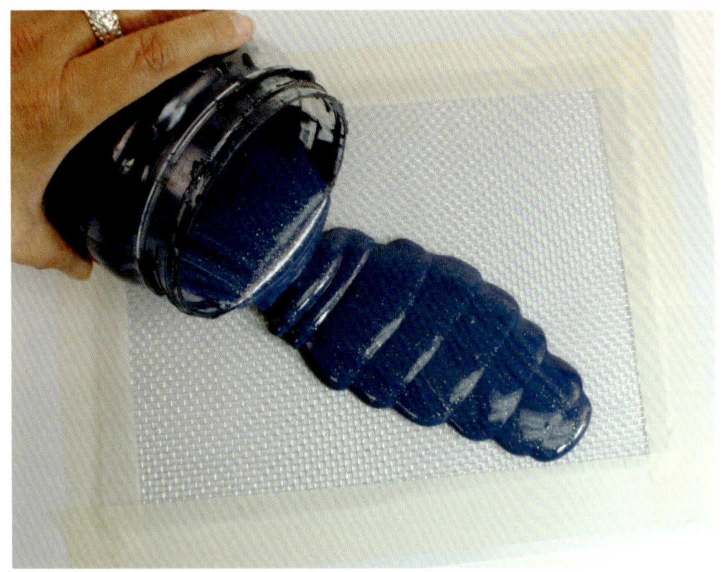

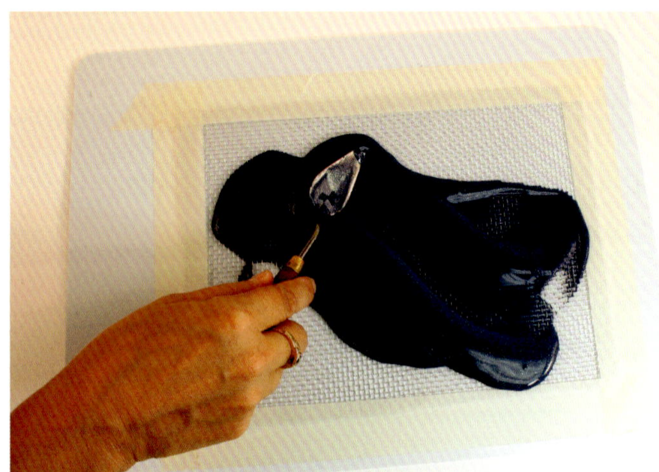

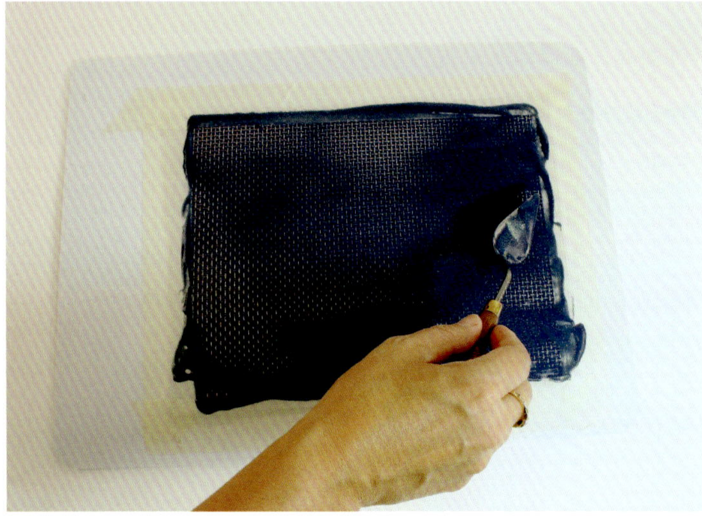

CREATING ARMATURES

When acrylics are used as a sculptural material, they can be reinforced in a number of ways with easy-to-find materials. Wire mesh is strong, malleable, and easy to cut. Metals used with acrylics should be non-reactive with water (to avoid rust). You can also use plastic-coated wire, testing first to see if the acrylic will adhere well to the plastic.

For invisible support, wires or mesh can be sandwiched between two layers of acrylic with a light application of acrylic medium to keep it in place. If you are going to be molding the acrylic, do not use metal armatures that have a "memory" unless you want the piece to spring back to its original shape regardless of how it is manipulated. Very delicate skins can be almost invisibly reinforced with fishing line or sheer fabrics.

THE TIME FACTOR

Producing three-dimensional works with acrylic mediums often demands that you use a significant quantity of mediums, and these take a lot of time to dry. Of all the factors involved in creating acrylic sculpture, patience is the most important.

Structures built up in layers require that you allow the appropriate amount of time to elapse between the application of layers. This process can take anywhere from days to months, and curing times must be respected for the best results. Liquid

TOP Tinted and sparkly self-leveling gel being poured onto craft mesh atop a nonstick palette.

CENTER AND BOTTOM Smoothing out the tinted, sparkly self-leveling gel over the craft mesh atop a nonstick palette.

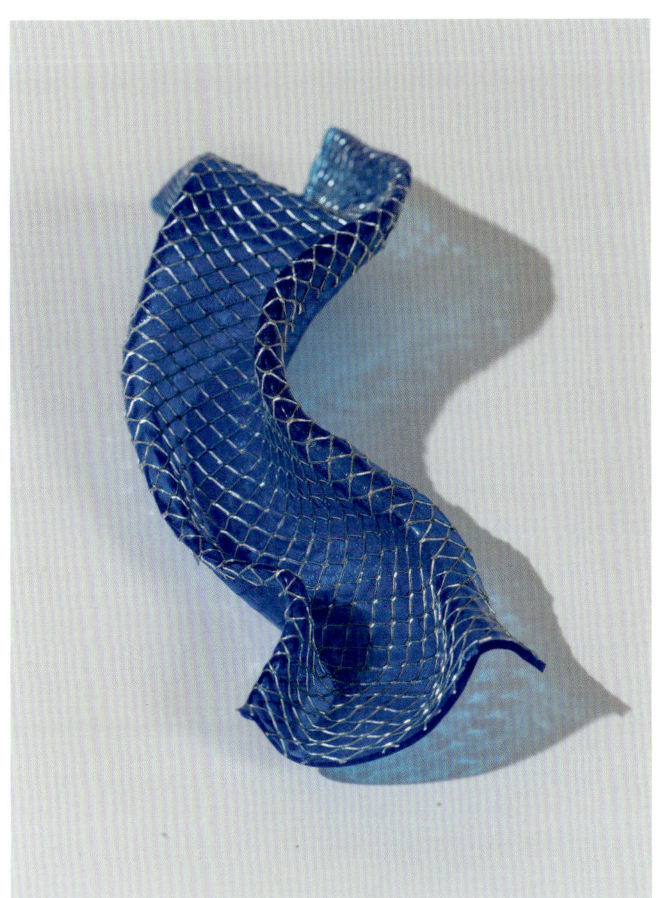

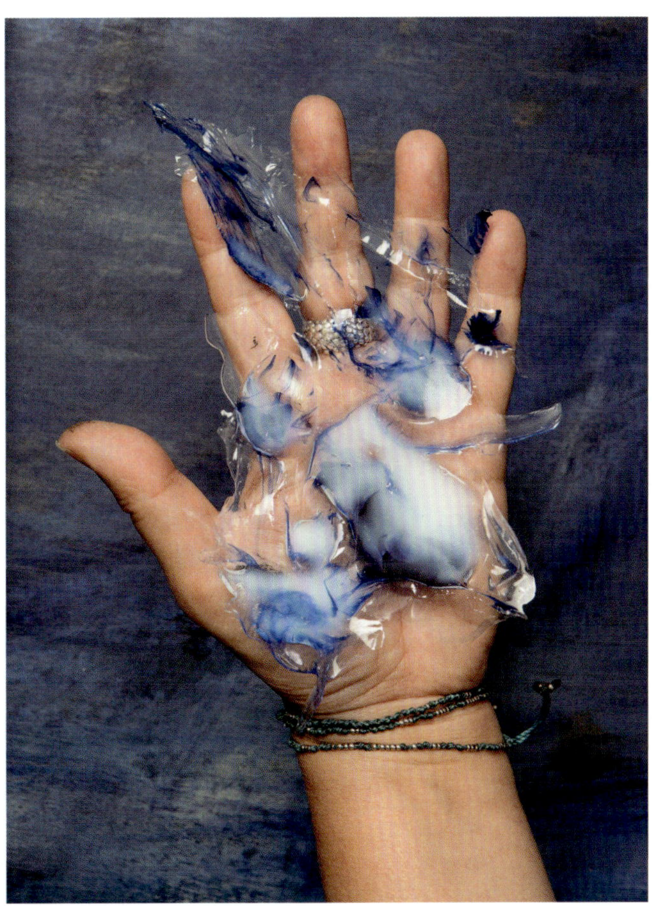

ABOVE, LEFT Acrylic medium reinforced with nonreactive craft mesh. Uniformly strengthened acrylic skin can be sculpted into any desired shape.

ABOVE, RIGHT This piece of acrylic gel loosely mixed with color was peeled off a nonstick palette. Notice that while most of the gel has dried clear, some areas (up to three-quarters of an inch thick) are still milky, because water remained trapped under the dried outer layer.

mediums and top coats dry to the touch within an hour or less. Gel mediums in films upward of one-quarter inch can take from a several hours to a couple of days to be dry enough to overpaint. The mediums that take the longest to cure are those that are very dense, like modeling pastes and those containing a lot of particulates. Self-leveling gels and pouring mediums should also be left to dry untouched for one to three days. Improperly cured layers can result in pockets of unclarified medium, weakness in the paint film, and increased water permeability. Here are some tips for getting it right:

1. A simple rule for building layers: Let each layer dry before adding the next. Take your time and build *slowly*.
2. If you add a wet layer over a layer that has not yet clarified, the milkiness of the acrylic will remain in the lower layer. If the color is opaque, it will be more difficult to determine

whether the paint is truly dry. If it is still cool to the touch, wait. The best advice is to wait three days between thick layers—longer if the air is very humid. Keep your working area well ventilated whenever you are using large quantities of medium.

3. A straightedge tool is your friend when producing large swaths of acrylic. Large-format palette knives—especially the offset ones that are nine to twelve inches long—are terrific, but if you don't have one you can use any rigid straightedge to get an even coat of paint onto a large area. I use pieces of mat board, extra-long rulers, and drywall tools for this purpose. My newest tool is a length of weatherstripping that I found at a hardware store. It's nice and soft and leaves a very light pinstripe of texture. If you want the tool to remain usable, make sure to wipe the excess paint off immediately after use.

BELOW This piece of weatherstripping is my new favorite straightedge tool. I love the subtle pinstripe of texture it produces.

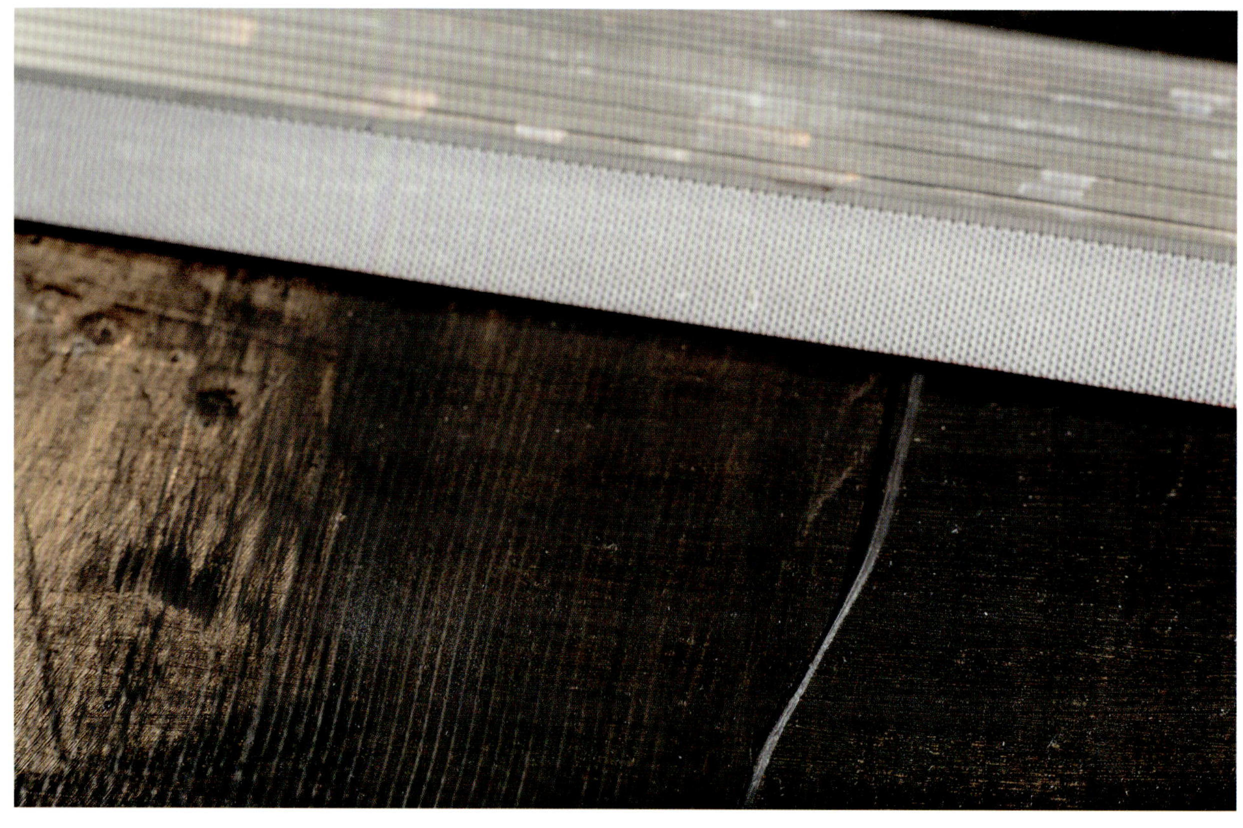

AN ACRYLIC SKIN PAINTING

Among the tools I use most often in my studio are my large (48 × 48 inch) nonstick palettes. These provide the base on which I begin many of my paintings. I apply numerous layers of acrylic, primarily of tinted gloss gel and self-leveling gel, over several weeks. Sometimes, as in my painting *The Conversation*, I use other materials, such as graphite and charcoal, for accents and details. When I am sure that it is dry enough, the entire painting is peeled from the palette and either stretched onto a frame, or suspended from a dowel.

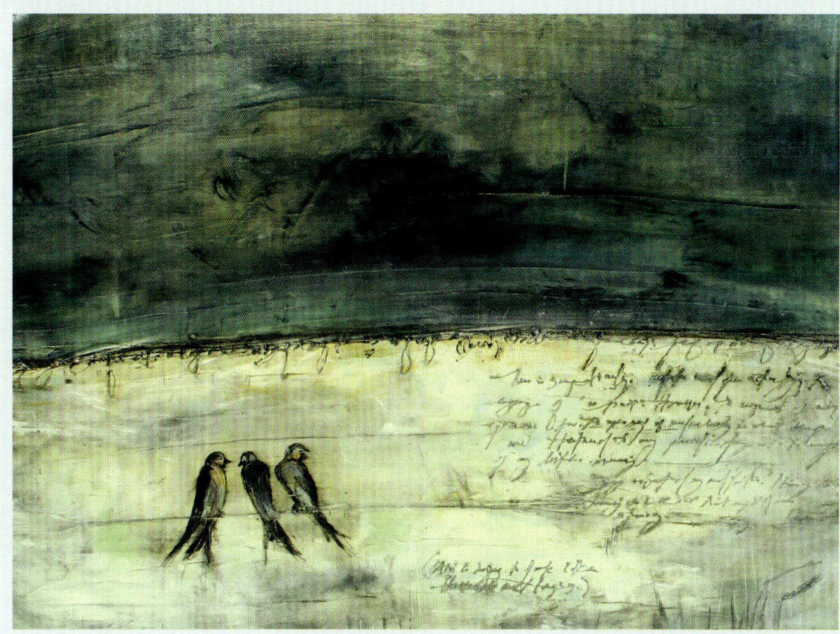

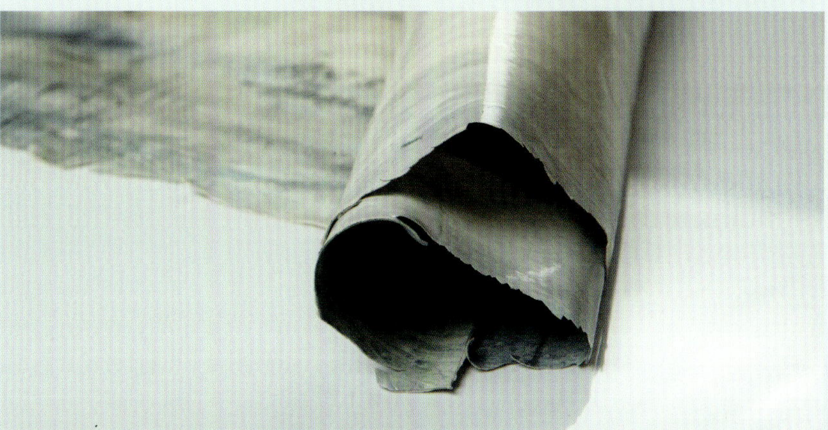

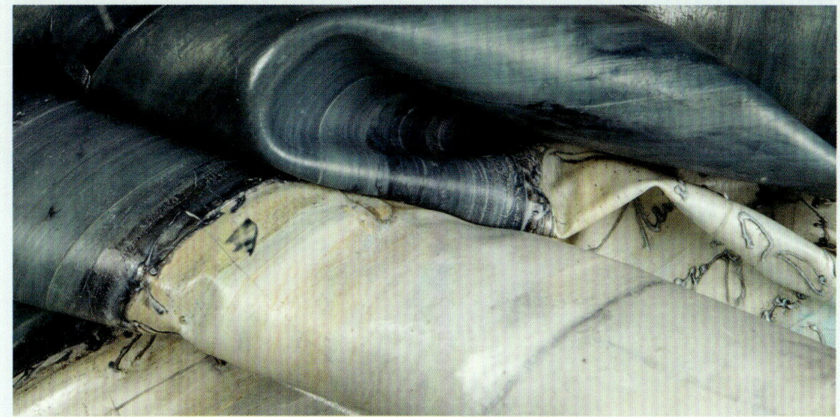

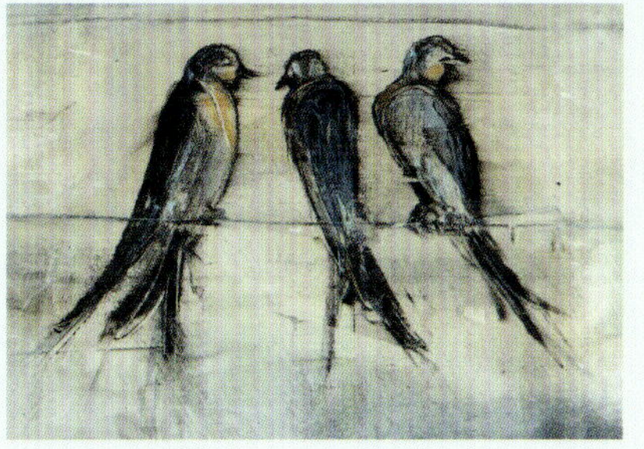

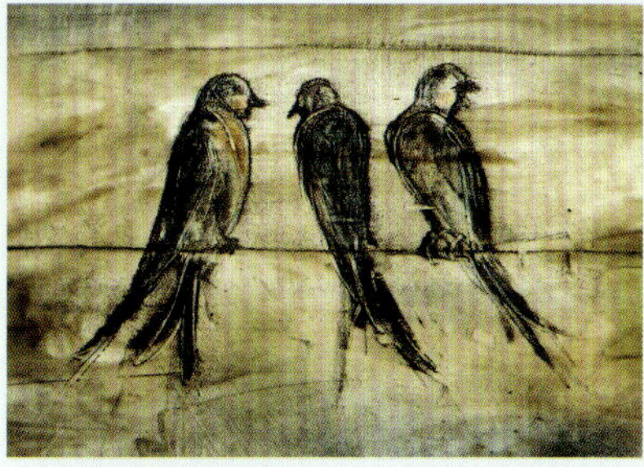

TOP **Rhéni Tauchid, *The Conversation,* 2013, acrylic skin with charcoal and graphite, 42 x 36 inches (107 x 91 cm). Private collection, Toronto.**

RIGHT, ABOVE I lightly rolled this acrylic skin painting to show the thinness of the paint film and the contrast between the glossy underside and the matte surface. Be warned, though: doing this for longer than a few minutes would result in a fused skin! You can, however, store an acrylic skin by rolling it with freezer paper or parchment to prevent it from sticking to itself.

RIGHT, BELOW The acrylic skin painting lying in folds.

BOTTOM, LEFT TO RIGHT A detail of the painting showing a charcoal drawing of swallows on a line over a matte medium layer. • The same detail, backlit.

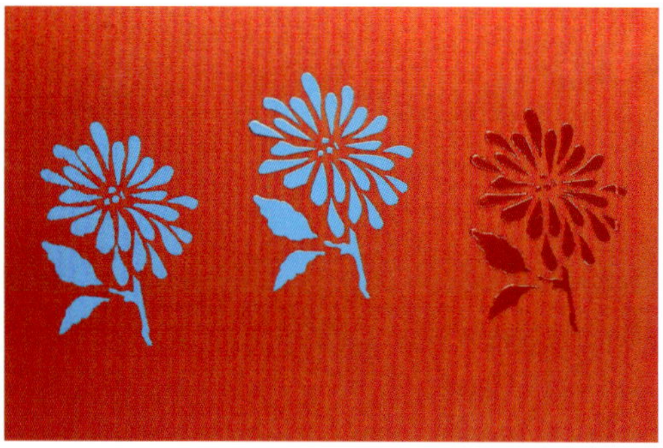

STENCILING

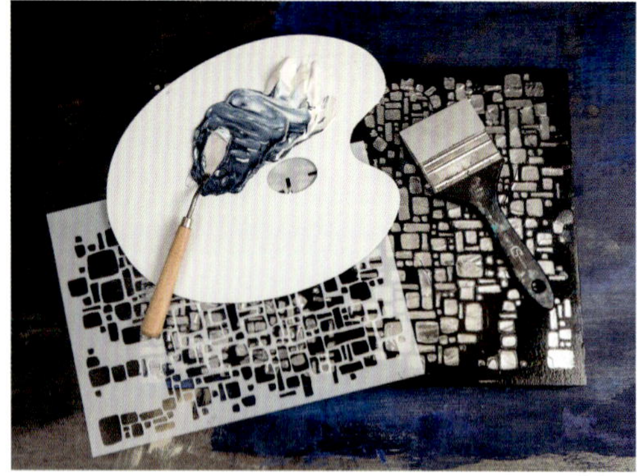

TOP Exploring the possibilities of stenciling. From left to right: pure color stencil, gel + color stencil, untinted gel stencil.

ABOVE The makings of a stenciled surface.

BELOW Applying a graduated blend of colors directly onto the stencil achieves varying tones.

The word *stencil* tends to evoke images of spray-painted graphics or the precious floral borders of 1980s decor. But when used with acrylic mediums, templates for repeated mark-making can be more inspired than insipid. Stencils give you sharp edges that are extremely effective at creating bas relief designs. Some mediums work beautifully for this type of application: choose a medium (smooth or rough) that holds texture, has a high-viscosity, and won't collapse or smooth itself out.

Celebrate the ornamental, decorative patterning possibilities using machine-made or hand-cut stencils and working with mediums you have on hand. For thin stenciled effects, use mediums with a low viscosity and a short rheology. For higher-relief stencils, choose high viscosity mediums with short rheology. Avoid using long-rheology, high-flow mediums like self-leveling mediums, pouring mediums, and very thin liquid mediums, however. These will not stay within the confines of the stencil and will creep under the cut edges; when the stencil is removed, the forms will flow into each other.

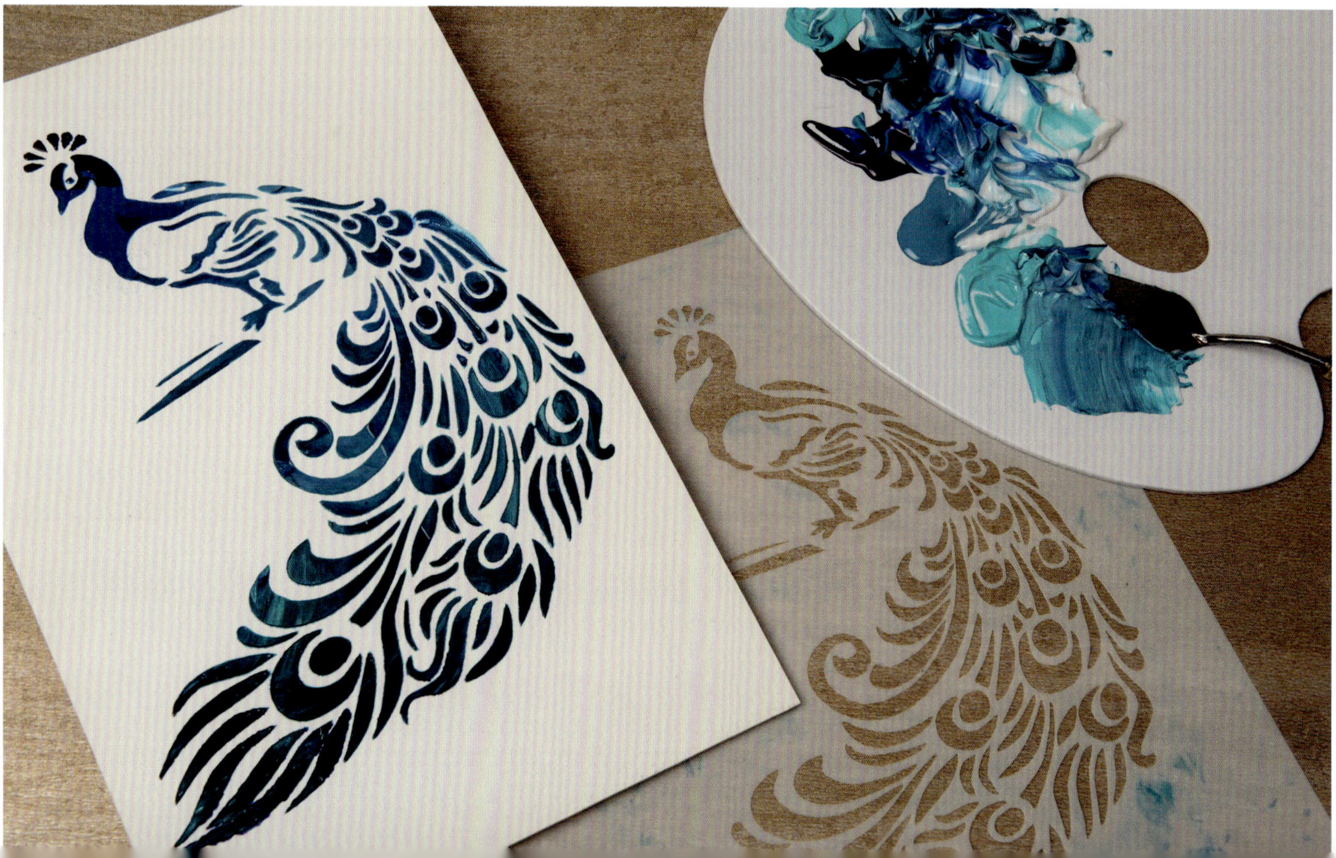

Rhéni Tauchid, *Never Been to Marrakesh,* 2015, acrylic on canvas, 36 x 36 inches (91 x 91 cm). Collection of the artist.

We tend to think of stencils as producing repeated, clearly delineated shapes. Here, layers of overlapping and intertwined stenciled shapes, produced with thick gel, create an area of profuse texture—a lavish abundance that generates a calmness. The layers below reveal deeply saturated blue and purple tones, the result of more than a dozen layered tinted glazes.

PROCESS

USING STENCILS

The trick to a crisply edged stencil is a light touch. Make sure that the surface is dry, and lay the stencil on it. Use a wide, flat tool to spread the medium over the stencil (unless you want brushstrokes in your stenciled medium). Pushing the paint too firmly will cause it to creep under the edges of the stencil, as will going back and forth over the stencil. A single pass, not too thick, is all that is needed. As soon as the stencil is fully covered, lift it off in one motion. Excess medium that has escaped into the "clean" spaces can be removed with a small color shaper or the tip of a palette knife. Once the raised stenciled medium is dry, it can be painted and enhanced like any other texture. To keep your stencil clean and reusable, remove it immediately after use, wipe off excess acrylic, and rinse clean with water.

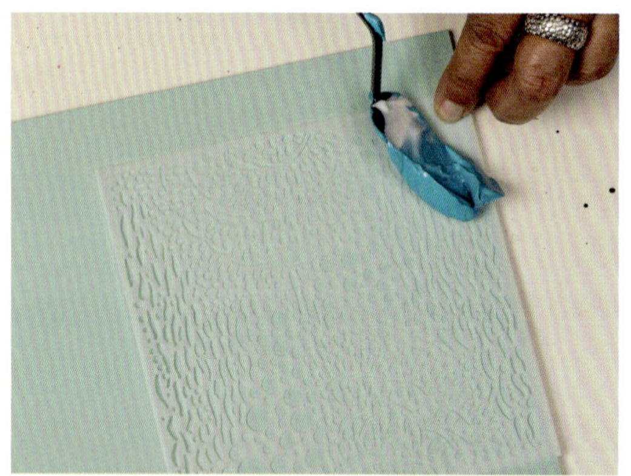

1
Use a palette knife to apply the medium (here, a soft gel tinted with cobalt teal) to the stencil.

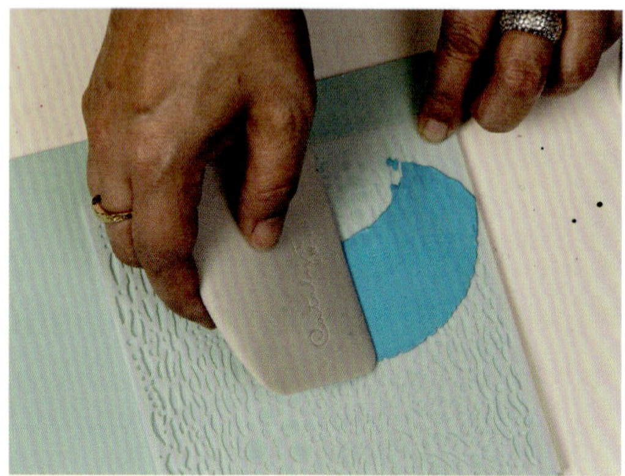

2
Spread the medium with a wide sweep, applying just enough pressure to fill the stencil without pushing color underneath.

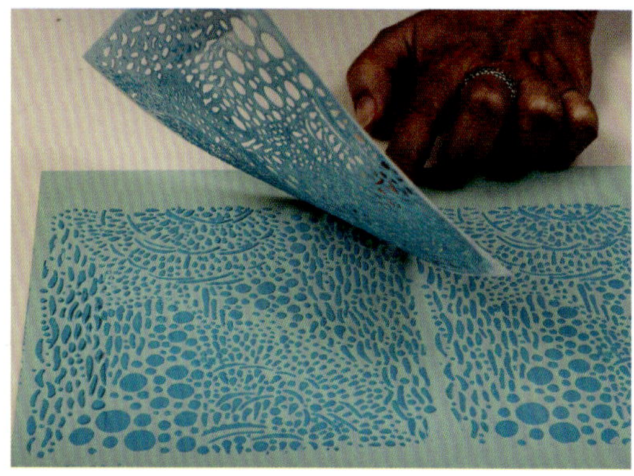

3
Remove the stencil immediately after the application.

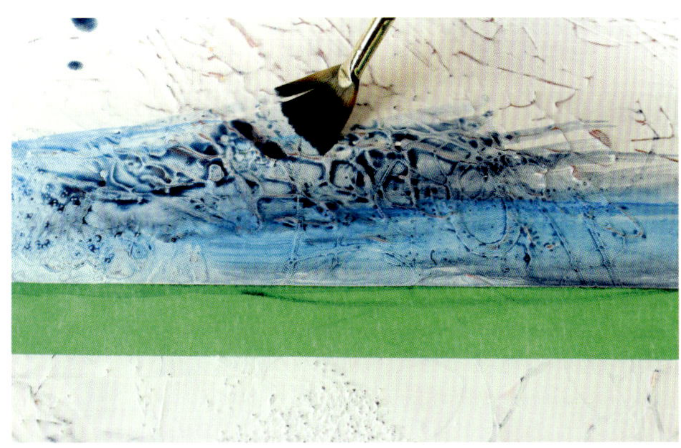

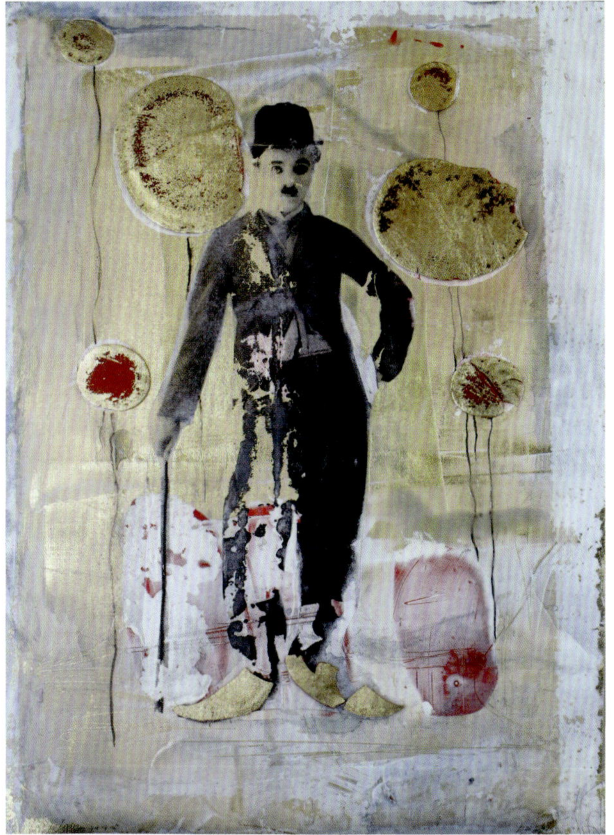

TOP Applying thin washes of color to bring out texture on a gessoed surface. An area has been masked with tape to create a horizon line.

BOTTOM **Rhéni Tauchid, *Charlie's Ghost mit Luftballons,* 2017, gold leaf, permanent marker, pouring medium, and transfer on Sintra foam board, 8 x 10 inches (20 x 25 cm). Collection of the artist.**

This piece was an exercise in using leftovers and printed material. The support was first painted with liquid matte medium and covered in gold leaf. Pools of red-tinted pouring medium were peeled off the gold-leafed surface when it was not quite dry, and the pieces were then inverted and collaged back onto the support with gel. The image of Charlie Chaplin was transferred onto the surface and then partly rubbed away to reveal the underlying gold leaf. Finishing details were added with permanent marker and a loose application of tinted matte medium.

MASKING

You can also use masking tape or masking film to create a crisp, hard line. Tape is ideal for marking horizon lines or blocking off areas. If your medium is viscous and has a short rheology, you can pull the tape off while it is still wet. If you are using a long-rheology medium, however, it's best to wait until the medium is dry to the touch before removing the tape. But this can be tricky: if you wait too long, the tape may tear away the dried medium that overlaps the tape. If you have missed the "dry but still soft" window, you can run the edge of a blade lightly along the edge of the tape for a clean line.

TRANSFERRING IMAGES

The process of transferring printed images onto acrylic paintings is very simple. There are two different ways to do it, and several different mediums can be used. If you are reluctant to sacrifice a drawing to a painting without the guarantee that it will work out, simply make a copy and preserve the original. Personal photographs or any printed works that are in the public domain are ideal starting points for transfers.

The first part of the process is to create a reproduction of the work. Print the work on a laser printer or copier, since these use waterfast inks. Stay away from jet printers or any other printers that use water-soluble ink, or you'll have a mess on your hands. Stick to black-and-white images unless you are sure that the colored printing inks are pigment based and lightfast (most are not). Use low- or medium-quality paper. In any transfer process, the paper is destroyed and discarded, so there is no benefit to using pretty or expensive paper.

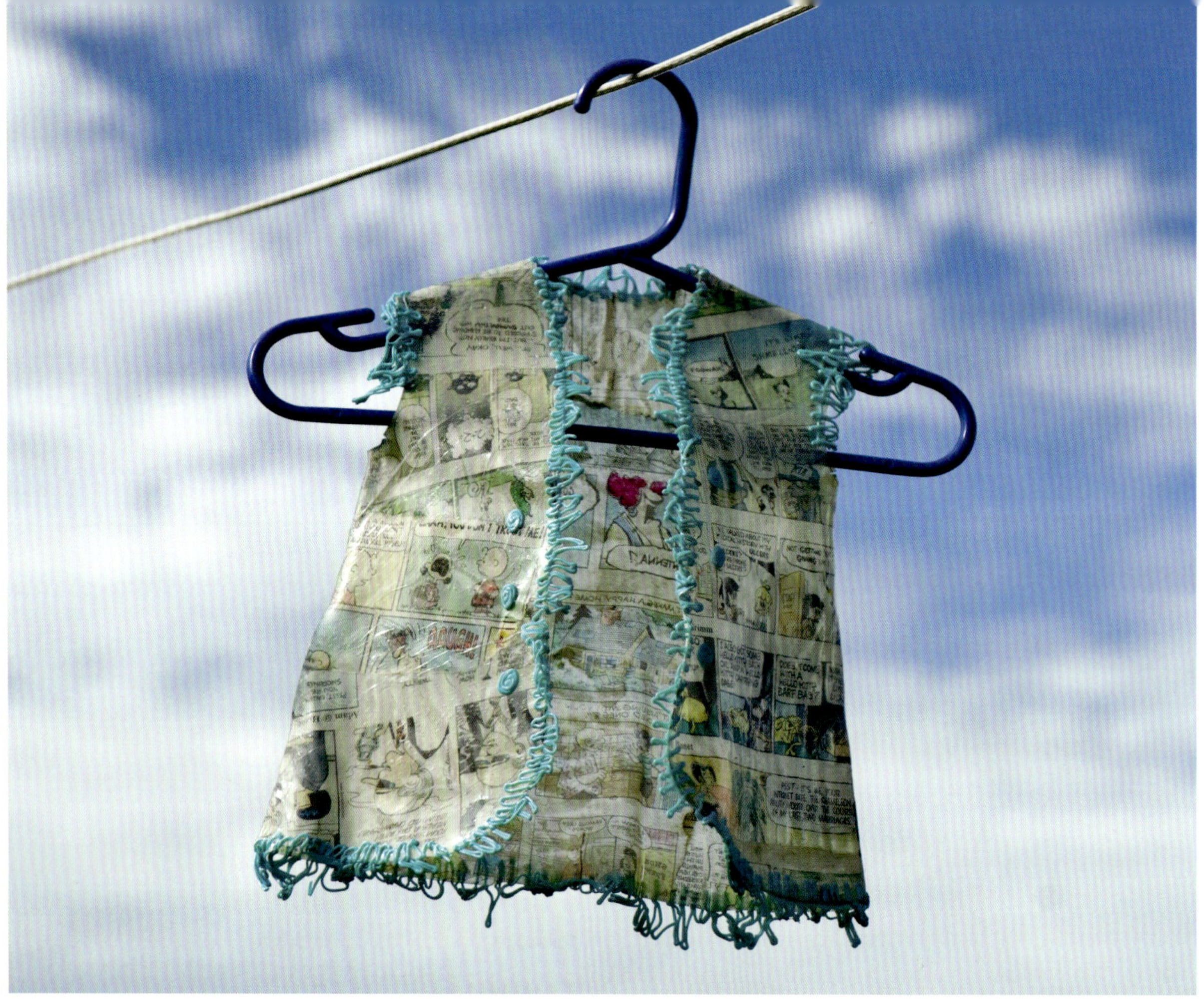

ABOVE Acrylic direct transfer (of newspaper comic strip) onto a child's vest with extruded fringe.

If you're doing an indirect transfer and the image you're transferring contains text, you will need to flip the image on your computer before printing; otherwise, you'll end up with an unreadable mirror image of the text.

Decide whether you want to use the direct or indirect transfer process. A *direct transfer* is one in which the acrylic medium is applied directly to the surface of the printed image and left to dry before the paper is removed. The result is an image embedded in a sheet of acrylic medium (an acrylic skin), unsupported by a substrate. When you transfer directly onto an acrylic paint film, you get an acrylic skin that you can then collage into your work. An *indirect transfer* results in a mirror image of the original—an inverted transfer that you adhere with medium onto a substrate. Paper, board, Plexiglas, and canvas are all suitable for this technique. Do be aware that the more porous the support, the more quickly the transfer will dry.

PROCESS

TRANSFERRING IMAGES DIRECTLY

In this simple example of the direct-transfer method, the drawing of the fish was scanned, then printed on a home laser printer. Excess paper was cut away. The transferred image was then incorporated into a composition on thick watercolor paper, adding a linear graphic element.

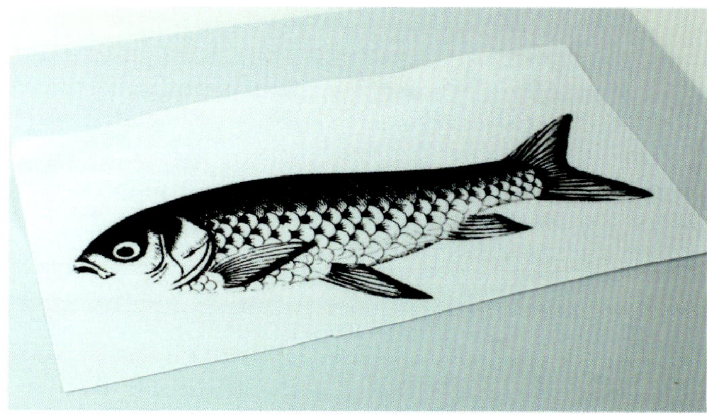

1
A black-and-white laser-printed image of a fish is ready for transfer.

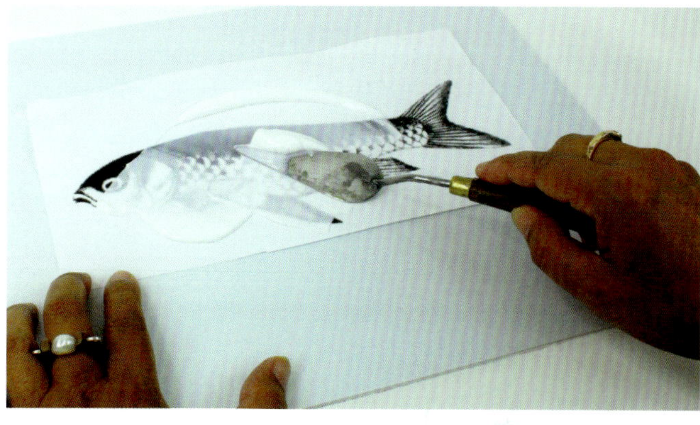

2
Hold or tape your printed image to a flat surface. Spread medium directly and evenly onto the image in a single, thick layer. Any clear drying medium will do. If using a thin medium, you can add an additional layer or two (when first layer is dry to the touch) to create a more robust paint film.

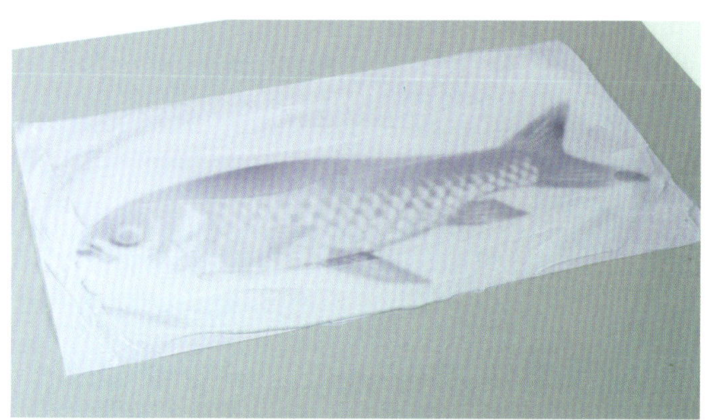

3
Make sure to cover the entire image with an even layer of gel. Allow the medium to dry completely. Thick applications should dry for 24 to 48 hours; thinner films will be ready in 12.

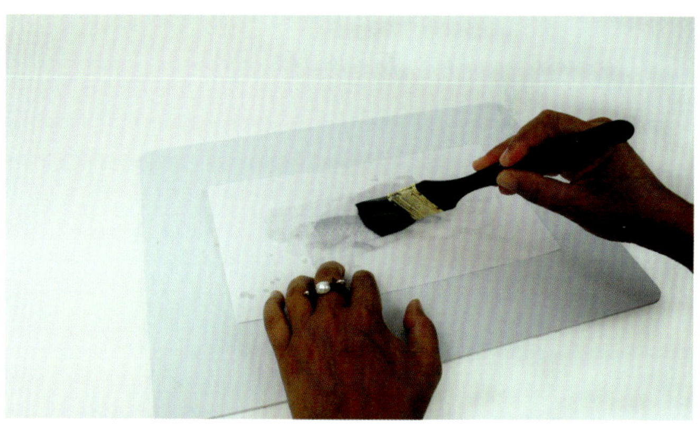

4
Once the gel has fully dried, soak the back of the paper with water using a very wet brush (as shown here) or a clean cotton rag saturated with water.

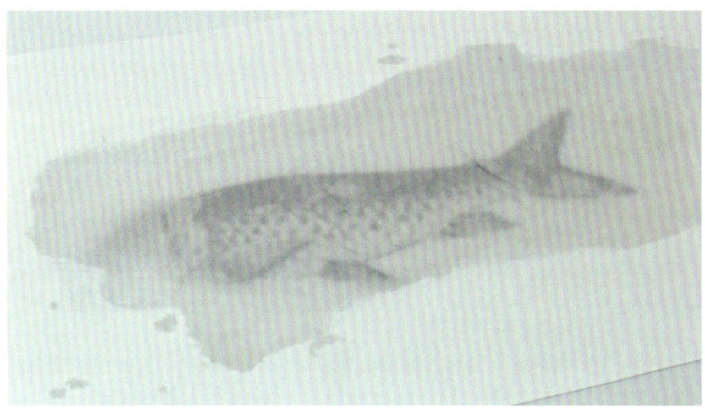

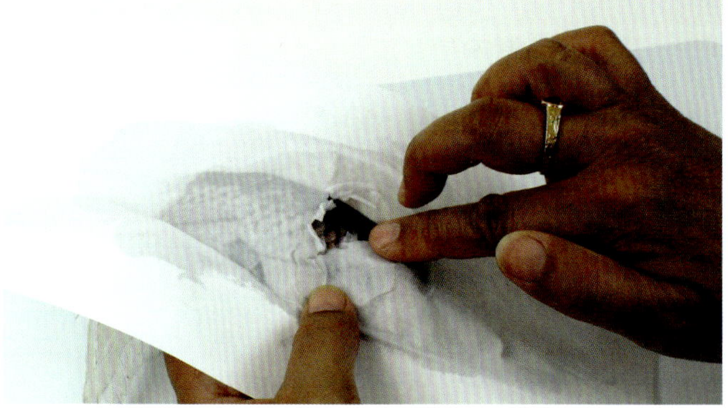

5
Soak the paper side of the image until the paper becomes soft and translucent. Here, the paper side of the transfer is completely saturated.

6
Then, using a circular motion, rub the moistened paper off the transferred image. You can use your fingers (as shown here) or a lightly dampened cotton cloth. Remove as much paper as you can. A very thin layer of paper will remain embedded in the medium, leaving a translucent residue.

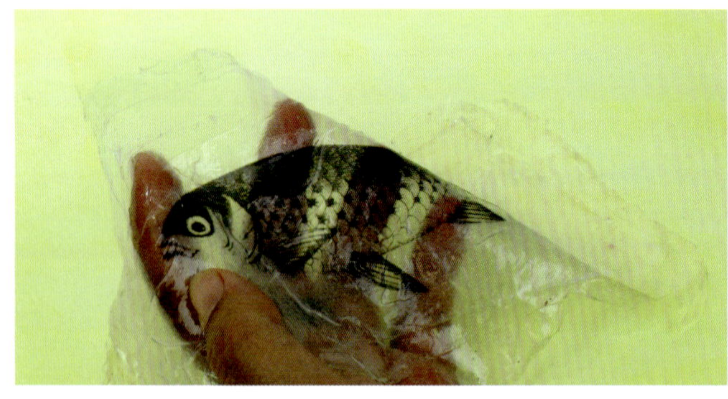

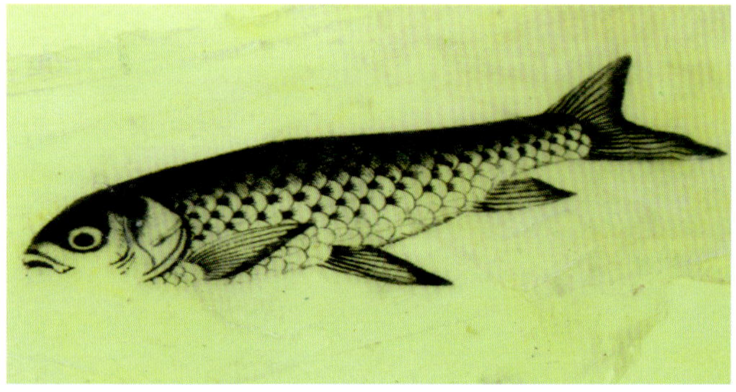

7
The process may cause your medium to become cloudy as it absorbs water, but the water will eventually evaporate, leaving you with a clear film. Here, the transfer is now ready to be fused to the support.

8
Adhere the transfer to the painted surface using a thin application of medium.

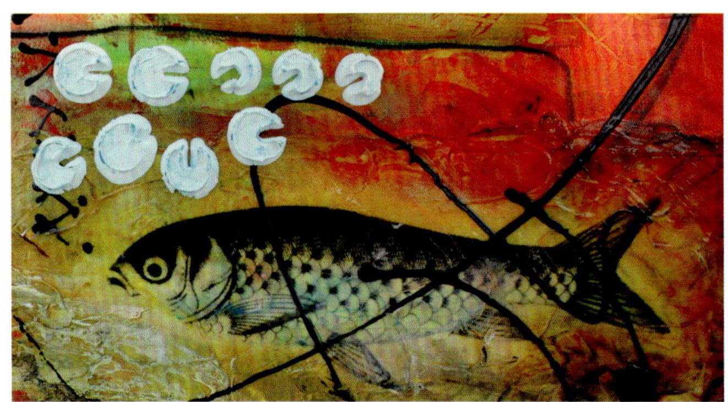

9
The transferred image of the fish has been incorporated into the painting.

PROCESS

TRANSFERRING IMAGES INDIRECTLY

Aside from your own drawings, photographs, and other graphic materials, images that are in the public domain are a great choice for acrylic transfers. (Be very wary of using copyrighted materials in your artwork.) The uncopyrighted drawing of a bicycle used here was available for download from a trusted site. Its clean lines and charming profile worked well in the transfer, and imperfections were easily touched up with a fine-line permanent marker.

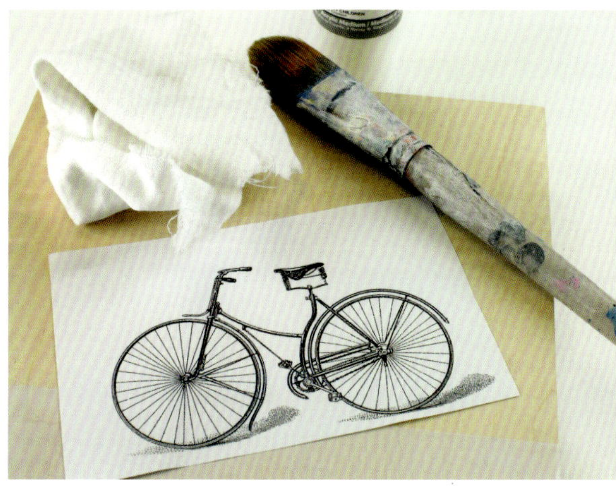

1
The black-and-white laser-printed image of a bicycle is ready for transfer onto a raw wood panel.

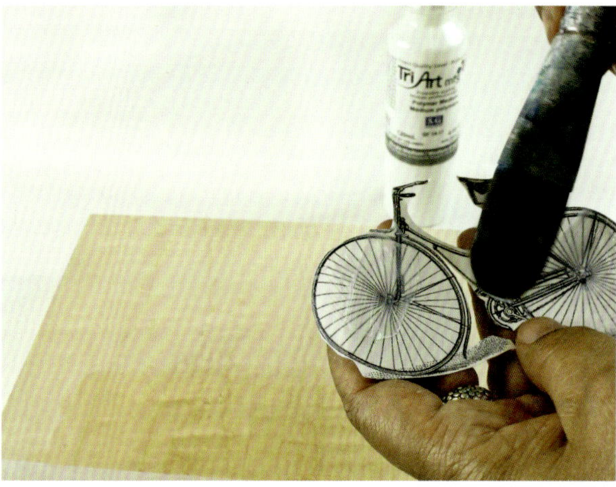

2
Apply semigloss polymer medium to the face of the image, making sure to spread the medium all the way to the edges of the image.

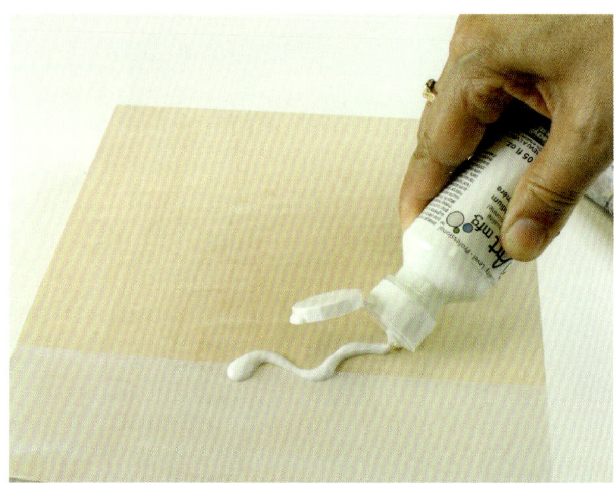

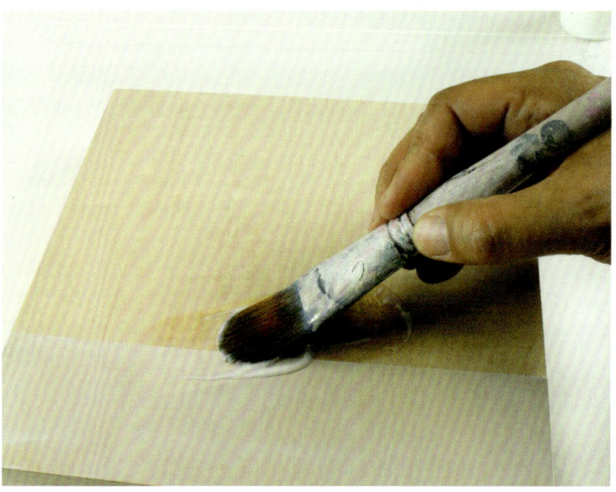

3
Apply semigloss polymer medium to the surface of the panel, as well. Be sure to do this quickly, especially if you are covering a large area, as thin mediums can dry quite rapidly. (If you work in a very dry environment, consider using a thicker medium for moisture retention.)

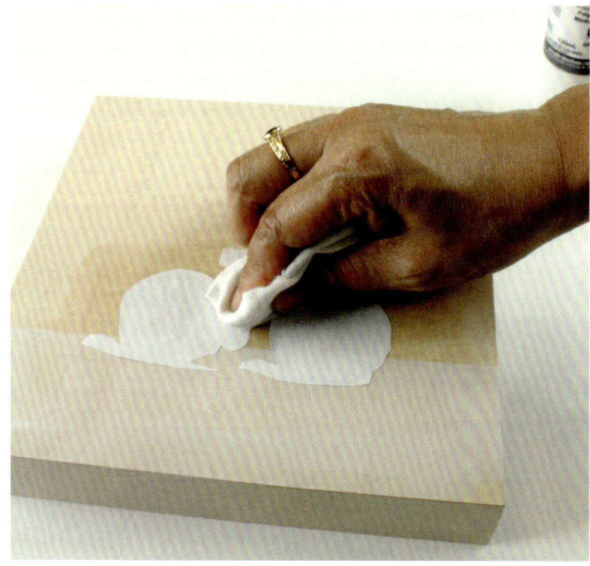

4
Place the image facedown on the surface. Fuse it to the surface, making sure there are no air pockets. You can use a flat tool or brayer (not shown here) to smooth out the back of the image, working from the center outward, to remove any air bubbles and ensure that the image is completely stuck to the medium.

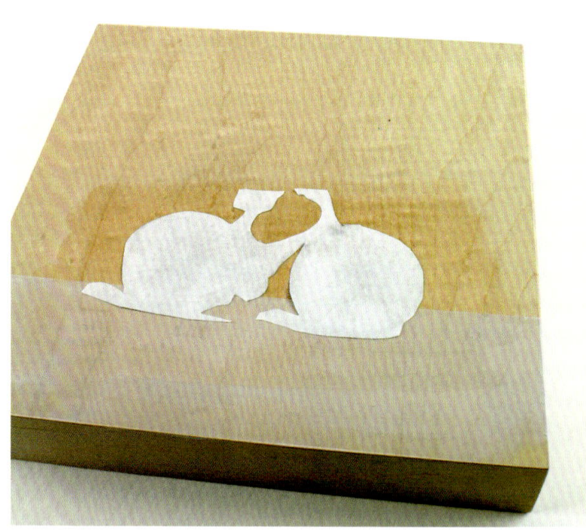

5
Let the medium dry completely before the next step. Thick applications should dry for 24 to 48 hours; thinner films will be ready in 12.

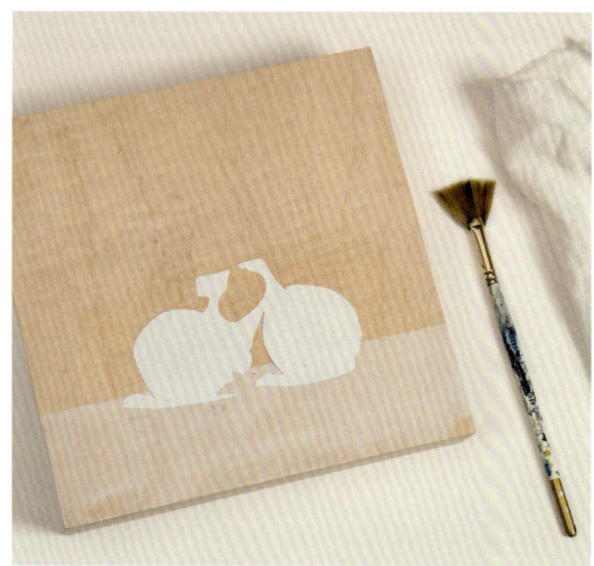

6
The transfer is now dry and ready to be revealed.

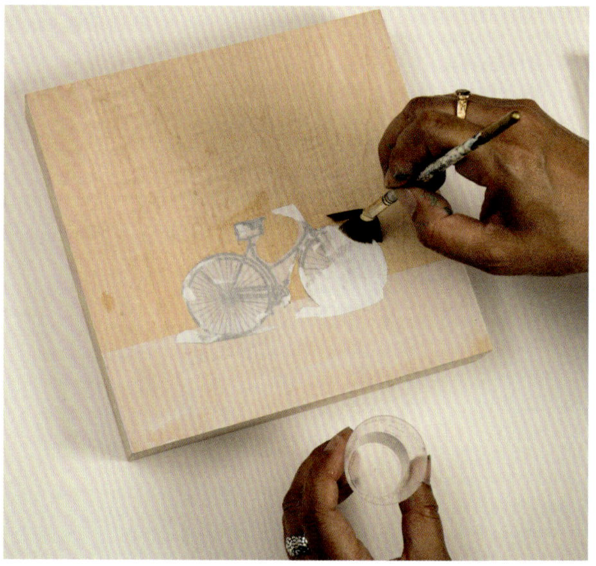

7
Saturate the paper with water until the paper becomes soft and translucent. You can use a wide, flat brush (as shown here) or a clean cotton rag to soak the paper side of the image.

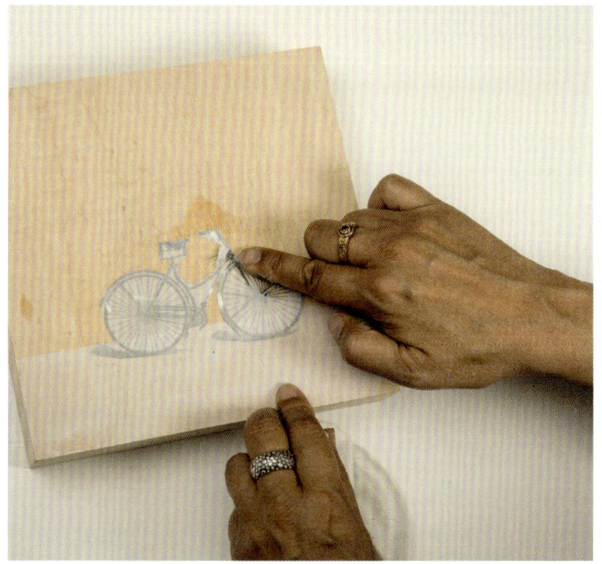

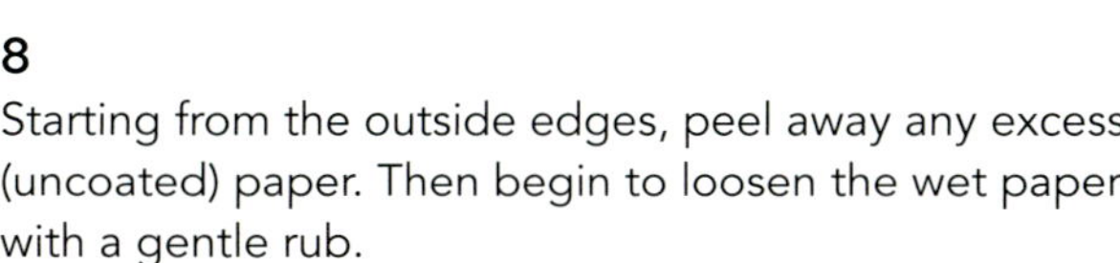

8
Starting from the outside edges, peel away any excess (uncoated) paper. Then begin to loosen the wet paper with a gentle rub.

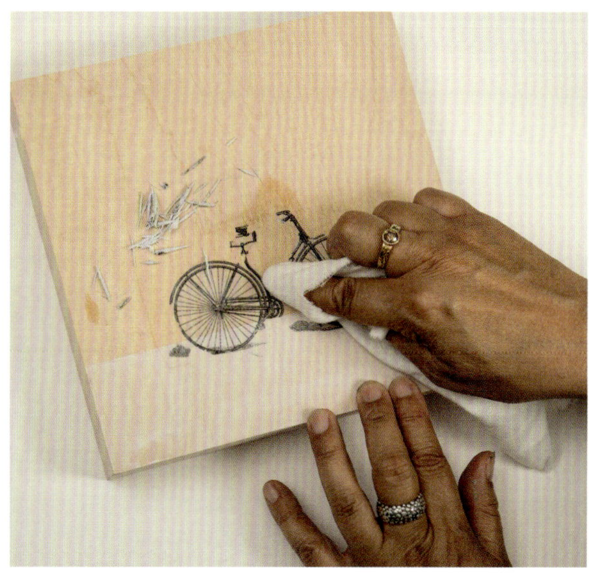

9
Using a circular motion, rub the moistened paper off the transferred image. You can use your fingers or a lightly dampened cotton cloth. Remove as much paper as you can. A very thin layer of paper will remain embedded in the medium, leaving a translucent residue. (For a more distressed look, you can scrub away at the image with an abrasive pad while the image is still wet or with sandpaper when it is dry.)

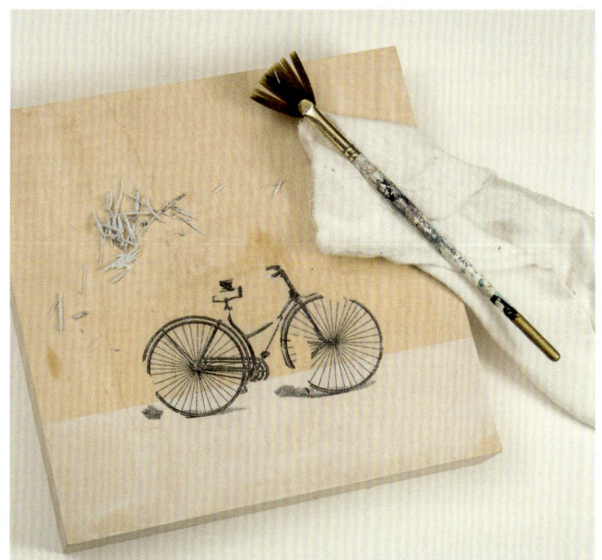

10
The paper has been completely removed, leaving the ink image embedded on the panel. Areas where the medium dried too quickly (failed adhesion) or were rubbed too vigorously during paper removal can be filled in later. If the transfer looks a bit dull, you can bring out the blackness of the ink by applying a thin layer of gloss or semigloss liquid medium over your transfer.

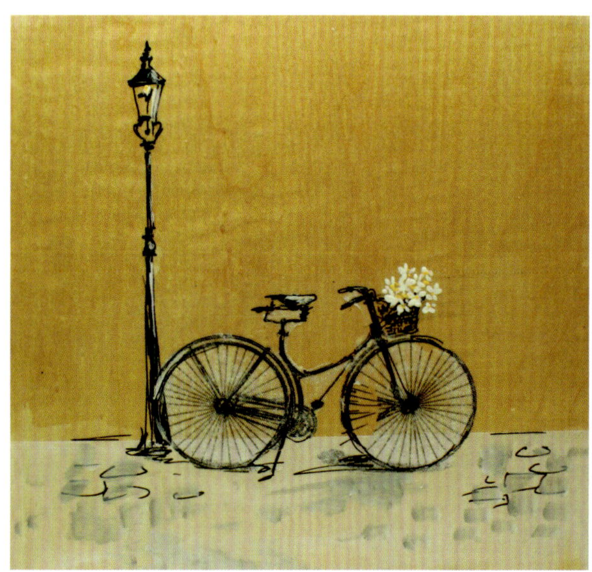

11
The final image. Using pen and ink and some acrylic color, areas were filled in and details were added to the composition to complete the picture.

CHAPTER 10

SPECIAL USES FOR MEDIUMS

ACRYLIC MEDIUMS PLAY MANY ROLES, on their own or in combination. We've examined how they affect various processes and enhance other materials. In this chapter we take a closer look at some of the jobs that acrylic mediums have been specifically designed to perform: extending color, controlling luster, changing viscosity and rheology, creating washes, controlling relative coverage, glazing, and creating thick, lustrous layers on the surface of a painting.

Acrylic mediums were designed for a range of specific jobs.

OPPOSITE **Lori Richards, *Pink Pond*, 2016, acrylic on self-leveling gel ground on panel, 36 x 48 inches (91 x 122 cm). Private collection, Kingston, Ontario, Canada.**

To mimic the physical feel of painting on Mylar, a liberal layer of self-leveling gel was applied to the surface as the ground for this painting.

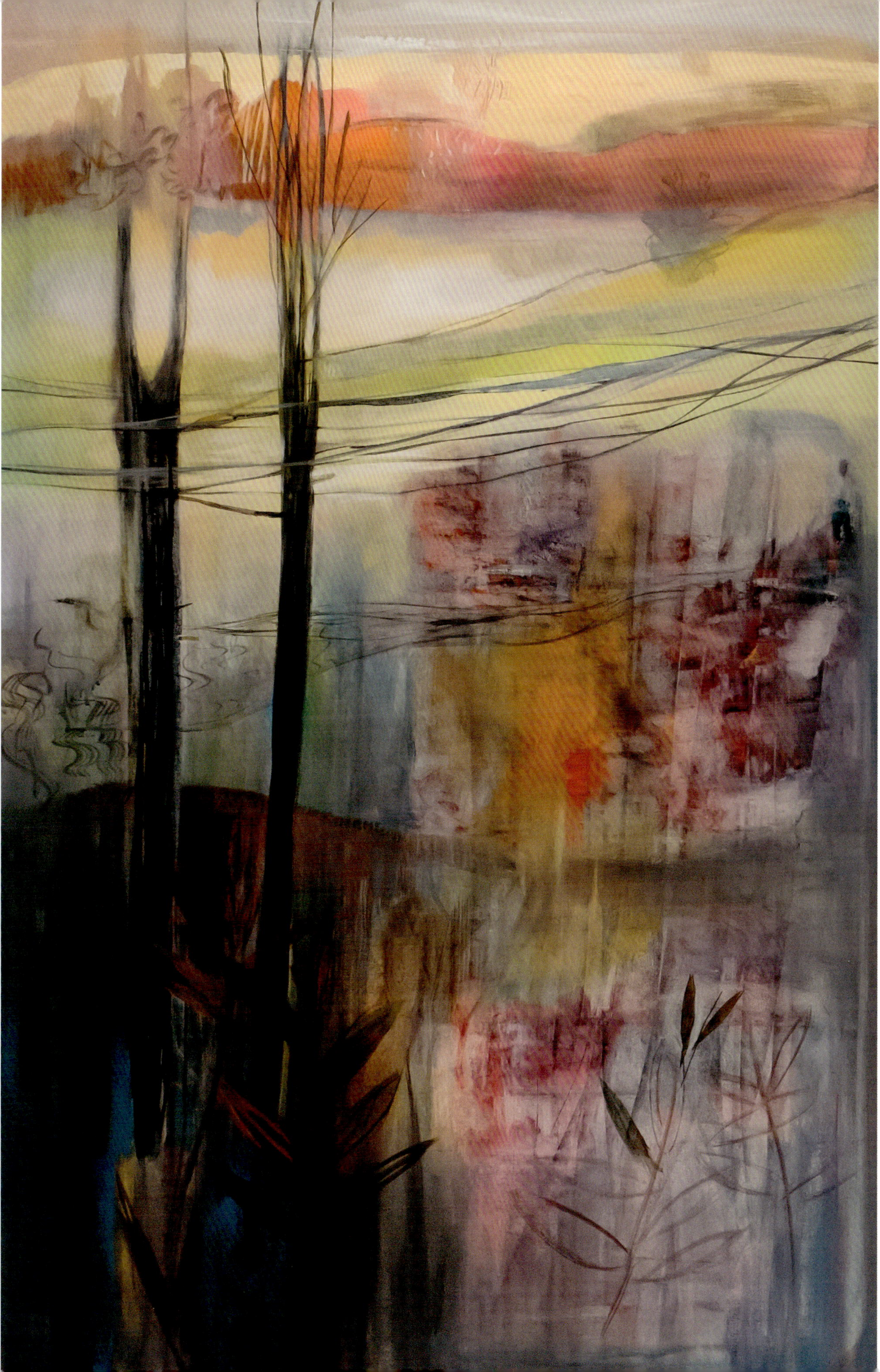

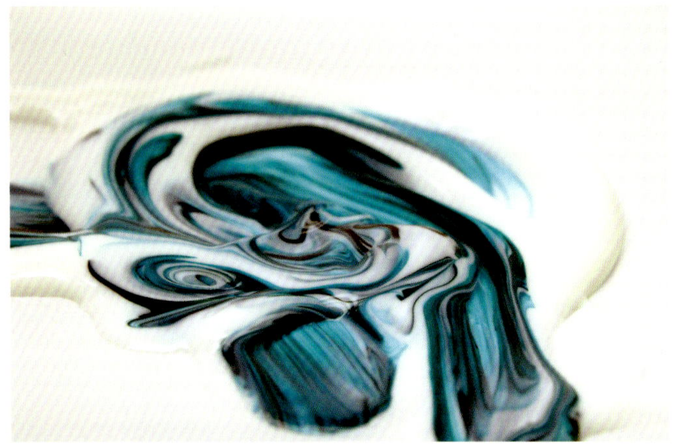

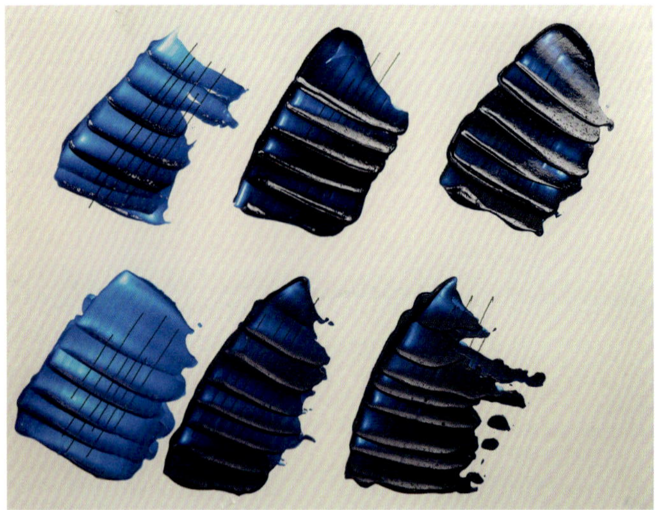

TOP Loosely mixed phthalo turquoise liquid acrylic and matte polymer medium.

BOTTOM Gloss gel (top row) and matte gel (bottom) mixed with phthalo blue green shade in increasingly saturated tints. From left to right: 90%, 50%, and 10% gel to color.

EXTENDING COLOR

Mediums can boost and stretch your colors. In fact, increasing the volume of an acrylic color is one of the more common purposes that gel and liquid mediums are used for. Adding medium to a color dilutes the color, but unlike diluting with water, it increases the durability of the paint film because it adds more acrylic binder to the color. So although the color becomes less saturated, the paint film becomes more flexible and transparent, and the overall tensile strength of the film is given a boost. The color also begins to show its undertone, and the added acrylic brings light and depth to the accumulated paint.

EXTENDING COLOR WHILE MAINTAINING VISCOSITY

To maintain viscosity while extending color, match the medium to the paint format:

- With high-viscosity (heavy body) paints, use gel medium.
- With liquid acrylics, use liquid medium.
- With acrylic inks, use low-viscosity polymer medium or airbrush medium.

For optimum color "pop," use a gloss medium, but be aware that this will remove the matte quality of some colors. For example, professional-grade ultramarine usually dries matte but will dry glossy with the addition of a gloss medium. (This topic is further explored in the "Controlling Luster" section, opposite.)

EXTENDING COLOR WHILE CHANGING VISCOSITY

Through the judicious use of mediums, you can extend acrylic colors while changing their properties. A liquid acrylic can be made dense and viscous with the addition of a gel medium. Conversely, a high-viscosity color can be made thinner and less viscous if extended with a liquid-format medium.

Mediums can also be used to extend each other. This may seem obvious, but it's remarkably widely overlooked. Remember, mediums play well together. Think of them as cheeses: if you run out of cheddar while making a cheese sauce, you can throw in some Gruyère or some Gouda. The rich sauce that results may even have a better flavor (thank you, international cheeses!) and an improved texture. For a range of aesthetic or practical reasons, you might want a medium that's different from what you can get straight from the jar. Say you want a thick gel medium whose texture when dry will be a bit softer and droopier; in that case, adding some liquid medium will accomplish what you want without compromising film integrity or transparency. Or suppose you want to extend your modeling paste without compromising on viscosity; if so, adding some thick gel—being careful not to add so much that the paste is rendered translucent—should do the trick.

CONTROLLING LUSTER

Mediums can also be used to change the luster of an acrylic color or medium, yielding a surface that is super glossy or dead matte or anything in between. You can manipulate luster on your painted surface with a large array of mediums, using varying methods. Knowing what kind of result you're aiming for is the first step in deciding what to use.

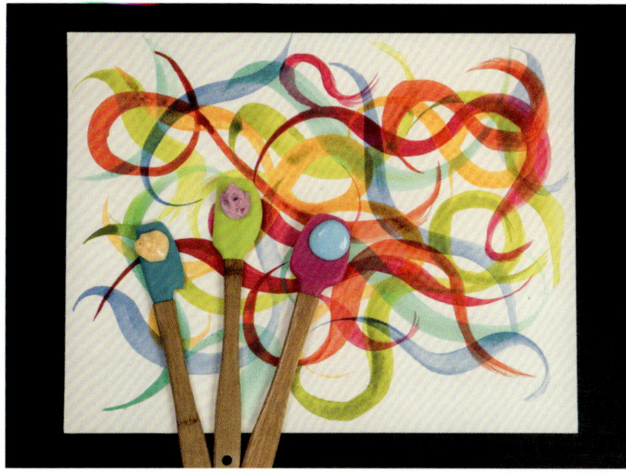

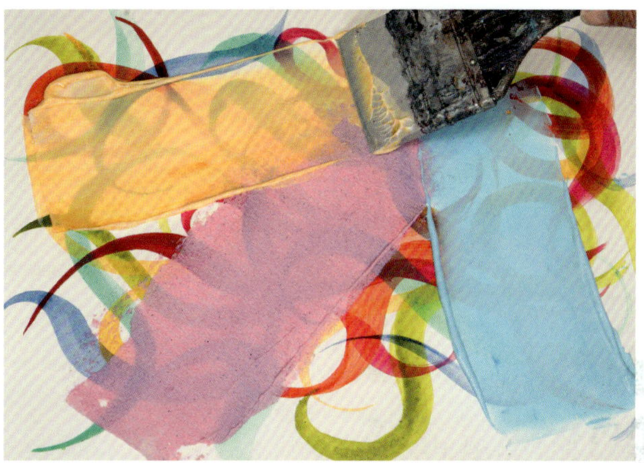

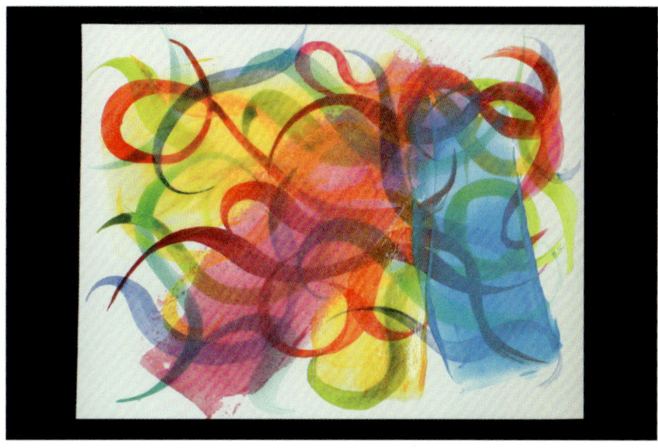

TOP Three thick glaze mixtures, in varying colors with various mediums, ready to be applied.

CENTER Testing out different mediums with varying lusters and textures will help you understand how mediums affect your paint layers.

BOTTOM Dried gloss gel with arylide yellow medium, nepheline gel with quinacridone magenta, and matte medium with phthalo blue glazes on ribbons of color.

APPLYING MEDIUM OVER A PAINTED SURFACE

Mediums can be used over a painted surface to increase or decrease gloss. Many mediums come in three formats: gloss, semigloss (sometimes called satin), and matte. The most obvious way to change the surface luster of your painting is to apply a thin layer of medium to the surface as your "last" layer.

For this purpose, use final finish, polymer medium, or any clear-drying liquid or self-leveling medium. The mediums that work best for a final finish layer are thin, hold little or no visible texture, and dry very clear. A thick medium can also work well as a last layer; for maximum clarity, don't apply it too generously. Gloss mediums will dry the most clearly, while semigloss and matte mediums will produce a foggier finish, lightly muting the underlying tones.

BELOW Matte self-leveling gel versus gloss self-leveling gel over stenciled pattern.

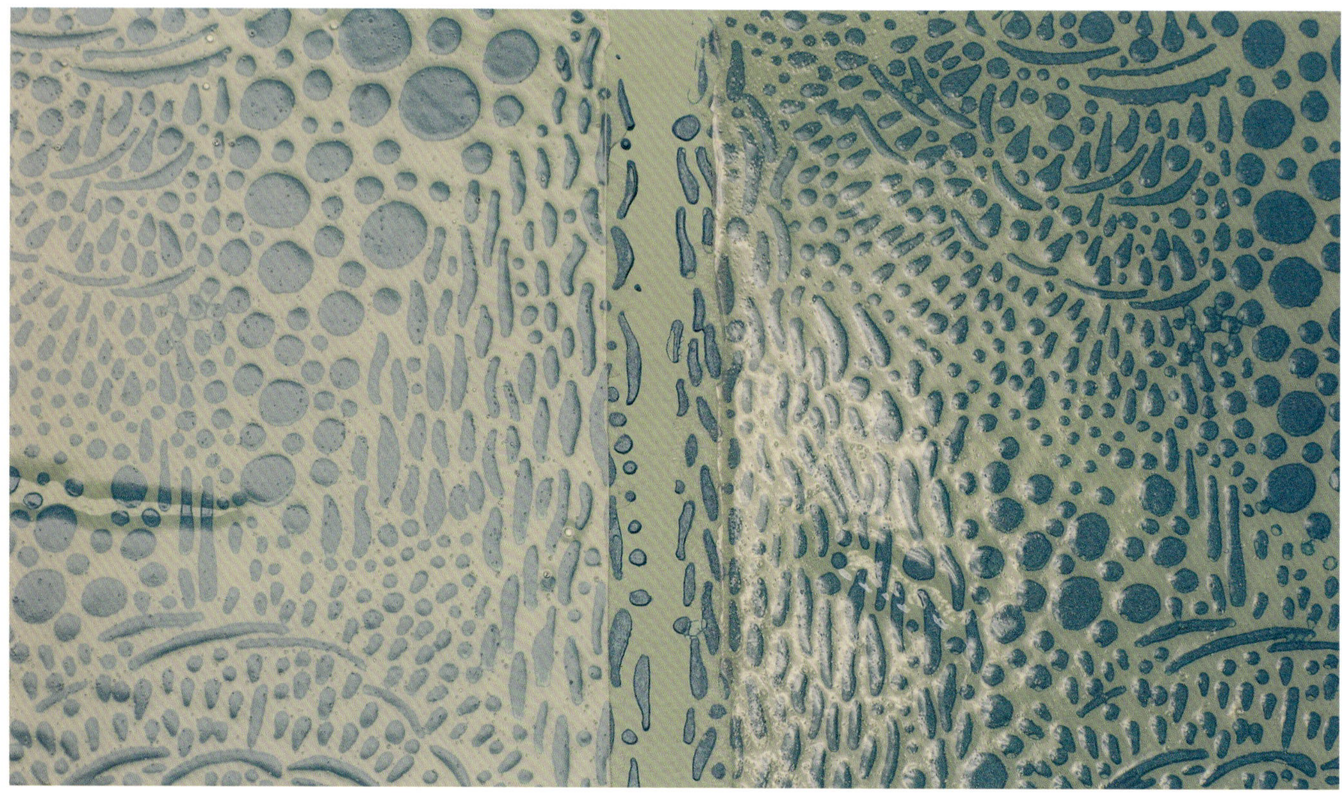

PROCESS

APPLYING A FINAL FINISH

A thick layer of clear-drying medium can do a great deal to enhance the color saturation, depth, and luminosity of a painted surface.

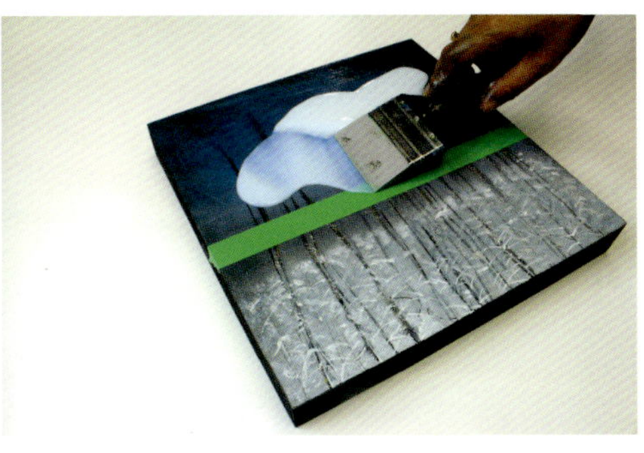

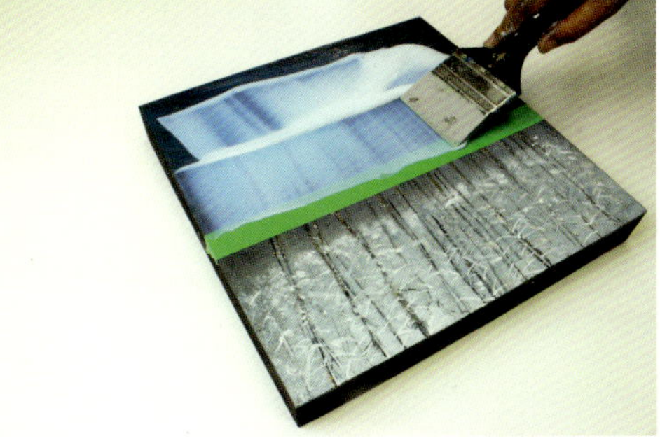

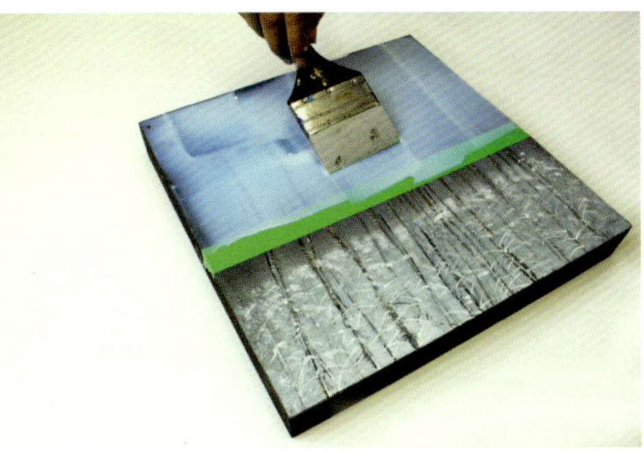

1
Applying a single layer of self-leveling gel gloss. A color shaper is an ideal tool for spreading the gel.

2
Work quickly, pushing the gel in one direction, to cover the full area.

3
Go over the area in a perpendicular direction to smooth out the gel.

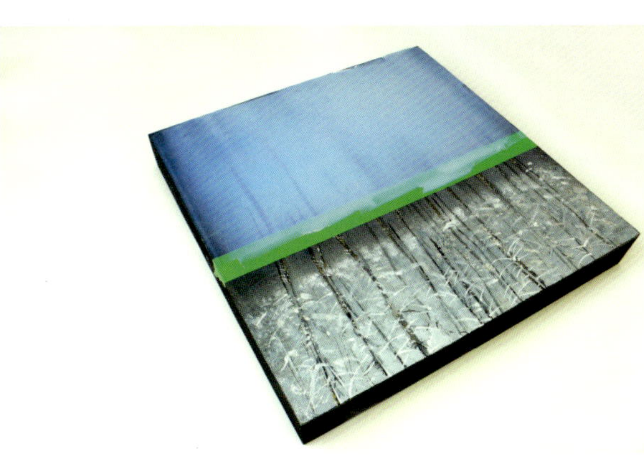

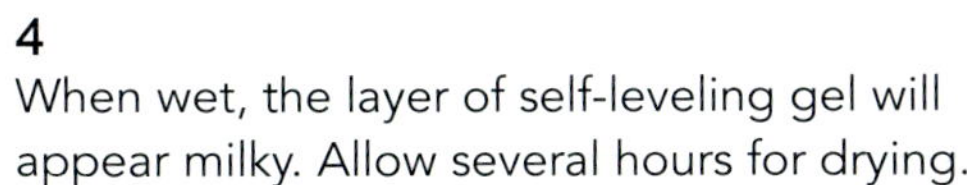

4
When wet, the layer of self-leveling gel will appear milky. Allow several hours for drying.

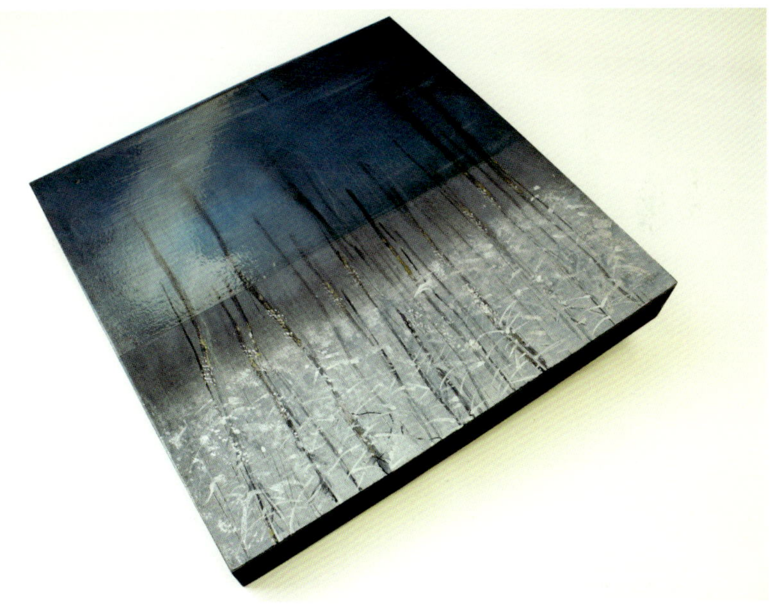

5 (right)
Once dry, the high gloss of the gel layer provides a nice contrast to the flat, matte surface.

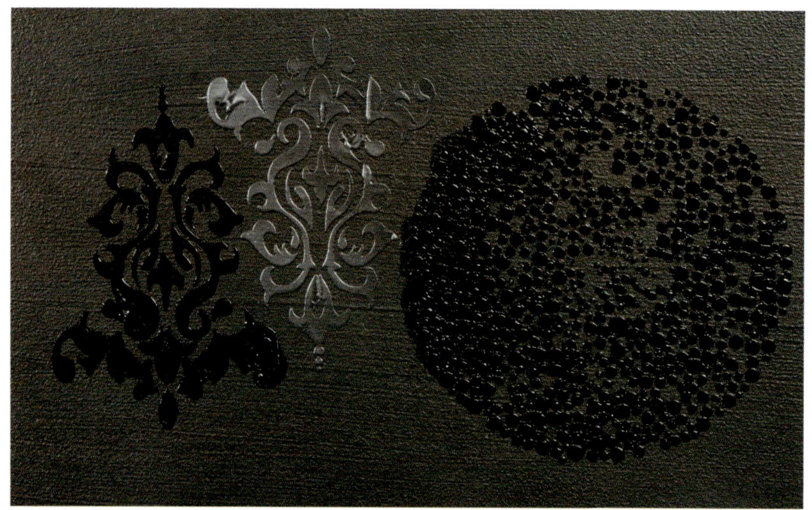

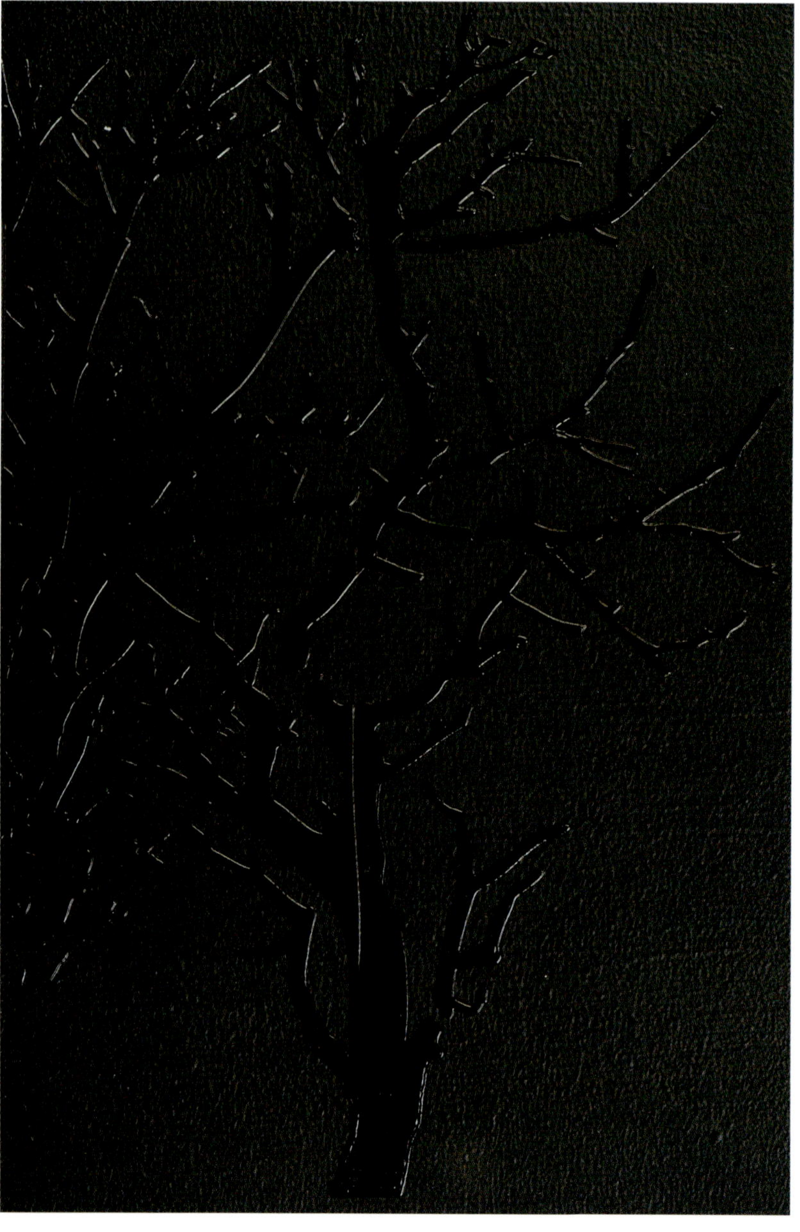

DEEPENING MATTE COLORS OR SURFACES

Is your painting not "popping"? Then it's time to get your gloss on. A matte surface can make colors appear a little dull, but you can change the lackluster appearance of those colors by adding a gloss layer. Dark colors, in particular, will appear richer, deeper, and more saturated by this simple addition.

REDUCING REFLECTIVITY

Decreasing gloss to reduce the reflectivity of a painting is an easy fix with the addition of a semigloss or matte layer. This treatment also makes photographing your artwork a whole lot easier!

PRODUCING SUBTLE CONTRASTS

Playing with light reflectivity on the surface of a work can create a strong visual impact. By pairing opposing surface treatments (matte and gloss) side by side, you can significantly alter a painting's appearance without changing anything about the other components of a painting—composition, color, or application style.

TOP Contrasting stenciled elements: Stenciled onto a black semigloss surface, gloss gel (left and right) and matte gel (center) take on dramatically different appearances.

BOTTOM A thin application of gloss gel (through a stencil) over a matte black surface makes the black deeper.

MIXING MEDIUM WITH COLOR

A superficial coat of matte medium, unless very thin, can give a painting a lightly veiled appearance. If you want to avoid this—that is, to make your colors very matte but saturated and uniform looking—mix matte medium directly into your colors as you paint.

This more subtle approach—which can also be used for glossy and semigloss surfaces—requires some premeditation. If your intention is to define luster without desaturating your color, keep the proportion of medium to color low. It does not take a great deal of medium to alter the reflective property of a color, so only use enough medium to change it but not so much that it renders the color too hazy, translucent, or transparent.

It is also useful to know whether individual acrylic colors are inherently matte or glossy. Generally, synthetic organic colors (like phthalocyanines, quinacridones, and pyrroles) are more glossy than inorganic ones; the most matte pigments are ultramarine, heavy-metal (cobalt- and cadmium-based) colors, and opaque iron oxide colors. Plan accordingly.

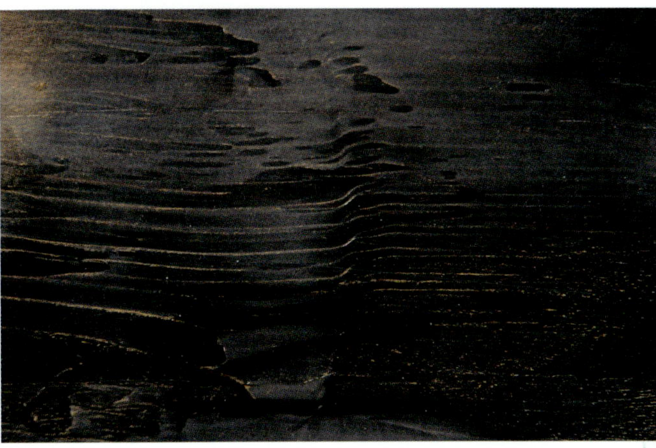

ABOVE Dry-brushed iridescent gold on layers of matte gel that were applied with a large palette knife. The translucent matte gel gives the surface a frosted appearance.

CHANGING VISCOSITY

Acrylic paints come in two basic formats: high viscosity (heavy body) and liquid. Adding medium can reduce or increase the viscosity of an acrylic color (or even another medium). This can expand a color's format range significantly to suit every application technique.

Of course, you can reduce the viscosity of a color—liquid or high viscosity—with water. But if you want to maintain flexibility, gloss, and durability, you must use an acrylic medium. Need to thin a thick color? Use a polymer medium. Thicken a liquid? Use a gel medium.

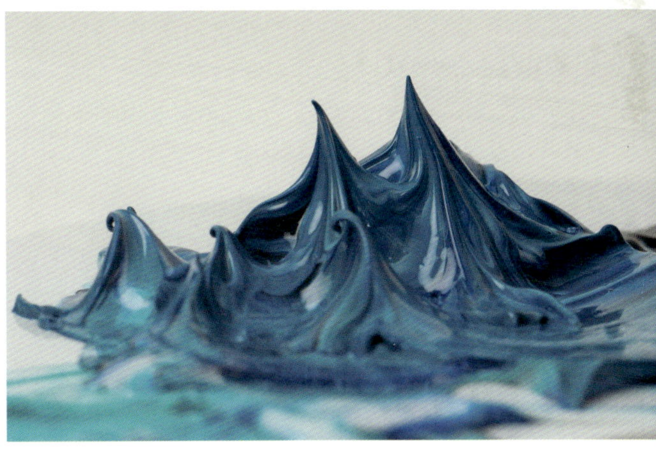

ABOVE Turquoise peaks and curlicues.

REDUCING VISCOSITY

The impulse to thin a heavily viscous color with water is natural, but it's not always a good choice. It is a good choice when you want to reduce viscosity to create a wash or increase flow. But thinning with water can also

- Increase the porosity of the paint film
- Decrease the paint film's tensile strength
- Reduce gloss

If tensile strength, water resistance, and gloss retention are concerns, reduce viscosity by adding a low-viscosity medium instead of water. Because polymer mediums (regular or low viscosity) are substantially thinner than high-viscosity colors but still quite a bit more viscous than water, they will blend with the color more consistently, giving you a smoother mixture with superior permanence.

Whether you thin with water or a medium, you will decrease the color's saturation, rendering it more transparent. This change will be most apparent in colors with larger pigment particles (and thus lower pigmentation), such as ultramarine, cobalt-based colors, umbers, and siennas. Colors with tiny pigment particles, such as quinacridones and phthalocyanines, will retain a higher degree of saturation when thinned.

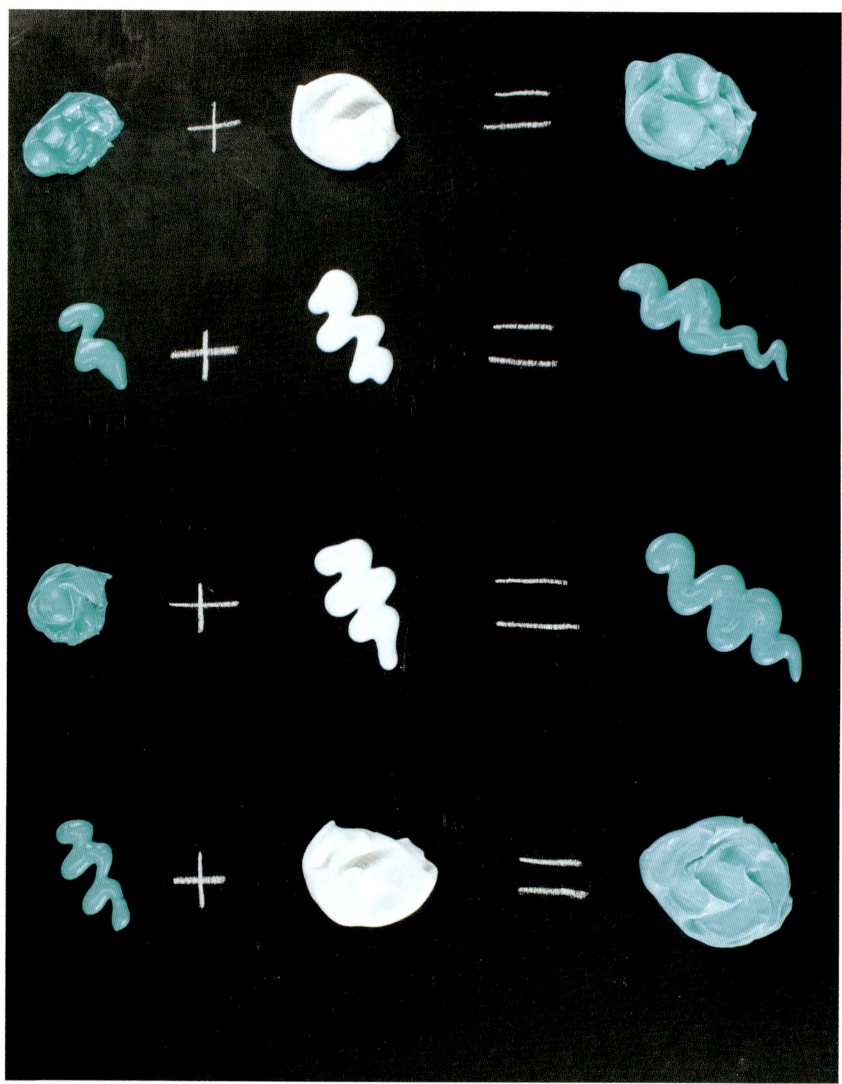

ABOVE Changing viscosity. From top to bottom: high-viscosity color + gel = thick paint; liquid acrylic + liquid medium = thin paint; high-viscosity color + liquid medium = liquid paint; liquid color + gel medium = thick paint.

INCREASING VISCOSITY

If you want to bulk up a color, there are several ways to do it. The simplest way—one that makes the paint more substantial but does not change its rheology—is to add a high-viscosity or heavy gel. Adding gel to a liquid color will make it look and feel like a high-viscosity color but with added flexibility, tensile strength, and transparency.

Clear modeling paste can also be used to thicken your color and increase its density. The components of modeling paste (which are absent from gel mediums) give such mixtures serious weight and strength, although the matting agent and calcium carbonate present in most modeling pastes do reduce a color's intensity, making it significantly more matte. Paints thickened with clear modeling paste make an excellent colored, textured ground.

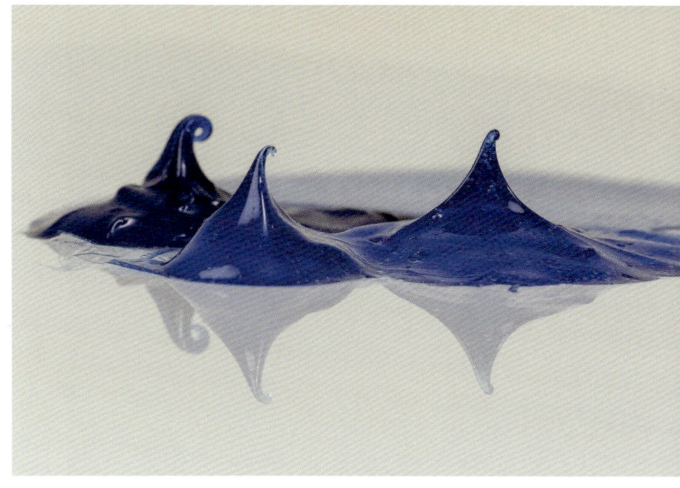

ABOVE Blue curlicues of color: low-viscosity gel tinted with phthalo blue.

CHANGING RHEOLOGY

You can change a color's rheology from short to long, increasing flow while decreasing viscosity. Self-leveling gels and pouring mediums produce a dramatic change in a color's rheology. These mediums are strong enough to break down all the texture-holding capabilities of a color, rendering it gooey, stringy, and outright oozy.

The unpredictability of long-rheology mediums can be very freeing, though to some the lack of control can be a bit alarming. These mercurial materials demand that you let go of any aversion to spontaneity and embrace random results. I myself have experienced applying an identical—or so I thought!—technique to two paintings with wildly different results. Why so different? Because there are always more variables than we know about when using materials of such a capricious nature.

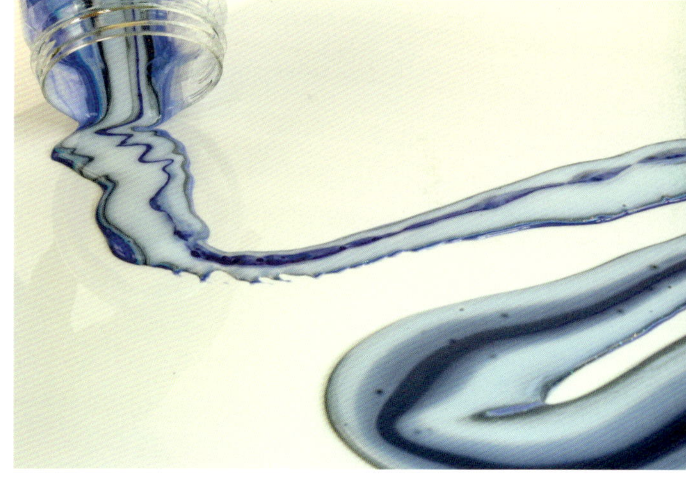

ABOVE Tinted self-leveling gel flows and pools on a palette.

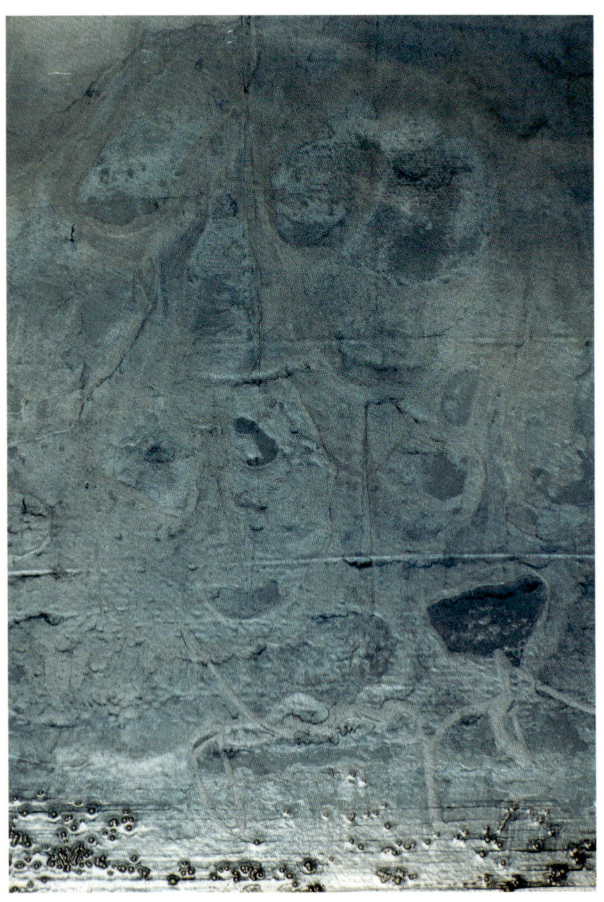

ABOVE In this detail from my painting *And the Stars Fell* (see page 237), watery washes and color separation lie under three layers of self-leveling gel.

CREATING WASHES

Water, the "first medium," is the one we automatically think of for thinning color. In many cases, it's not only the obvious choice but also the correct one. A wash of color thinned with water behaves very differently from a color thinned with a medium. Water-thinned color flows very freely and beautifully; at the same time, the color desaturates, and the paint loses all ability to hold any textural detail. When dry, the wash is flat and dead matte. What we can't see is the extreme decrease in paint film flexibility and integrity. The more water you add, the farther apart the acrylic molecules move, progressively weakening the bond between them. In this veil of diluted acrylic, the pigment barely holds on to the surface. Water can also disrupt paint colors that are made of mixed pigments, spreading the individual pigment particles away from each other.

CONTROLLING RELATIVE COVERAGE

Mediums can also be used to increase or decrease the opacity of an acrylic color or another medium. Mediums vary in clarity and coverage: Transparent mediums, which dry to a very clear and transparent film, are ideal for creating a clear, transparent layer and for increasing the transparency of a color. Semi-opaque mediums dry clear, though with a slightly reduced clarity. Opaque mediums dry totally white or colored.

To control relative coverage, you must understand the transparency/opacity of both the color (pigment) and the medium. Remember that each pigment is unique. Getting the right ratio of color to medium may take some experimentation —take notes!—but here are the basic principles:

1. Use an opacifying medium to opacify color or medium.
2. Use matte medium (any viscosity) to partially opacify color.
3. Increase transparency by diluting color with medium. (Changing viscosity will have no effect on transparency.)

WASHES VERSUS GLAZES

There is a marked difference between a wash and a glaze. A wash will coat a surface but remain thin and matte, while a glaze can be thicker and thus more saturated in appearance; it can be glossy or matte depending on the type of medium used. Glazes are also more durable and flexible than washes and are less likely to be easily marred. (For more on glazing, see page 210.)

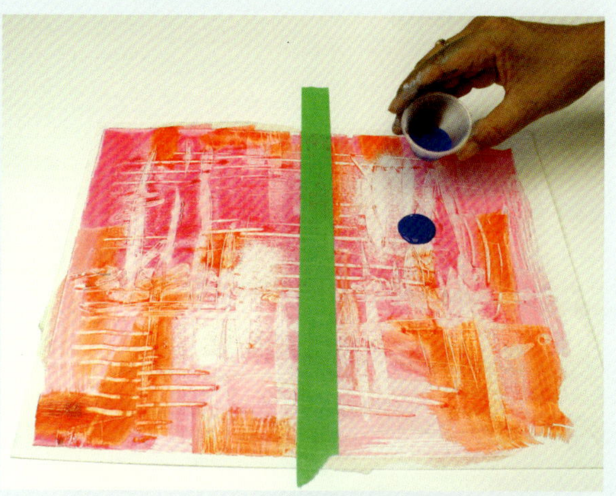

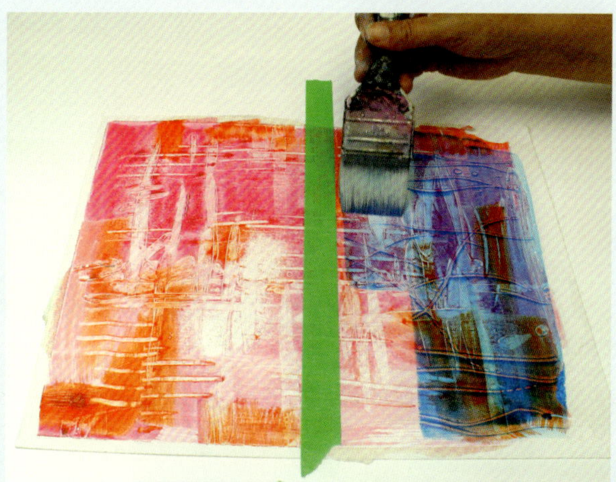

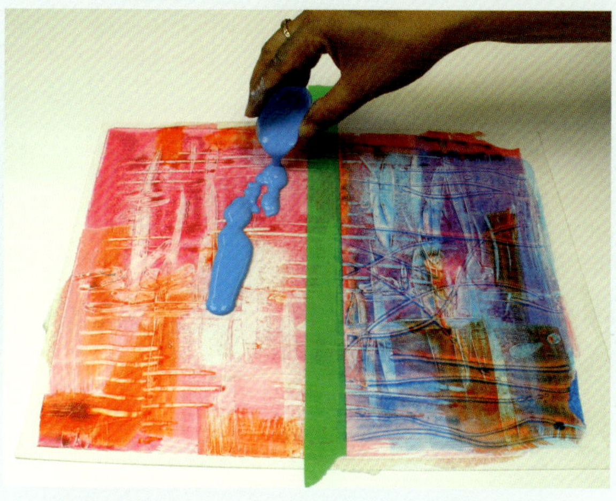

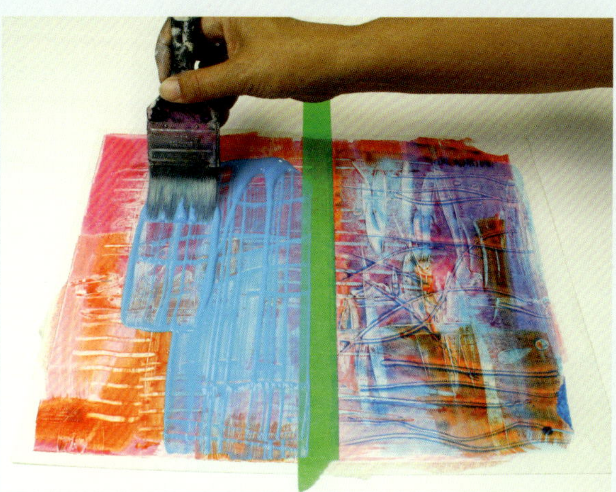

TOP ROW Applying a wash (color + water) to a colored ground.

CENTER ROW Applying a glaze (color + liquid medium) to a colored ground.

RIGHT The glaze (left) and the wash (right) have the same proportion of medium/water to color. Each unifies the background in its own way.

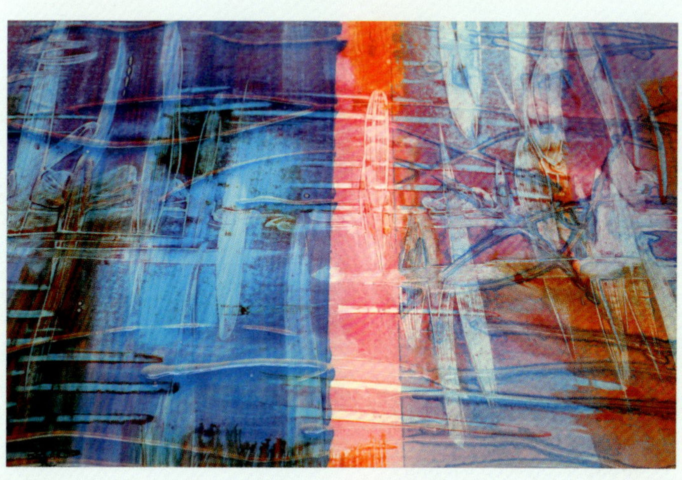

PROCESS

CREATING A SMOOTH COLOR FIELD

There are several acrylic-painting techniques for producing a soft, diffuse color field. In the process shown here, a thick medium and semi-opaque colors are used in combination to simultaneously dull the underpainted color and give it a more uniform appearance. Tinting the self-leveling gel medium with semi-opaque colors changes its relative coverage from transparent to semi-opaque.

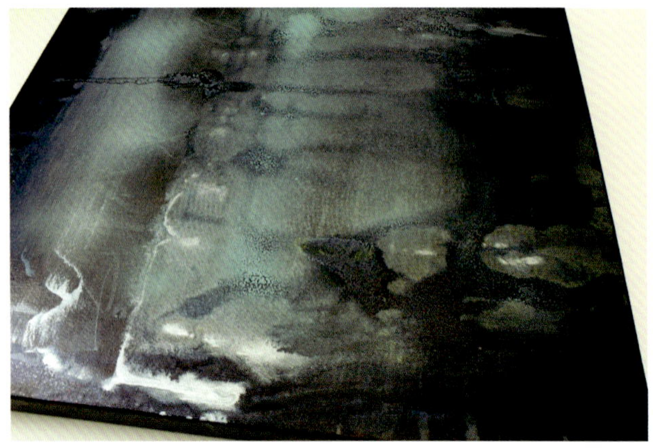

1
The surface before the application.

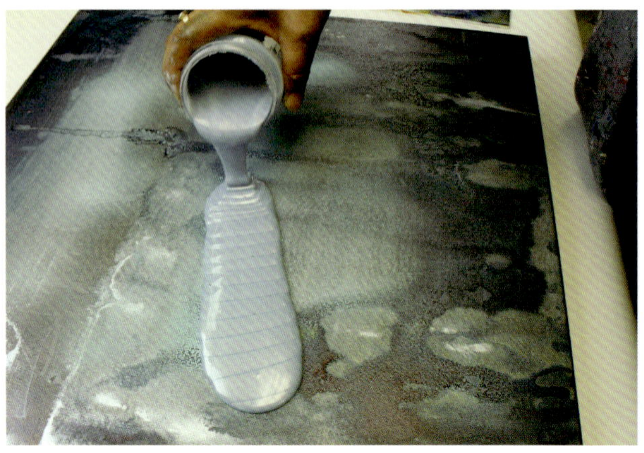

2
Layering tinted self-leveling gel over the painted surface. Self-leveling gloss and matte gels were combined to create a semigloss medium, then suffused with zinc white and Payne's grey, both semi-opaque colors.

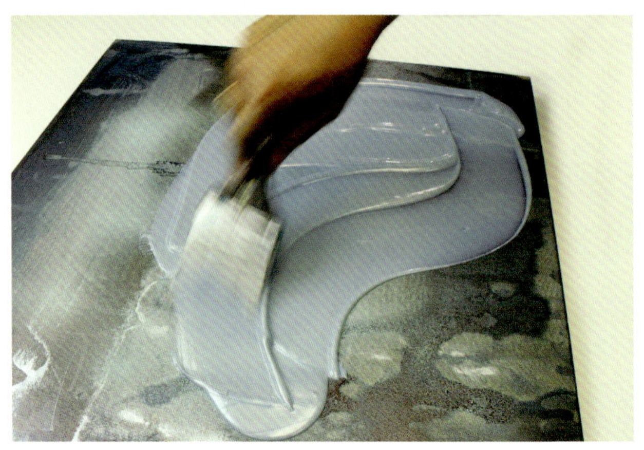

3
When covering a surface with self-leveling gel, it is essential to move quickly.

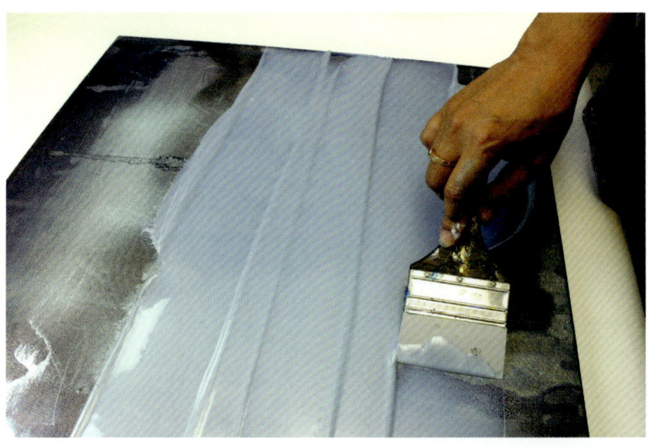

4
Spread the gel as evenly as possible over the surface. The ridges will gradually level out and disappear.

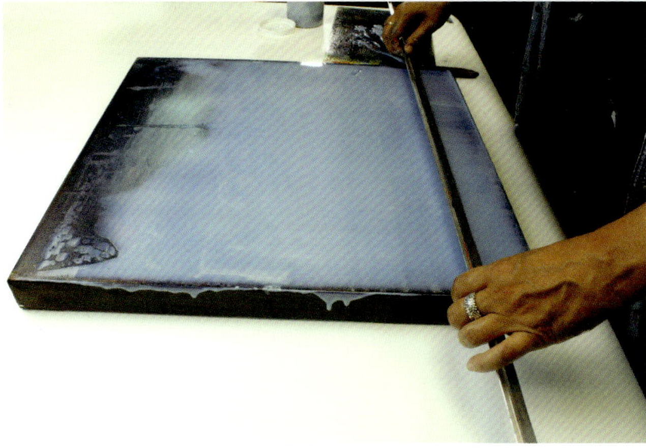

5
Use a long straightedge for the final smoothing.

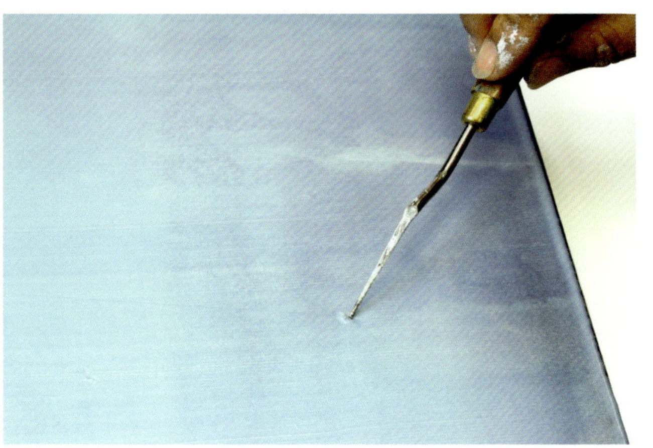

6
A small piece of dried paint, a hair, or any other errant speck can mar the smooth surface of the gel. Remove it as quickly as possible while the gel is still wet. As air bubbles rise to the surface, they can be popped with a pin or sharp tool.

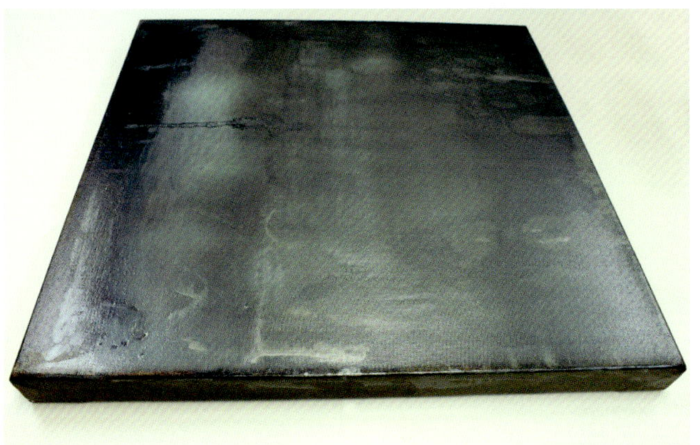

7
The layer of semigloss self-leveling gel tinted with a semi-opaque color mixture has now dried. The added layer softens and adds a luminous quality to the surface.

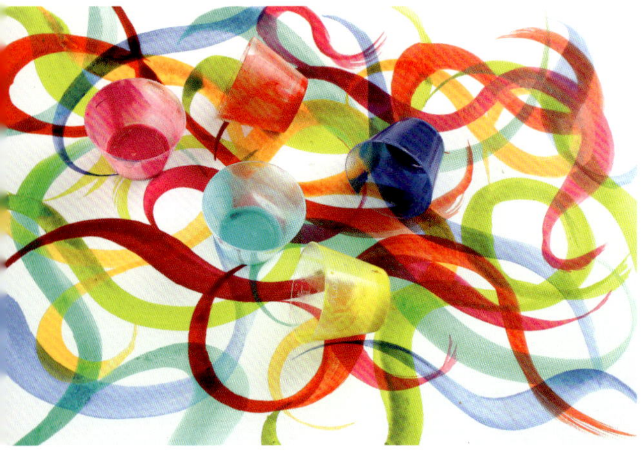

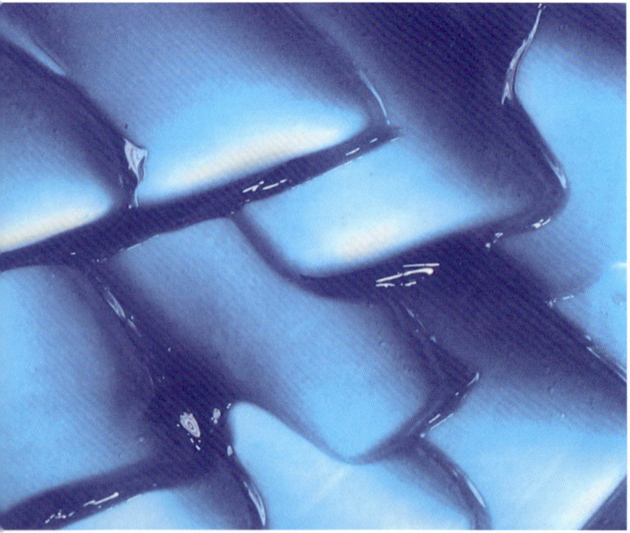

GLAZING

Glazing is one of the great strengths of acrylic mediums. These transparent vehicles are ideally suited to carrying small amounts of color that, when combined and layered, produce paintings with rich depth and subtle color transitions.

Professional-grade acrylic paint is really, really saturated. Used full strength, its bright undertones are obscured by the sheer intensity of the color. Mediums draw out the color's nuances—the bright juiciness of its essence. The light they bring ramps up the power of any color.

The transparent or translucent color of a glaze has a seductive quality, urging us to look deeper into a composition, hunting for the elusive color veils laid down layer over layer. The surface appearance of a painting rich with transparent layers carries a life and depth that is not present in an opaque application.

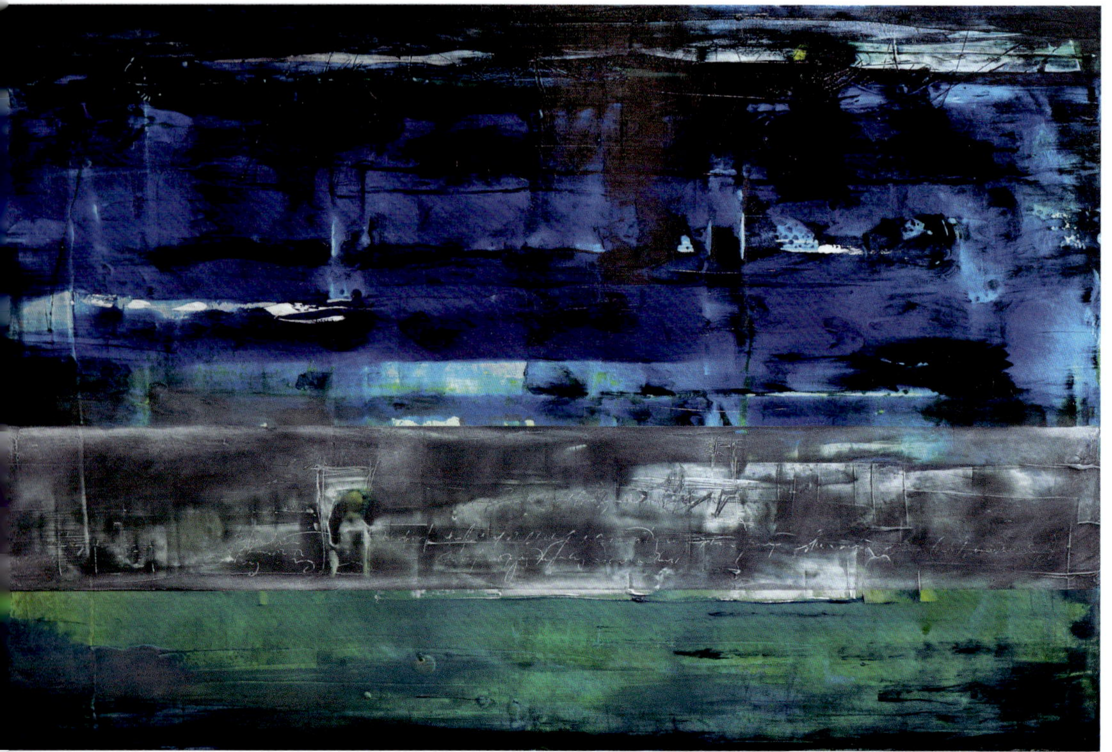

TOP Ribbons of colored glazes.

CENTER Gel glaze.

LEFT **Rhéni Tauchid, *Behind the Falls,* 2014, acrylic and charcoal on canvas, 40 x 60 inches (102 x 152 cm). Private collection, Kingston, Ontario, Canada.**

A half-inch-thick layer of gloss-gel glazes intensifies the aquatic tones in the top half of this painting. Below it is a band of matte gel with text texture scratched into it with a skewer. Along the bottom, a band of silky smooth self-leveling gel draws the viewer into its depths. There are ten to twelve individual glaze layers in this painting.

PROCESS

GLAZING

This step-by-step demonstration guides us through the multilayered glazing process that artist Connie Morris followed in creating a still life of pears.

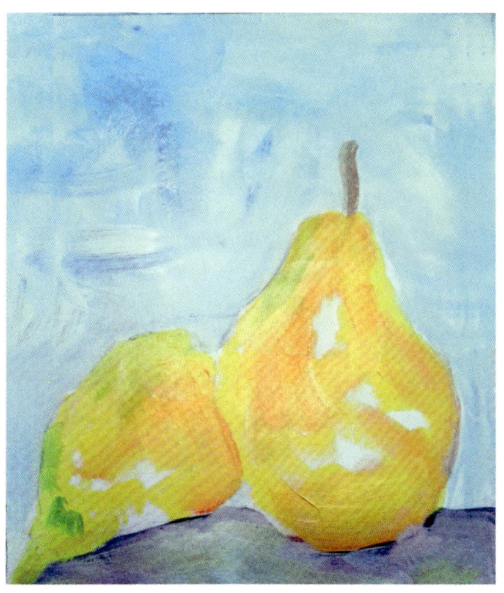

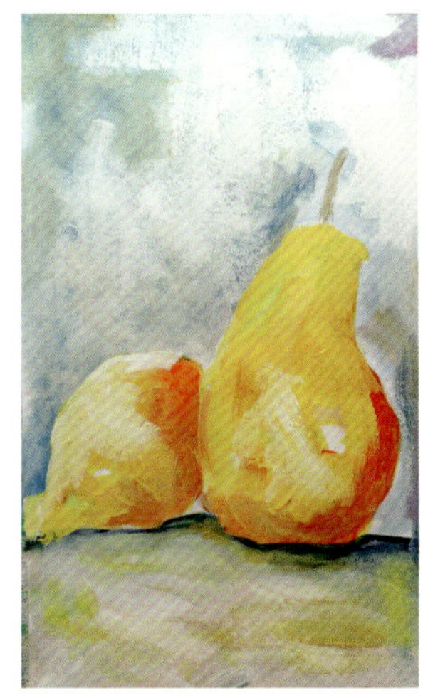

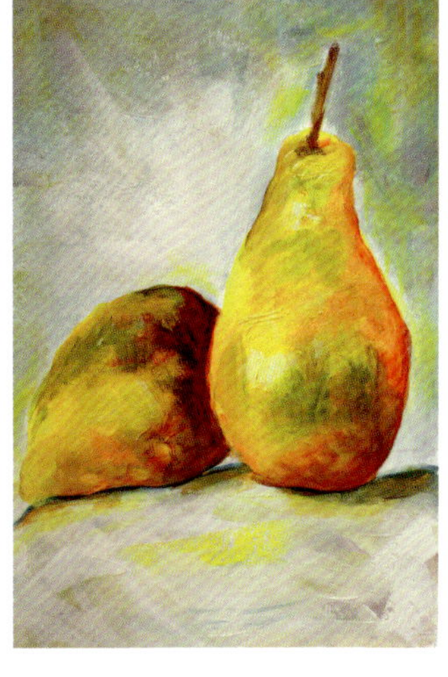

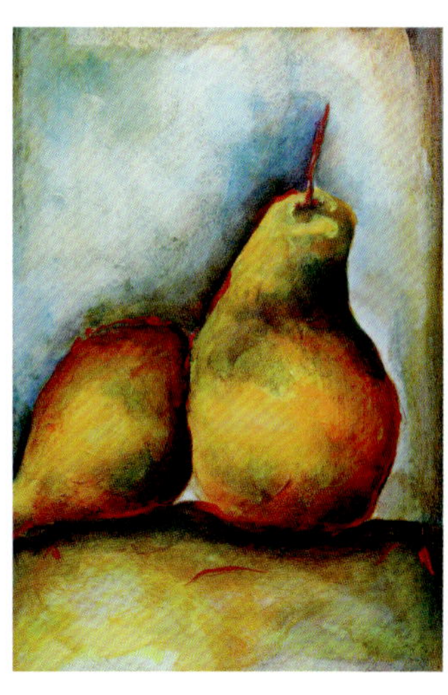

1
First, Morris locked in the composition with loosely applied color tints using liquid acrylic colors mixed with glazing medium.

2
Still using glazing medium, she started building up the surface with both transparent and opaque color overlays. Morris painted the background with white to open it up and to create a reflective ground for subsequent layers of color. Then she began adding darker colors to create form in the pears.

3
To further define the pears' form, she added more layers of transparent color with glazing medium, continuing to create contrasts of lights and darks (shadows). She toned down the colors by using a very small amount of transparent brown and glazing medium, giving the background a soft patina.

4
Finally, Morris blended and softened the color shifts. She darkened the shadows to enhance the forms of the pears. She finished by adding vibrant, delineating red strokes and giving the painting a final clear coat of untinted glazing medium.

Thin mediums can be tinted with color in any proportion to create glazes. The more medium, the more transparent and desaturated the color will be, though the addition of the medium brings light into the surface and creates a feeling of depth.

When creating a glaze, think about proportions and test how your glaze will look from wet to dry. The more medium you use, the whiter the wet glaze will be. In thin films this will not be not as apparent, but if you are applying a thick glaze layer, there will be a significant change as the medium clarifies and the color asserts itself.

It's good to have an understanding of color theory so that you don't muddy your work, but it's more important when glazing to understand the type of pigment you are using and its relative coverage. Semi-opaque and opaque colors become more opaque (to varying degrees) as you build up layers of glazes. If you're using very opaque colors, apply them in very diluted mixtures to avoid masking the underlying layer.

Glazes of color build gradually, giving you more control. One thick layer of color, even if it's very saturated, will not have the luminosity of the same color built up in glaze layers, which pull in the light and increase visual depth.

Additives such as retarder and retarder gel can be used to dual purpose in a glaze. When you are glazing a large area and need to keep a quantity of glaze wet, you can add retarder (up to 15 percent of the total volume) to the medium-paint mixture to increase the flow and keep the glaze from drying too quickly. The extended drying time can also make the brushstrokes in the paint film less prominent and visible.

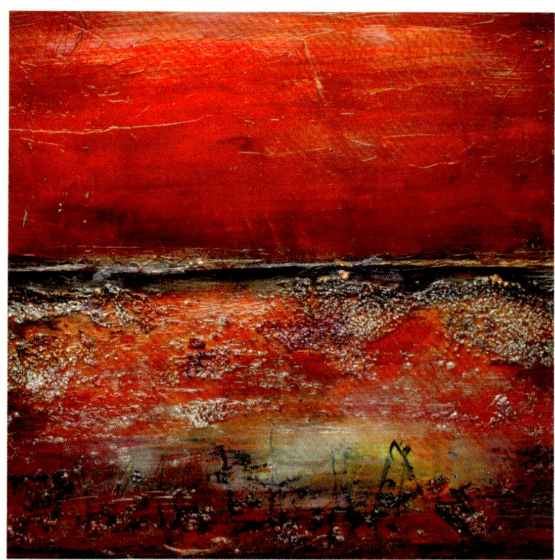

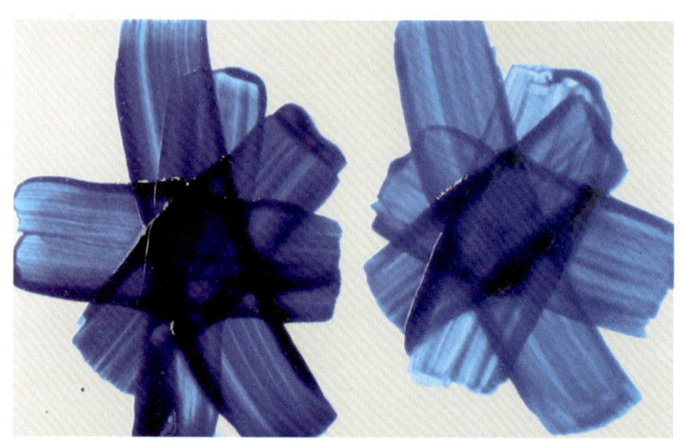

TOP An underpainting of extremely reflective color adds brilliance and luminosity to the glaze layers in this painting.

CENTER Saturated glazes (left) and dilute glazes (right). Glazes build up color one brushstroke at a time.

Tip: When glazing, imagine you're layering stained glass—building depth and chroma. If one piece of glass is frosted, it will obscure what is underneath. Be mindful of the transparency of your mediums when layering. Gloss medium will give you the clearest glaze layers, while matte medium will render glazes more translucent. Even if you prefer a matte finish, for increased clarity between layers it is prudent to use gloss mediums until the final layer.

PROCESS

TRANSPARENT LAYERS

For her painting *Twice As Much*, artist Lorena Kloosterboer built up layers of transparent color over a grisaille underpainting.

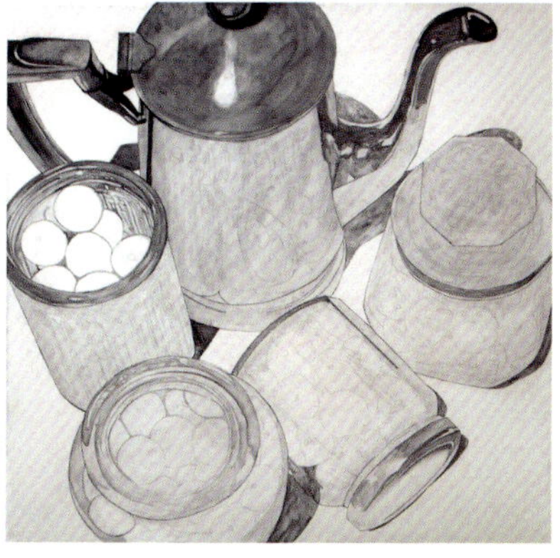

1
Tracing her outline onto a prepared panel, Kloosterboer then covered the objects with a highly diluted wash of Payne's grey, gloss medium, and water, leaving the background white. Continuing with Payne's grey, she added darker lines, areas, and cast shadows to build value intensity.

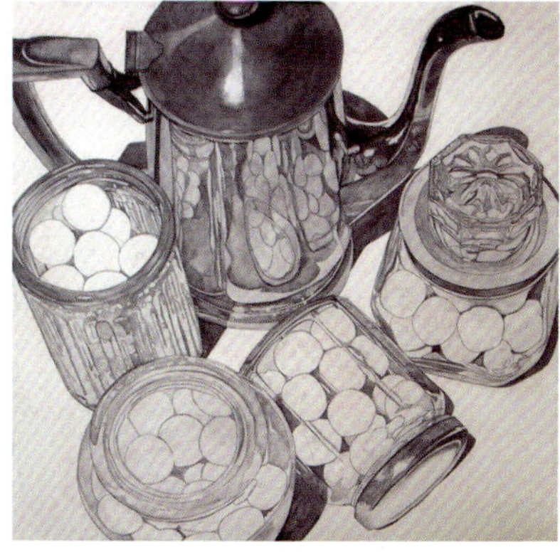

2
She outlined the shapes and deepened the darker areas with Payne's grey. Grisaille underpaintings like this establish all the values and provide a great base over which to glaze colors.

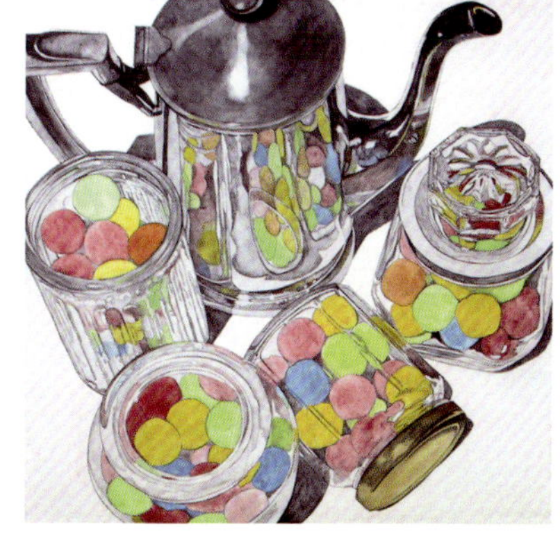

3
She added color slowly, building up transparent layers and assessing intensity and chroma as each layer dried. Working with transparent colors helps you to vaguely see the composition's lines, using them as guides. Kloosterboer keeps the background clean, concealing little smudges with titanium white.

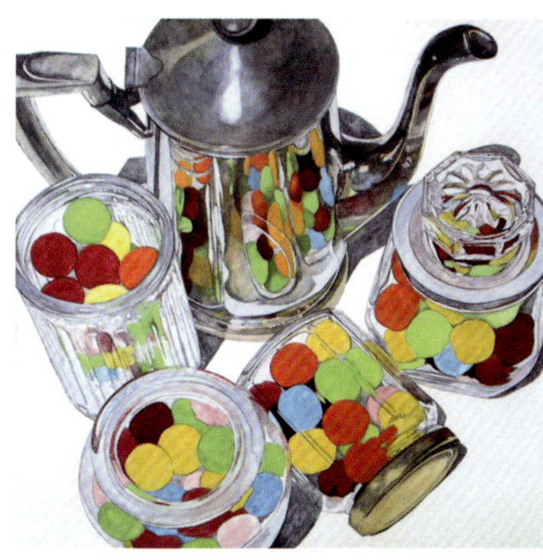

4
Kloosterboer added more glaze layers to intensify colors, suspending pigments in clear gloss medium to create wonderful jewel tones.

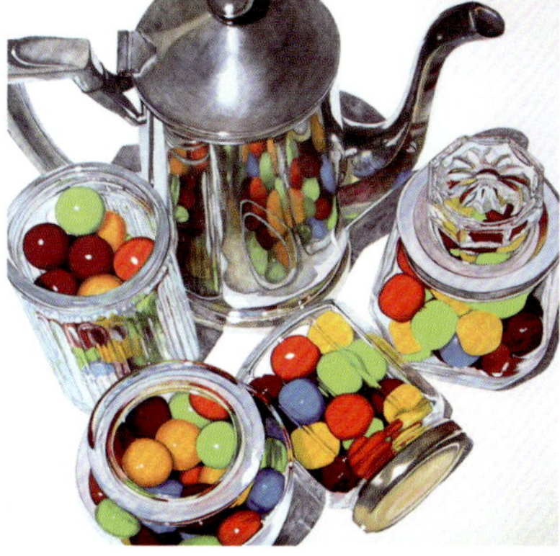

5
Small details are added with titanium white to suggest highlights and distortions in the glass. To imply three-dimensionality, shadows were added around the gumballs; a Payne's grey glaze added to the coffee pot gives it a tarnished look.

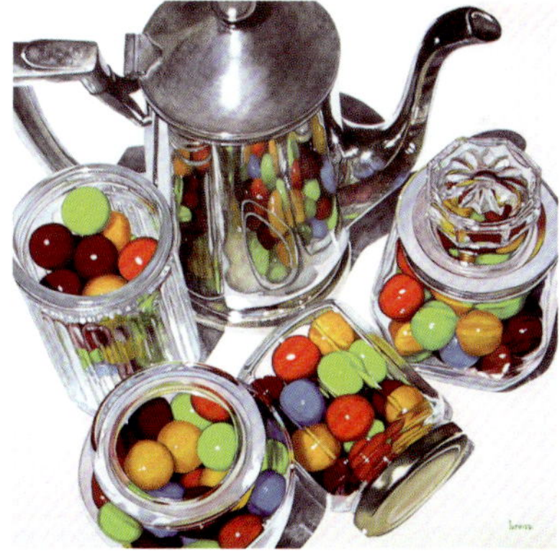

6
The finished painting. A larger image of this beautiful piece can be found on page 131.

CREATING THICK, LUSTROUS LAYERS

There is something very satisfying about spreading a large volume of medium over the surface of a painting. It's a bit scary, too, because you can never be 100 percent sure how it's going to turn out. But that can be part of the fun. If you are a stickler for control or are aiming for consistency in application, be sure to take note of the ratio of color to medium so that you can reproduce the effect. If mixing your color and medium in a sealable container, apply a dab of the mix to the lid so that you'll know what to expect when you next use it.

For this kind of application, use gel, self-leveling gel, pouring medium, or a top coat. There are various

BELOW, LEFT Rhéni Tauchid, *Meltwater Trench,* 2016, acrylic, graphite, and charcoal on canvas, 48 x 60 (122 x 152 cm). Private collection, Wolfe Island, Ontario, Canada.

The three bands of color in this painting were built with layers and layers of medium.

BELOW, RIGHT *Meltwater Trench* detail: Under a thick, clear laminate of self-leveling gel lie the remnants of the interplay of water and color over a slick gel surface.

BOTTOM, RIGHT *Meltwater Trench* detail: Between bands of glossy medium treatment lies a swath of dry media ground over a frosty white surface, where sharp charcoal marks add stark linear accents.

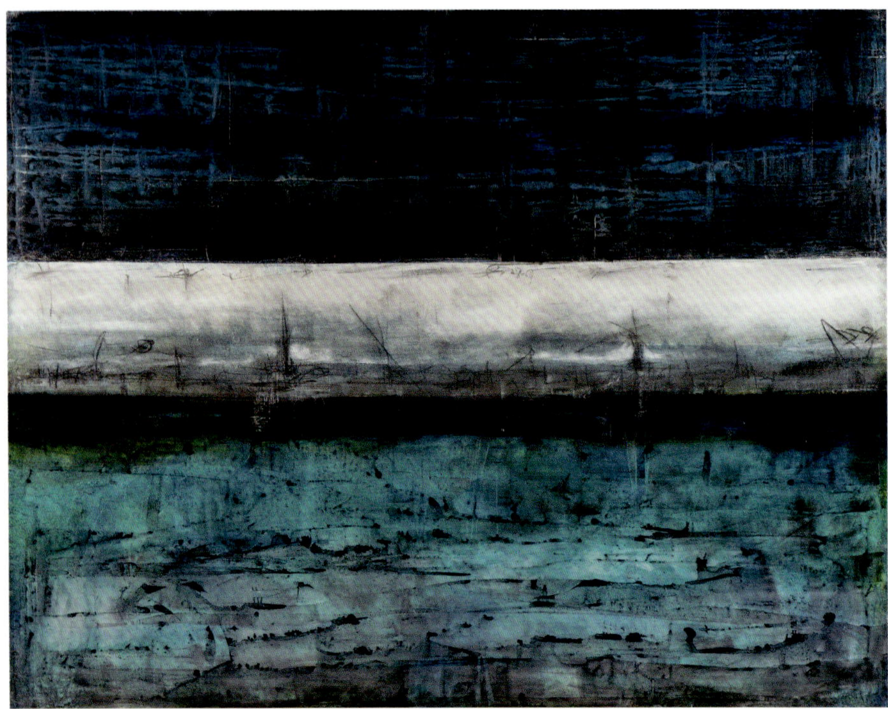

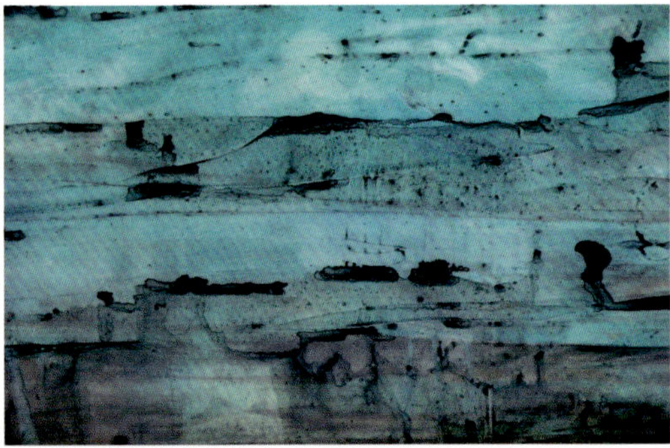

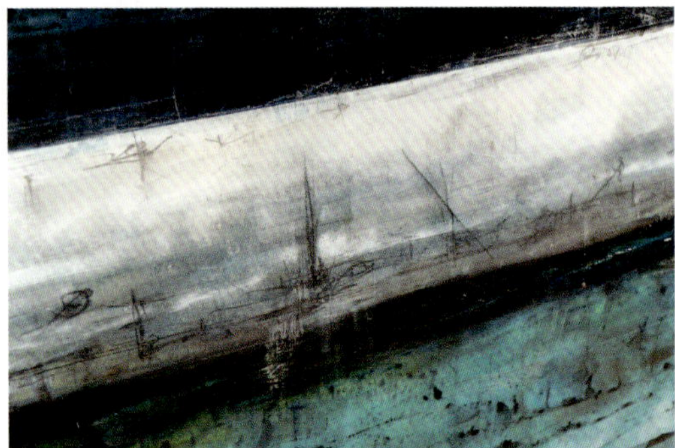

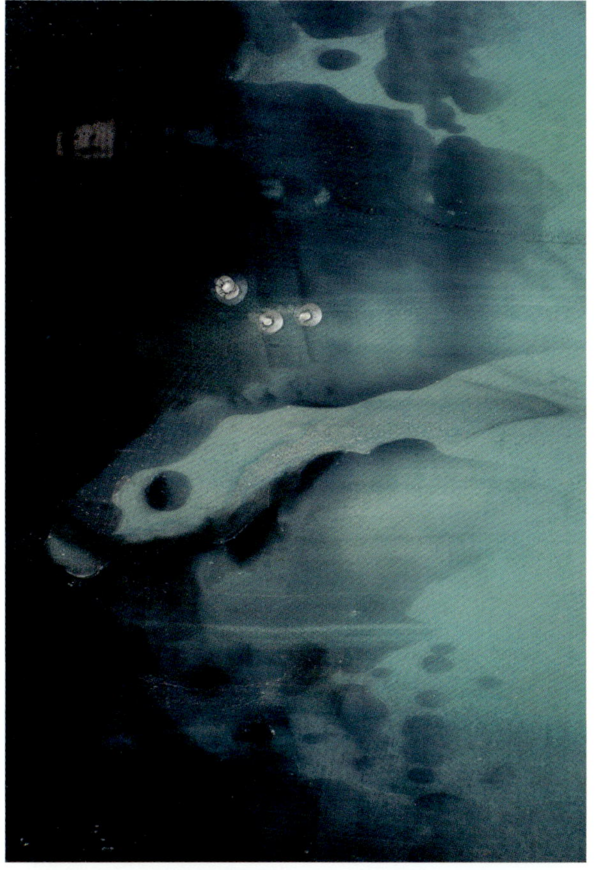

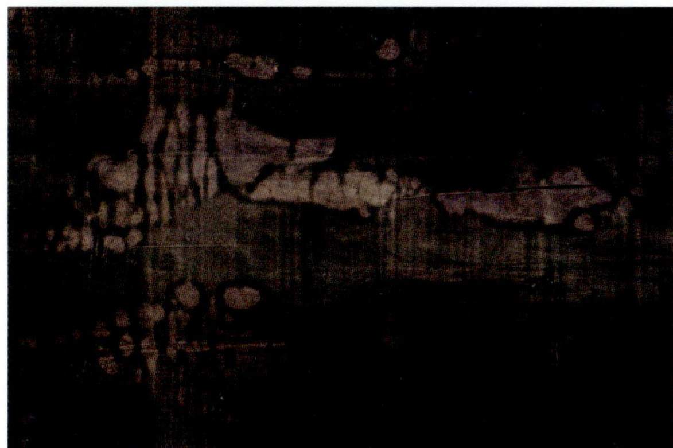

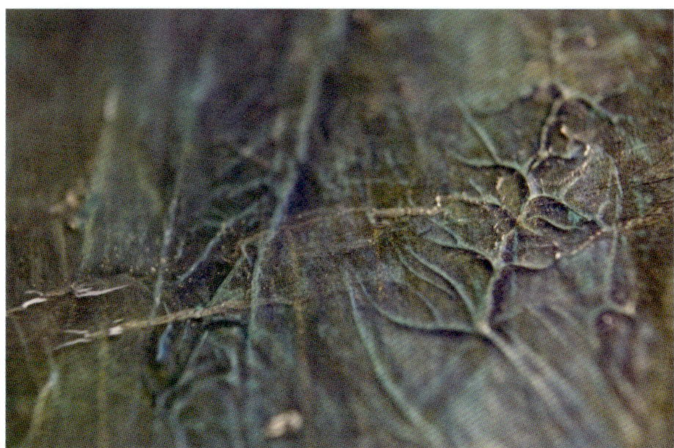

approaches to the application of thick layers of clear mediums. All require a lot of patience. Focus on the medium's viscosity and rheology and on the appropriate tools:

- *For gel:* Unless you are looking for a highly ridged surface, brushes are not very effective for applying large quantities of gel. A large spatula, palette knife, or similar flat-edged tool is more suited to spreading the gel over a flat surface.
- *For self-leveling gel and pouring medium:* These types of mediums can be susceptible to the accumulation of bubbles, which can be exacerbated by the use of a brush. Using a spreading tool—color shaper, spatula, or straightedge—will produce a smoother surface. Keep in mind, too, that gravity has a tight grip on these mediums; if your surface is not level, your layer of medium won't be either.

TOP This smooth, subtle area resulted from several layers of thick glaze in two colors (phthalo blue and Payne's grey).

CENTER Here, a layer of manipulated watery wash is trapped between layers of gloss gel. You get a peek at the underpaint.

BOTTOM This veined texture is highlighted with opaque color and enveloped in a layer of clear gel.

CHAPTER 11

EXPLORING COLLAGE, DECOUPAGE, AND MIXED MEDIA

Acrylics are essentially glue. It is their adhesive quality that binds the pigment to the support. Although we don't ordinarily think of paint as glue, all paints are. We identify paints by their binders—acrylic, oil, wax, gum arabic, egg yolk—and each of these binding agents acts as an adhesive. But of this whole group, the acrylic polymer emulsion binder is the most transparent and the best adhesive, making it ideal for collage, decoupage, mixed media, or any other type of gluing application.

Acrylic polymer emulsion binder is the best adhesive.

OPPOSITE Heather Haynes, *Norma*, 2017, collage and acrylic on canvas, 20 x 16 inches (51 x 41 cm). Collection of the artist.

In this work, a printed image of an old photograph was adhered to a painted acrylic ground with acrylic polymer medium, then sealed with several layers of glazing medium for both surface gloss and UV protection.

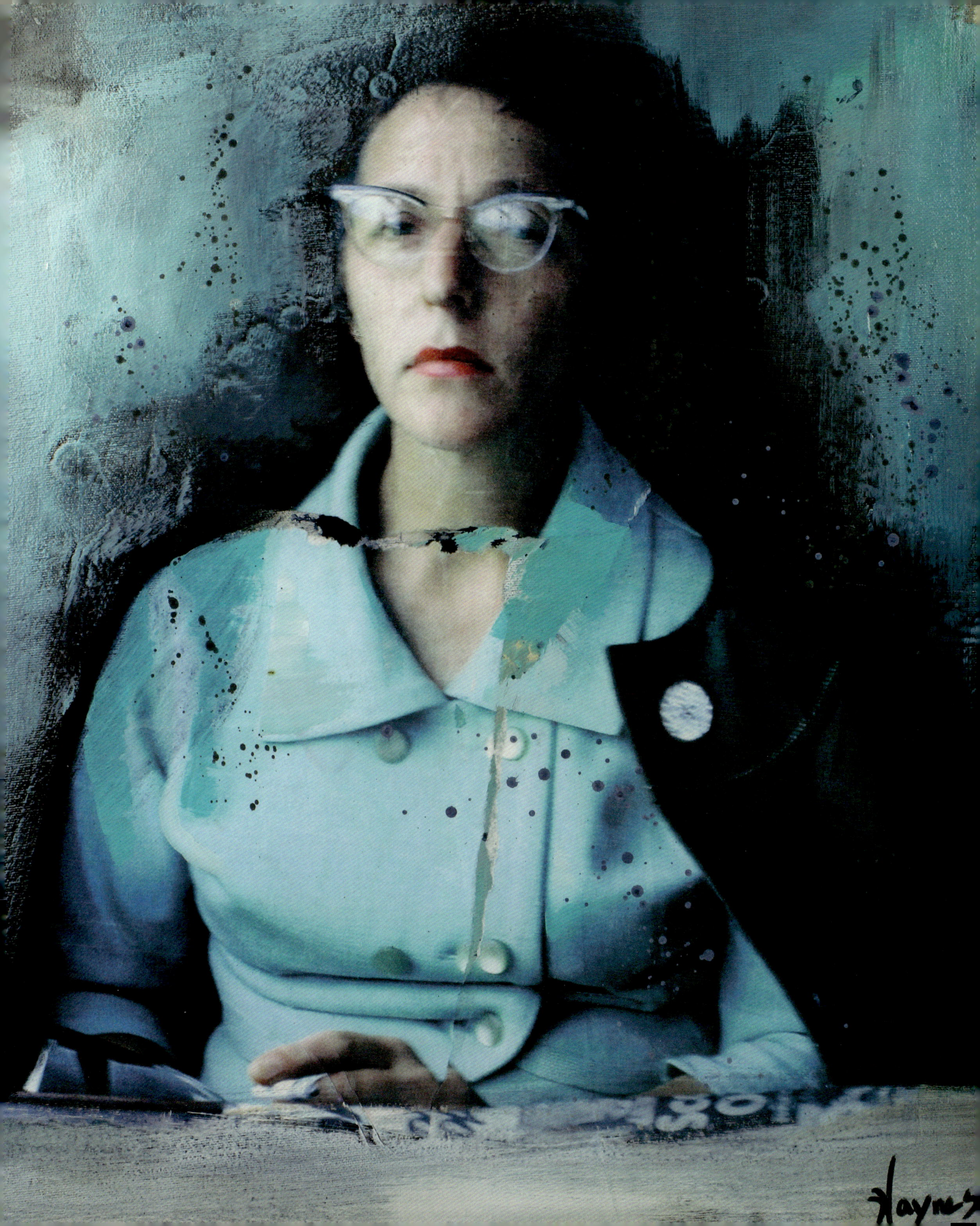
Haynes

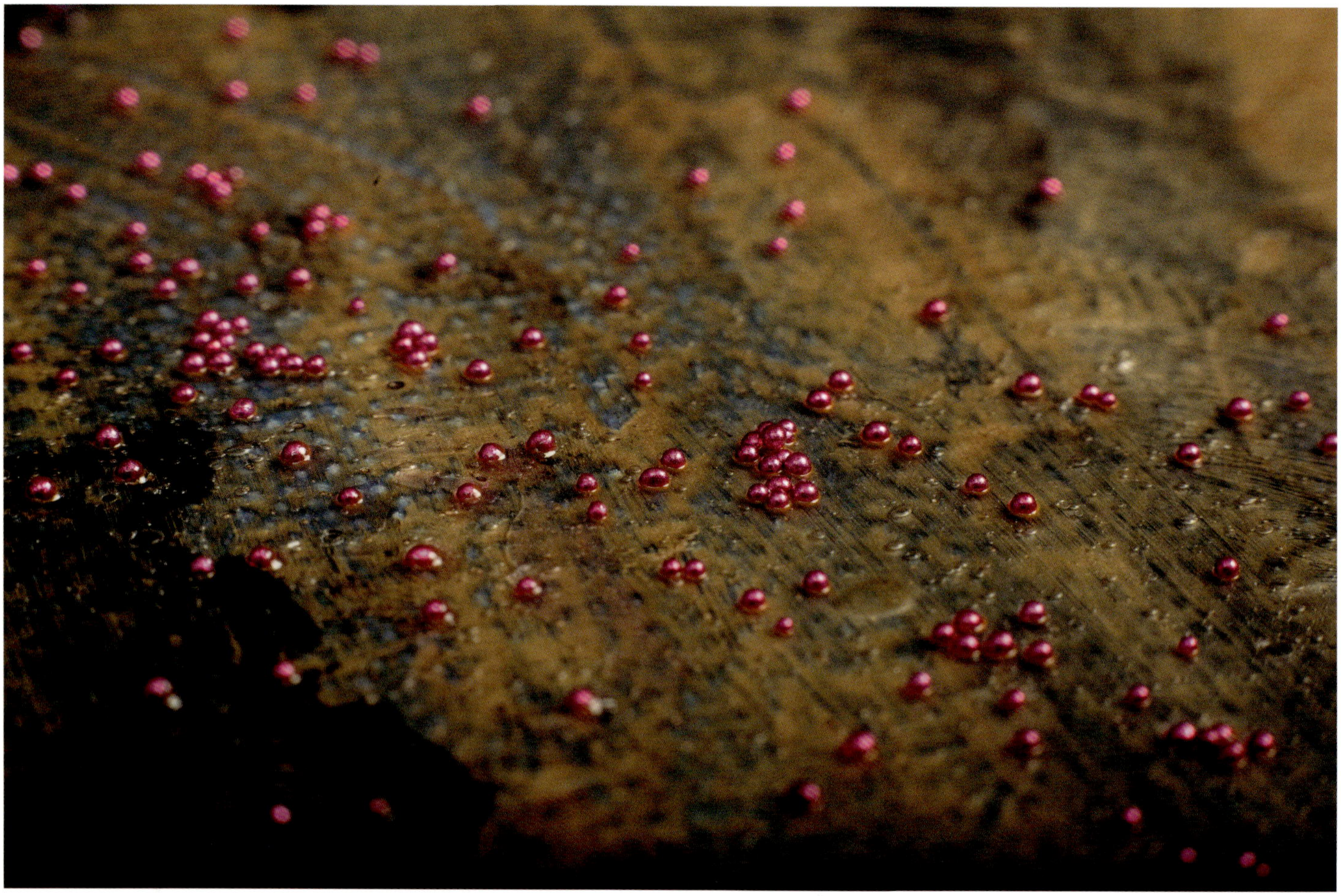

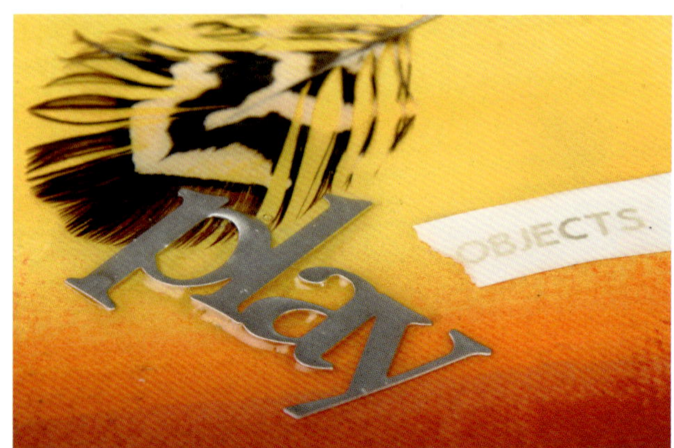

TOP A scattering of fuchsia metal beads provides a delightful textural color pop. The gel they were dropped into holds them to the surface.

BOTTOM Collage elements (feather, metal letters, paper) adhered to a surface with gloss liquid medium.

COLLAGE

If you want to place objects on your acrylic surface, there are details you need to consider. Beautiful papers, printed images, and photographs are natural partners to acrylic collage. Because of their adhesive quality and the clarity of their film when dry, clear mediums are most effective at laminating paper and fabric into acrylic compositions. The table on pages 220–21 provides information on various materials' suitability for collage.

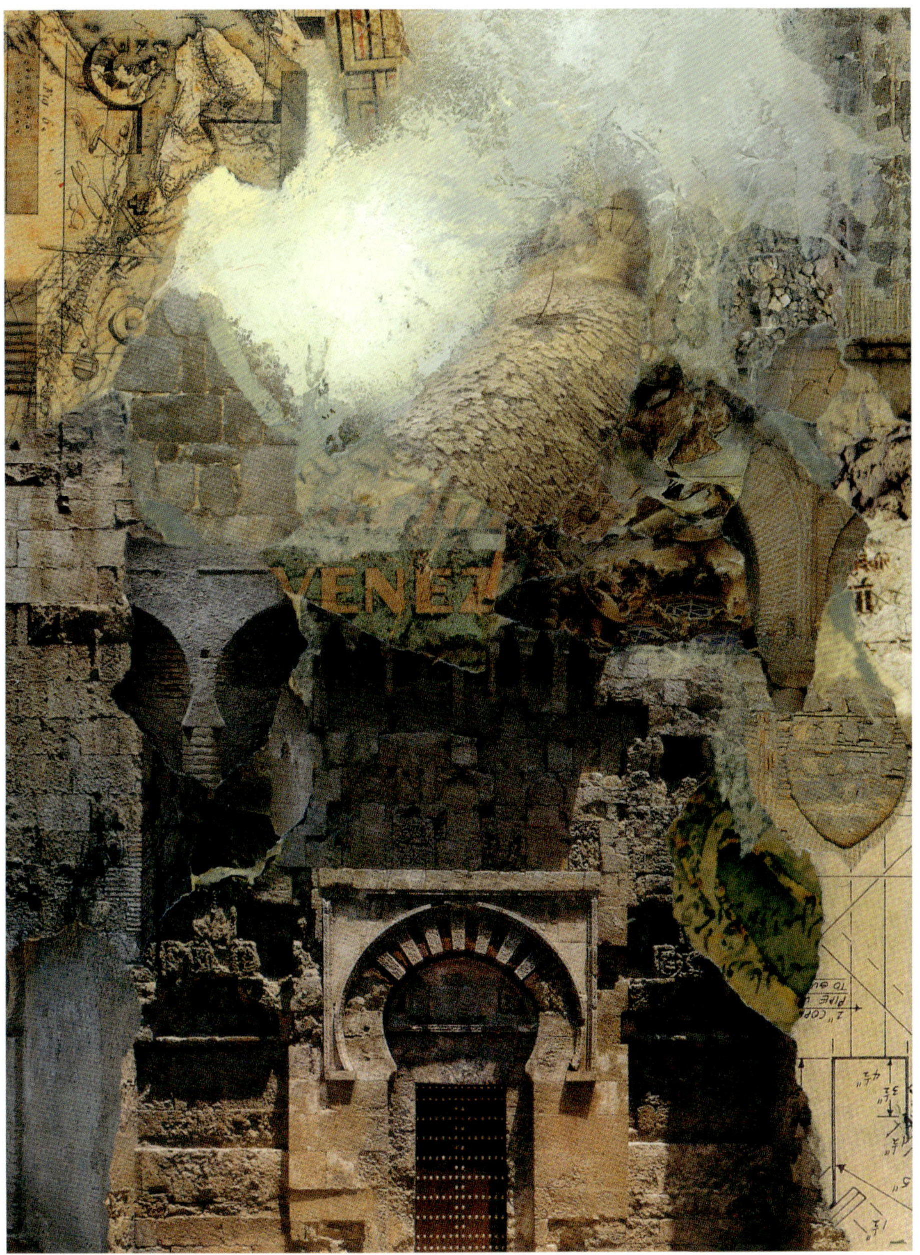

ABOVE **Nick Bantock, *Pharos Gate*, 2015, collage on artboard, 9 x 7 inches (23 x 18 cm). Collection of the artist.**

Created for Bantock's book *Pharos Gate* (the fourth volume of his Griffin & Sabine series), this work contains approximately sixty collage fragments attached with matte medium, together with acrylics (impasto and glaze) and chalk pastels bound with matte gel and finished with a 50/50 mix of matte and gloss acrylic varnish.

About his experience with acrylics, Nick Bantock says, "I've been using acrylics since they first appeared in the '60s. In fact, the paintings I fought with when I first arrived at art college employed powder paint mixed into a foul-smelling acrylic medium. The mix was grainy and clumsy and had lousy covering power, but unlike oil paint it didn't take forever to dry.

"Nearly fifty years on, acrylics are so much more sophisticated. However, I still try to play, experiment, and take risks with the materials as I did back then. To my mind, the great gift of plastic paints and mediums is their potential to go as far as we push them."

COLLAGE ELEMENTS

MATERIAL	ADHESION TO ACRYLIC PAINT	POTENTIAL PROBLEMS	NOTES
BISQUE-FIRED CLAY	Excellent	Bisque-fired clay is quite porous, and will absorb paint readily.	Apply clear top coat to dry paint to improve surface resilience.
FABRIC: COTTON, SYNTHETIC, OR BLENDED	Excellent	Complete saturation with acrylic may cause discoloration. Wash prior to painting to remove any sizing that will inhibit paint absorption.	
FABRIC: NATURAL WOOL, FELT, LEATHER, OR SUEDE	Excellent	Some leather may require cleaning prior to collage work.	
FIBERS: NATURAL AND SYNTHETIC THREAD OR YARN	Excellent	None	Using threads or yarns in an acrylic composition is a great way to provide extra structural strength, particularly with large-scale works or sculptural forms.
FIBERS: UNTREATED ORGANIC	Good to excellent	Some organic fibers—hair, feathers, fur—may be oily and require cleaning before application. Test fibers prior to use to ensure that any dye will not bleed into the acrylic layer.	
GLASS	Good	As oil is commonly used to cut glass, any glass pieces should be washed completely with mild detergent and water and allowed to dry before use.	Abrading glass lightly will improve adhesion. (Glass beads do not require abrasion.)
GLAZED CERAMIC	Good	None	
HANDMADE PAPERS	Excellent	Colored papers may not be lightfast. Test papers prior to use for potential bleeding of colors.	
MARBLE DUST	Excellent	High concentrations of marble dust will create a more brittle paint film. Marble dust will settle to the bottom in lower-viscosity mediums due to its weight. Use with gel-type mediums or high-viscosity paints.	Marble dust is often used to add tooth to a surface and create a tough, dense paint film.
METAL LEAF	Excellent	Because metal leaf tears easily, apply with a soft gilding brush. Some synthetic metal leaf can oxidize in a thick application of acrylic medium. Test prior to use.	Apply a thin layer of acrylic medium to the surface, allow to partially dry, then float the leaf onto it, smoothing with a soft brush. Allow the medium to dry completely before sealing with more acrylic.
METAL PIECES	Good to excellent	Determine whether or not the metal is corrosive in a watery environment. If so, it will need to be sealed before adding into paint to prevent further corrosion. If not, go ahead and stick it on.	To improve adhesion, abrade the metal. Metal produced for crafts (mesh, florist's wire) can be collaged into an acrylic composition with no ill effects.
METALLIC POWDER (COPPER BASED)	Good	Metallic powder will oxidize (rust) in a watery environment, causing discoloration.	Once the paint has cured, oxidation will cease and the metal should remain as it is.

MATERIAL	ADHESION TO ACRYLIC PAINT	POTENTIAL PROBLEMS	NOTES
MYLAR GLITTER	Excellent	None	Glitter mixes beautifully with acrylic paints and mediums.
NEWSPAPER AND MAGAZINE CLIPPINGS	Excellent	There may be lightfastness issues. Items printed in black ink (carbon based) will have good lightfastness, but others will likely fade with prolonged light exposure. Colored clippings will discolor and fade unless printed with lightfast pigment inks.	The paper on which newspapers and magazines are printed is generally not archival and will yellow and grow brittle with prolonged light exposure.
ORGANIC MATERIAL (FLOWERS, SEEDS, LEAVES, SHELLS, INSECTS, ETC.)	Good to excellent	Plant and other organic matter will adhere best if *completely* dried. Any remaining moisture may interfere with the drying of the paint and cause poor adhesion and uneven drying. For best results, ensure that organic materials are completely encased with acrylic to inhibit bacterial or mildew growth.	
PLASTER	Excellent	When mixing loose plaster with acrylic, make sure it is well blended; this will improve adhesion and minimize cracking. Painting onto dried plaster must be done in stages due to the extreme porosity of plaster. Use a diluted polymer medium as a sealant; this will create a barrier for additional layers. For a tougher basecoat, apply a second layer of undiluted medium or acrylic gesso before adding the color layer.	There are different varieties of plaster: gypsum, limestone, and synthetic composites. Always test before applying to artwork.
SEA SAND	Poor	The saltiness of sea sand will cause adhesion problems. To remove most of the salt, place the sand in a fine strainer or cloth and rinse repeatedly with fresh water. Allow to dry thoroughly before use.	Once cleaned, sea sand can be used to create granular textures.
SILICA SAND (CRAFT)	Excellent	Silica sand is potentially hazardous if inhaled.	For best adhesion, mix with acrylic medium prior to application.
SNAKESKIN, INSECT WINGS, ETC.	Excellent	Make sure material is free of moisture and dirt to ensure uniform adhesion. Very colorful insect wings can be sprayed with a fixative to prevent marring of the dusty color coating.	These types of collage materials can be quite delicate. A good method is to apply the object to a thin layer of wet medium, allow to dry, and brush a second layer of medium over top, effectively encasing it in acrylic.
STONE	Excellent	Remove any moss, salt, or dirt before painting.	
UNGLAZED CERAMIC AND STONEWARE	Excellent	None	
WOOD PIECES	Excellent	Sap is the only potential impediment to good adhesion between wood and acrylic. Seal with shellac-based sealant.	

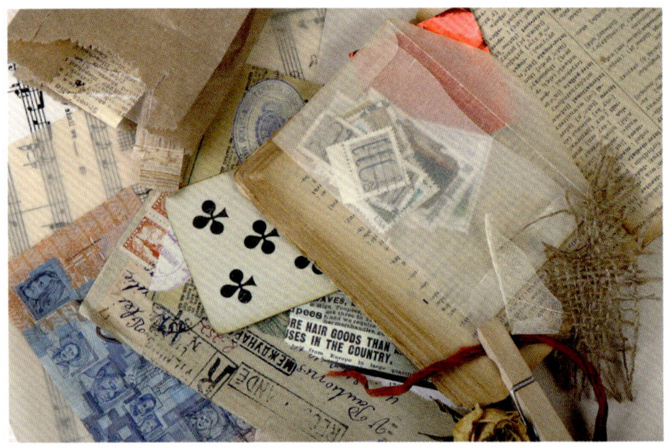

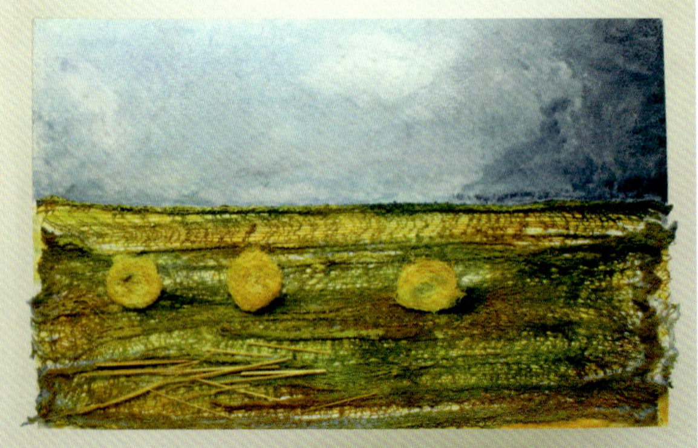

TOP Collected paper ephemera, ideally suited to acrylic collage.

BOTTOM Here, porous cheesecloth saturated with matte polymer medium has been collaged onto a panel. Once dry, this medium-soaked loose-weave cloth looked like cultivated fields. The "hay bales" were made with curls of soaked cheesecloth. After drying, the textured surface was painted with washes and *alla prima* color.

Tip: One method for giving printed paper a distressed look is to soak it in solvent and then manipulate its surface. But beware: even if the paper feels dry and free of solvent, it may cause serious issues with film formation when adhered to an acrylic surface.

When creating a collage using acrylic medium as glue, take the following into consideration:

- *Weight.* The heavier and more substantial the material you are gluing onto your composition, the heavier the medium needed to hold it.
- *Shape of object.* If the collaged object is irregular in shape and does not contact the surface evenly, use a thicker medium to hold it in place.
- *Porosity.* Acrylics absorb into porous objects and supports. If both the object and surface are porous, the bond will be stronger than if one or the other (or both) are not, and thus a thinner medium should be sufficient. Where both object and surface are not porous, a thicker medium will ne needed to create an enveloping seal.
- *Water fastness.* Is the object you are gluing susceptible to water damage? If you are not sure, test a small area of the object to see if it changes when water is applied.
- *Colorfastness.* Although acrylic resin offers a small amount of UV protection, you should be wary of incorporating objects that can fade with light exposure. Things like flower petals and naturally dyed yarns, for example, should be treated with a layer of UV-stabilizing medium prior to gluing.
- *Clarity and luster of paint film.* When gluing collage material or covering it completely, be aware of how the surface luster of the medium you're using will affect the appearance of the collaged material and the surrounding area. For example, if you use a glossy medium to glue an object onto a matte surface, chances are that some of the gloss will show. Remember, too, that matte medium will make a surface look flat and lusterless.
- *Adverse effects.* Materials that react poorly in acrylic collage include those that contain or are coated with oil or wax. Some inks repel acrylic, causing it to crawl on the surface.

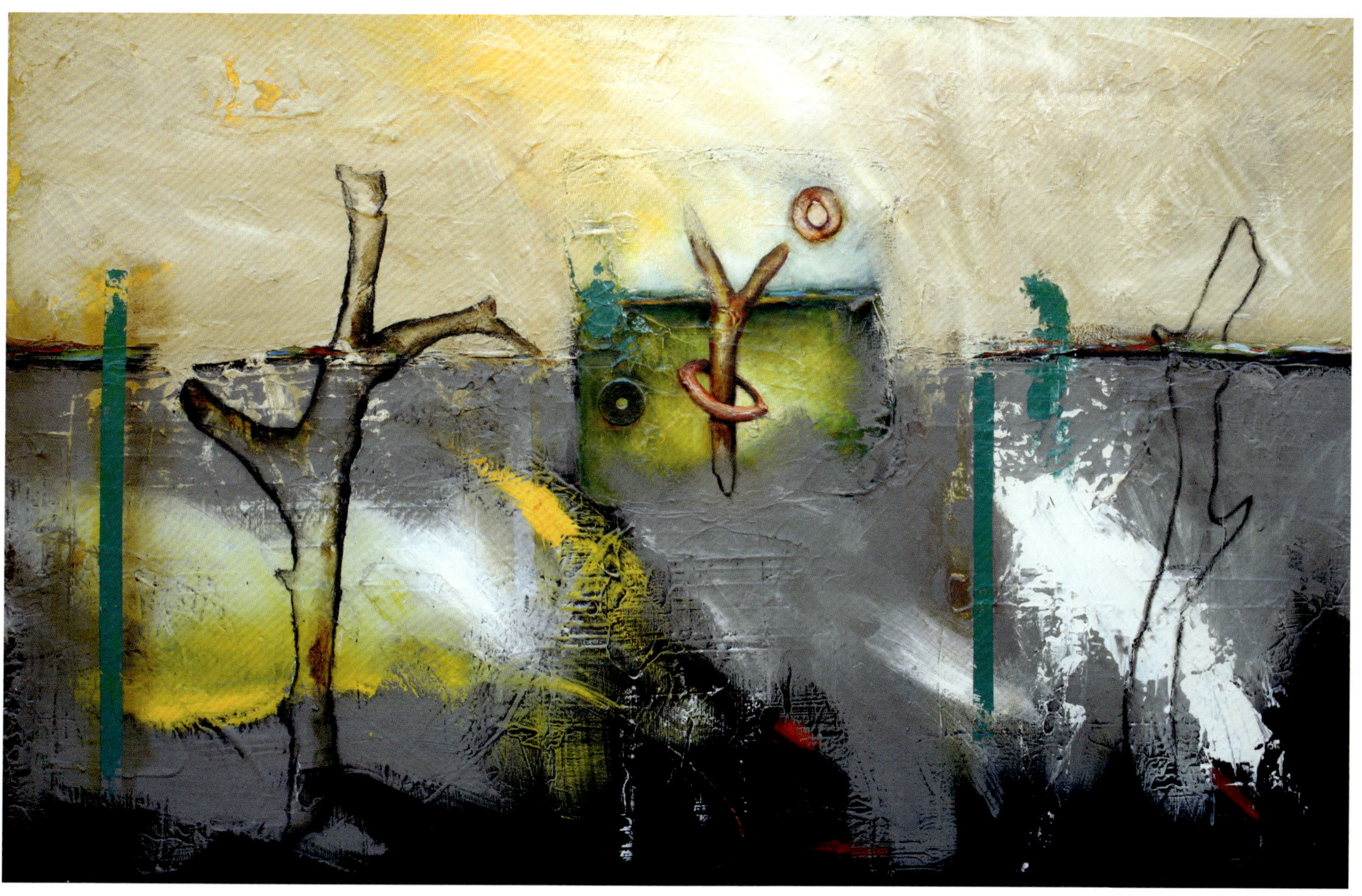

Connie Morris, *3 Sticks with 3 Rings*, 2017, acrylic and mixed media on wood panel, 30 x 48 inches (76 x 123 cm). Collection of the artist.

Artist Connie Morris describes the concept and process of 3 Sticks with 3 Rings*: "This piece was part of a series that explored the nostalgia of place, time, and memory and the duality of what was—or was perceived to be—perfect and timeless. It stylistically uses dualities in objects, materials, and process in an effort to try to capture a sense of visual memory. I started by applying modeling paste with a large palette knife to the surface in order to both seal the wood and create a textured ground. Thin layers of liquid colors mixed with gloss polymer were then applied. The result was an open landscape of subtly glazed areas of color with bordered landscape vignette. Gel medium was used to create a border for the vignette and to adhere the metal rings. I lived with this version for a while but struggled with the tightness that resulted from the same surface and the predictable and safe arrangement of elements. I later revisited it by painting over and obscuring areas with gesso and protecting passages that I wanted to still be revealed with gloss polymer. More texture was created with gel, then partially covered with color. I then used semigloss polymer to create a toothy ground in order to draw onto the surface with charcoal. The surface now has dual finished sheens, textured and smooth areas, thick and thin applications of color, and drawn, painted, and adhered objects. The result is a more evolved piece that far better expresses my creative and aesthetic objectives due to its spontaneous and reactive use of materials and mediums.*

"Acrylics have always been my primary medium for creative expression. It wasn't until I began to work for an art materials manufacturer and was exposed to the vast array of acrylic mediums available that I integrated them into my creative process. I have been able to learn the technical aspects as well as the many possibilities for creative expression, inspiration, and artistic control they can provide. Now, acrylic mediums are an integral part of my work. I feel I have only scratched the surface in my applications and will continue to be awed and surprised by what is possible."

DECOUPAGE

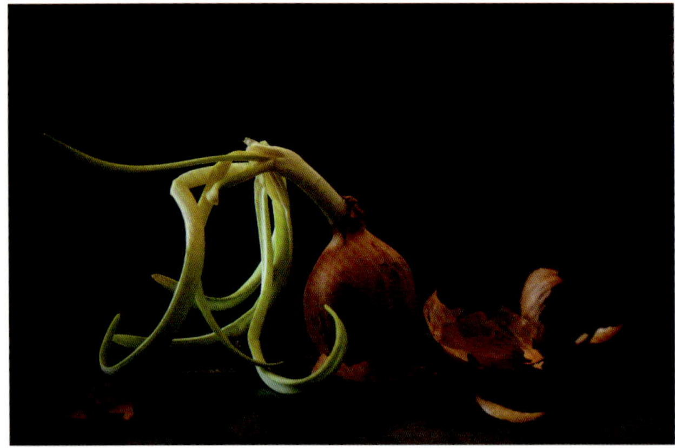

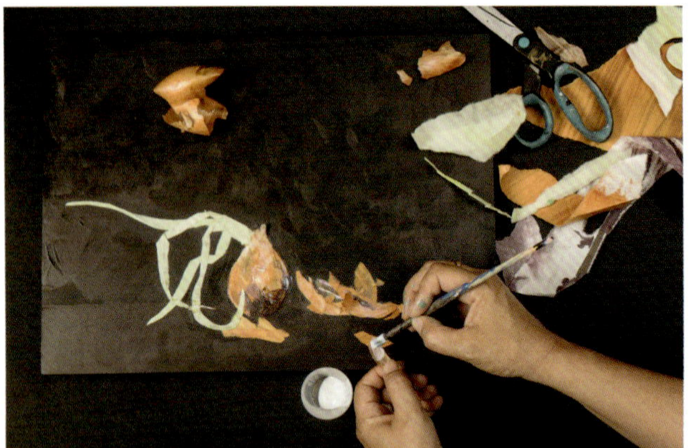

Decoupage—the "poor man's art" of gluing paper cutouts to other surfaces—has its roots in twelfth-century China. It was all the rage in seventeenth-century Italy, where artisans hand-colored prints and engravings of better-known artists' works to affix to furniture and decorative objects with lacquer. Decoupage reached a fever pitch in eighteenth- and nineteenth-century Europe. (The "Paper Mosaiks" of the eighteenth-century English artist Mary Delany—botanical images made of cut paper and glued to black backgrounds—are perhaps the best-known examples of the art.) Now, many centuries later, decoupage is still being widely practiced.

Although the concept of decoupage remains virtually unchanged and it's still a laborious process, the results are now far more durable and lightfast thanks to advances in materials. Any clear-drying acrylic medium, thick or thin, makes an excellent adhesive for paper. The more delicate the paper, the thinner the medium that should be used.

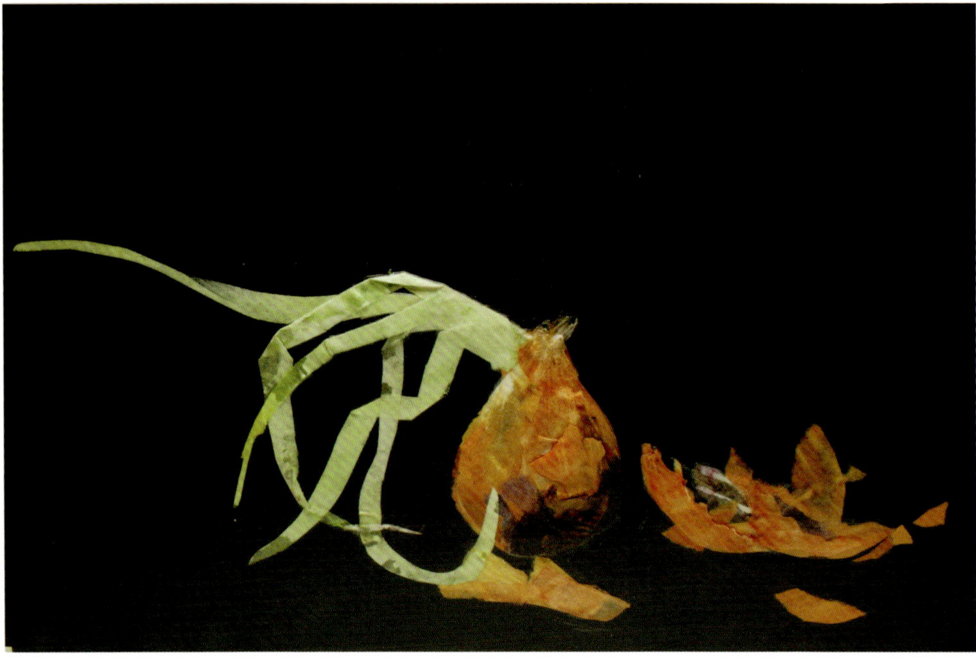

TOP **Suzy Lamont, *Take Root,* 2017, digital photograph, 16 x 24 inches (41 x 61 cm). Collection of the artist.**

Suzy Lamont's photo of an onion inspired my riff on the theme of collage and decoupage.

CENTER I soaked and stained sheets of rice paper with liquid acrylics, then cut, tore, and folded them to form parts of the composition. Pieces of onion skin add a harmonizing accent.

BOTTOM The collage elements give added dimension and texture to this simple composition made of torn paper and onion skins cobbled together with a clear-drying thin medium on a black gesso ground.

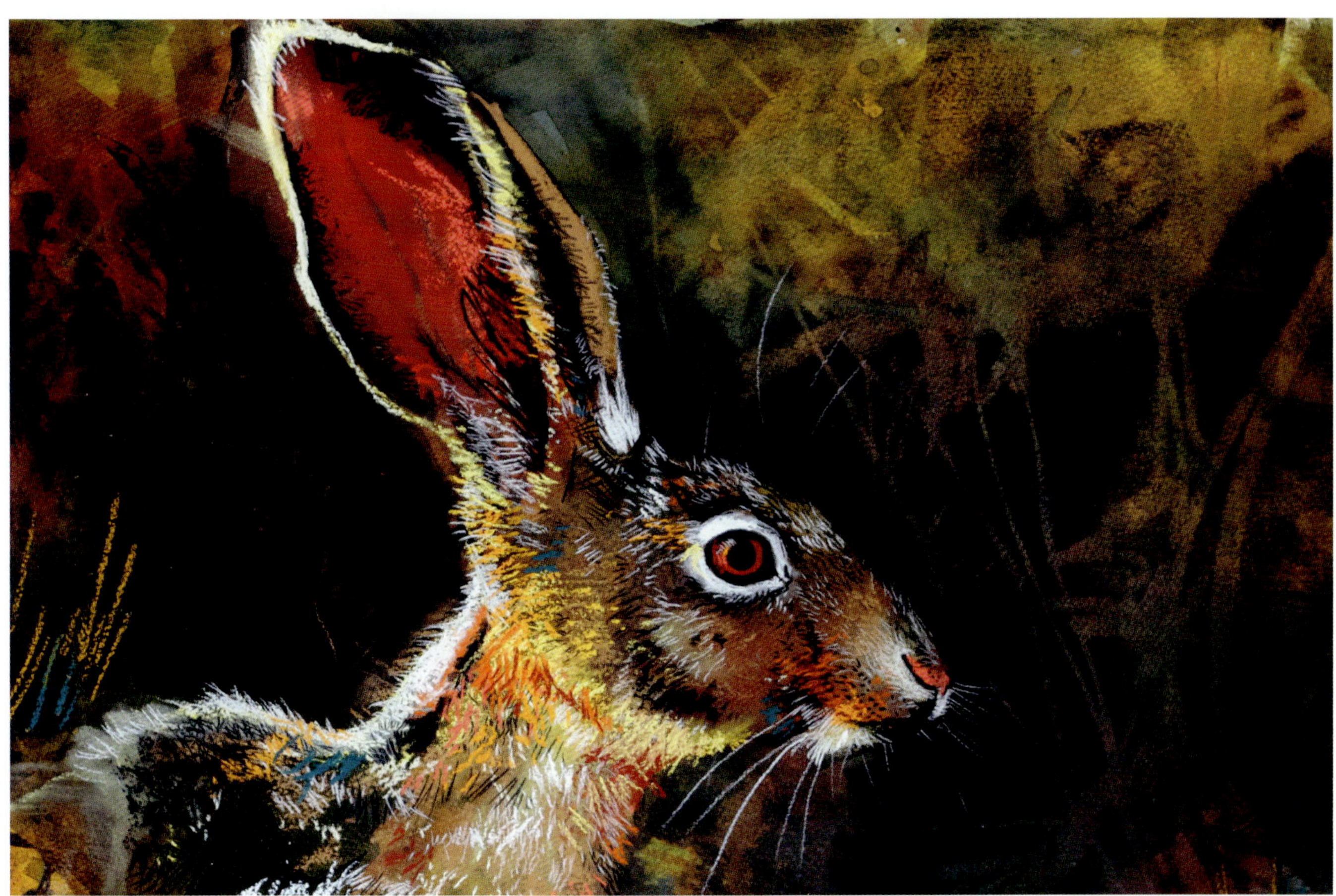

ABOVE **Sharlena Wood, *Jack Rabbit*, 2012, acrylic, charcoal, and soft pastel on 200 lb watercolor paper, 11 x 15 inches (28 x 38 cm). Collection of the artist.**

Artist Sharlena Wood built up watercolor-style glazes with transparent acrylic colors (golden orange, transparent pyrrole red, Indian yellow, Payne's grey) and drew details on dry media ground using charcoal powder, Unison soft pastels, and General's charcoal.

MIXED MEDIA

Mixed media works incorporate a variety of media: not just acrylics but other materials such as watercolors, pastels, oil paint, colored pencils, and so on. Acrylics are the glue binding these works together. They adhere very well to many different kinds of surfaces, and multiple materials adhere well to them.

Dry media ground is the unofficial mascot of mixed media work. It's matte and toothy, grips dry media, and holds them fast to the surface. This medium makes creating mixed media paintings with pastels, charcoal, and graphite a breeze. Although toothy, dry media ground is very uniform. It is also smooth, which allows for the removal of dry media with erasers. Using dry media in conjunction with acrylics allows you to be more precise—or simply to include a different type of mark-making into a work.

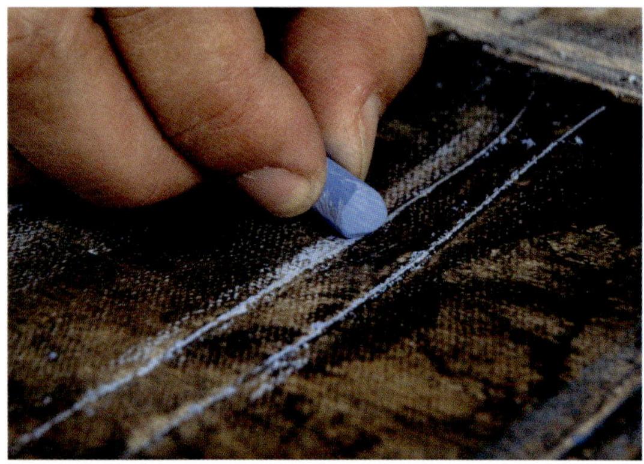

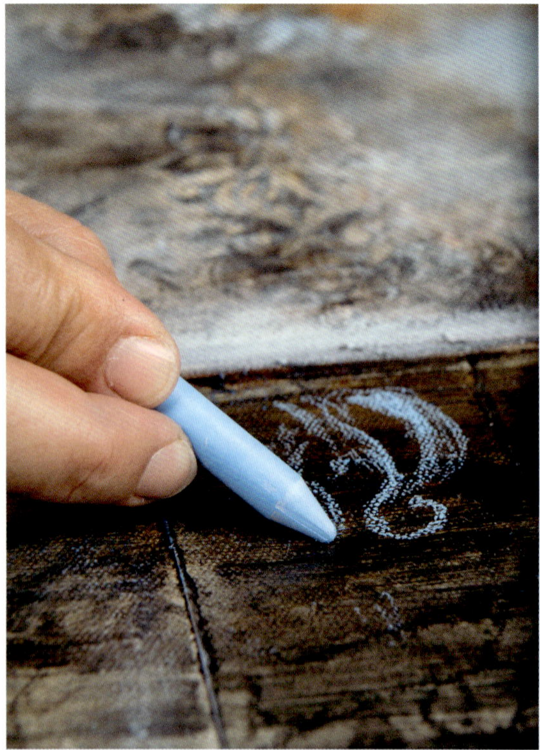

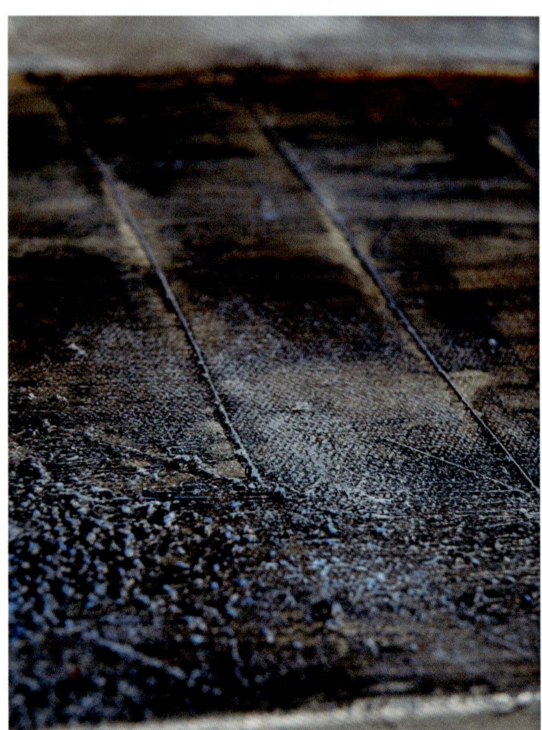

MIXED MEDIA DO'S AND DON'TS

WHAT TO DO:

- Know your materials. What is the base or binder of the material you are combining with acrylic? Is it oil, wax, gum arabic, clay? Each adheres differently to acrylics.
- Test first. This is the most important—not optional!—rule of thumb. If you can, wait a day or two to establish adhesion and compatibility between media.
- Plan your layers. If you want to add oil- or wax-based materials, leave those for last. Allow sufficient drying time between layers.
- To set loose or resoluble media, use an archival fixative spray in two to three light layers.

WHAT *NOT* TO DO:

- Do not mix oil- or wax-based paints with water-based paints. (You can layer oil- or wax-based paints over water-based paints, but not the other way around.)
- Do not layer water-based paints over oily or greasy surfaces. They won't adhere. (However, you can use oil- or wax-based materials as resists when layering with water-based paints.)
- Acrylic applied to a thicker application of watercolor can sometimes cause the fresh layer of paint to crackle. Test in a small area first! If cracks do occur (and you like the effect), let the acrylic layer dry and seal it with a thin layer of liquid medium.

TOP A sharp ridge of gel medium catches and holds soft pastel.

CENTER Watercolor crayons are great for drawing in creamy opaque color.

BOTTOM Soft pastel gently rubbed onto the surface brings out its granular textures. Hit with a light spray of fixative before adding another layer.

PROCESS

WORKING WITH MIXED MEDIA

Artist Sharlena Wood used a variety of media when creating her work *Phoenix*. Here are the steps she followed.

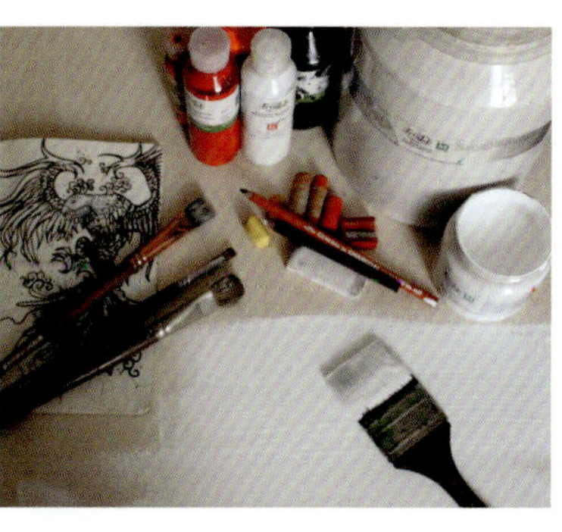

ABOVE For *Phoenix*, Sharlena Wood used some of her favorite materials and tools: a birch panel, Tri-Art acrylic gesso, Tri-Art dry media ground, Tri-Art liquid acrylics (for this piece she used carbon black, transparent pyrrole red medium, and Indian yellow), General's 6B charcoal pencil, General's white pastel pencil, white vinyl and kneaded erasers, General's soft lead pencil sharpener, Unison soft pastels, and, for blending, bristle brushes in a variety of sizes.

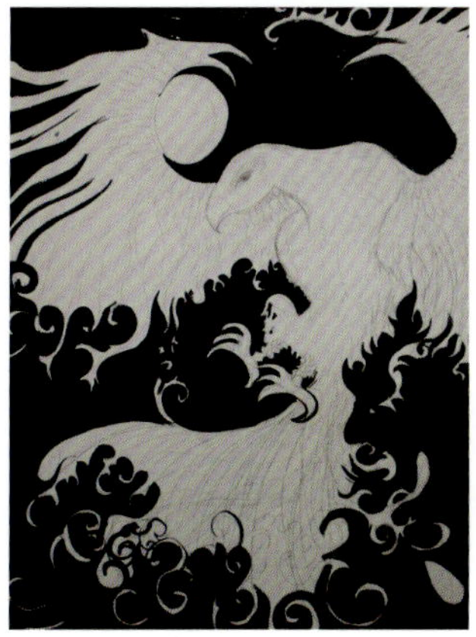

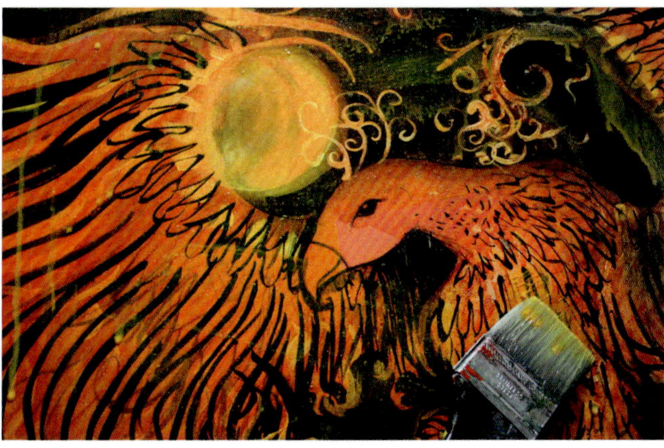

1
First, Wood blocked in the design with gesso and carbon black acrylic.

2
She then added glazes of transparent acrylic colors (transparent pyrrole red and Indian yellow).

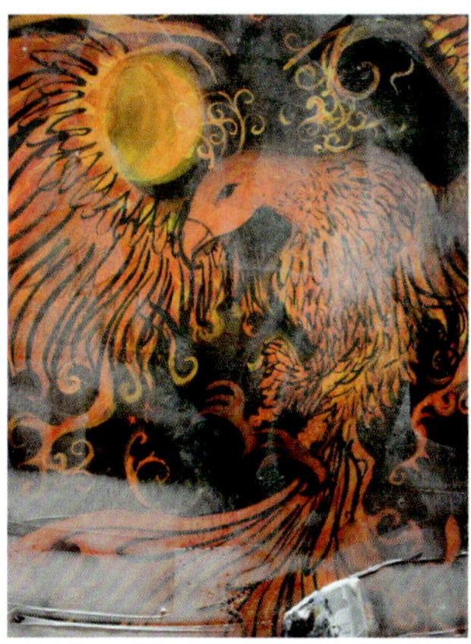

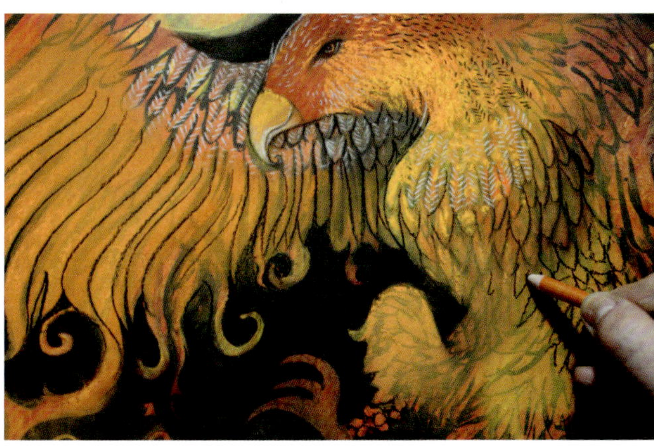

3
She applied a thin veil of dry media ground over the whole surface using a color shaper rubber brush tool.

4
Next, charcoal pencil details were added . . .

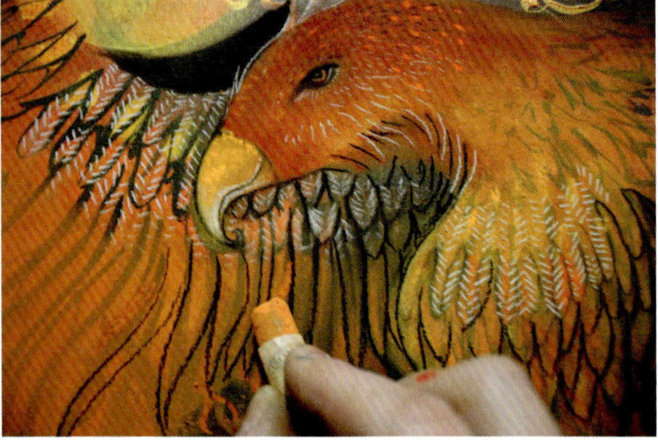

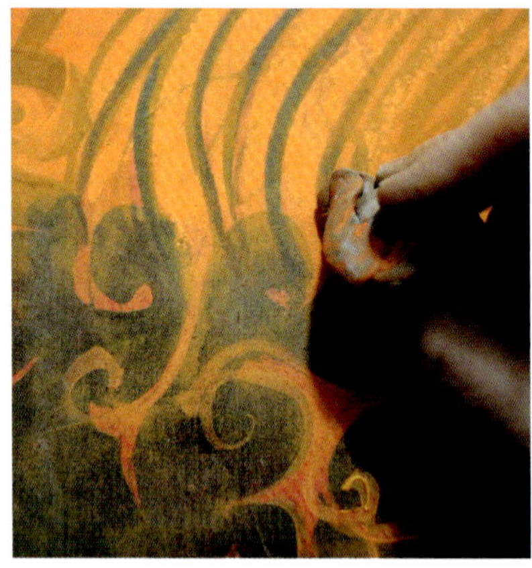

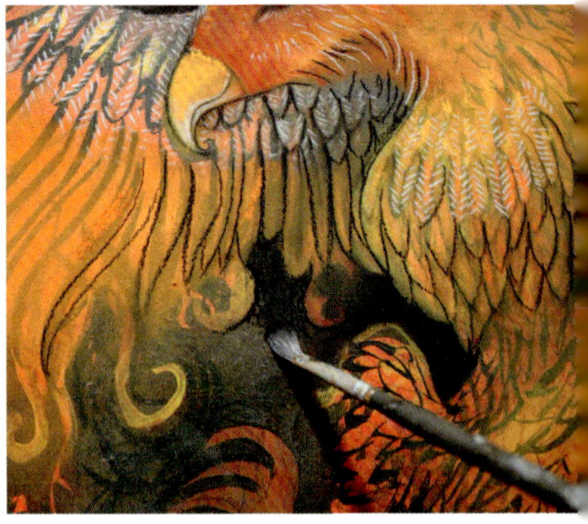

5
. . . followed by details in soft pastel.

6
Wood erased lines that were covered by too much pastel, allowing some of the paint beneath to show through.

7
She refined edges and dark areas with powdered charcoal, smudging with a dry bristle brush.

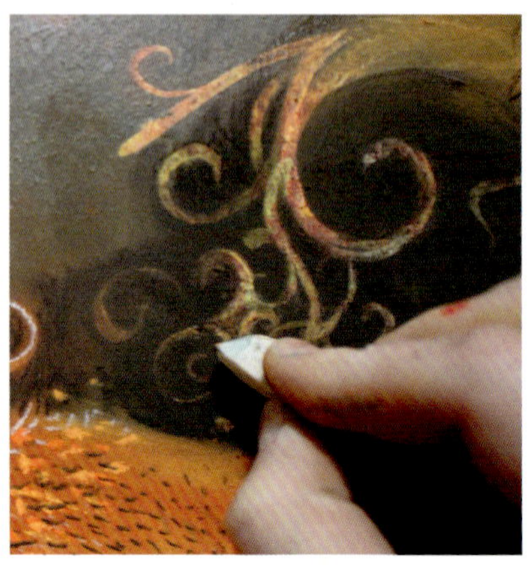

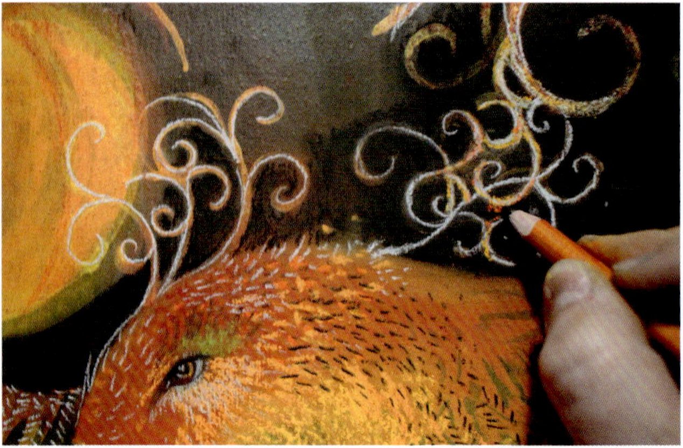

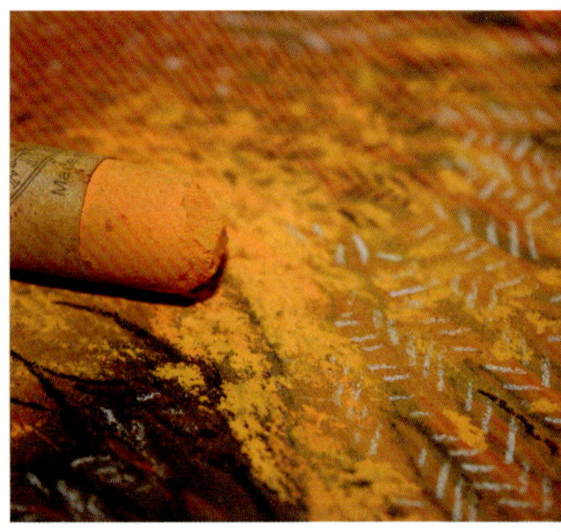

8
Wood used a white vinyl eraser cut into a small triangle to achieve fine lines and details.

9
Charcoal or pastel pencils are perfect for delicate line work. (Soft pastels are too large for this.)

10
Some final details were applied with soft pastel.

OPPOSITE Sharlena Wood, *Phoenix,* 2012, acrylic, soft pastel, and charcoal on birch panel, 24 x 18 inches (61 x 46 cm). Collection of the artist.

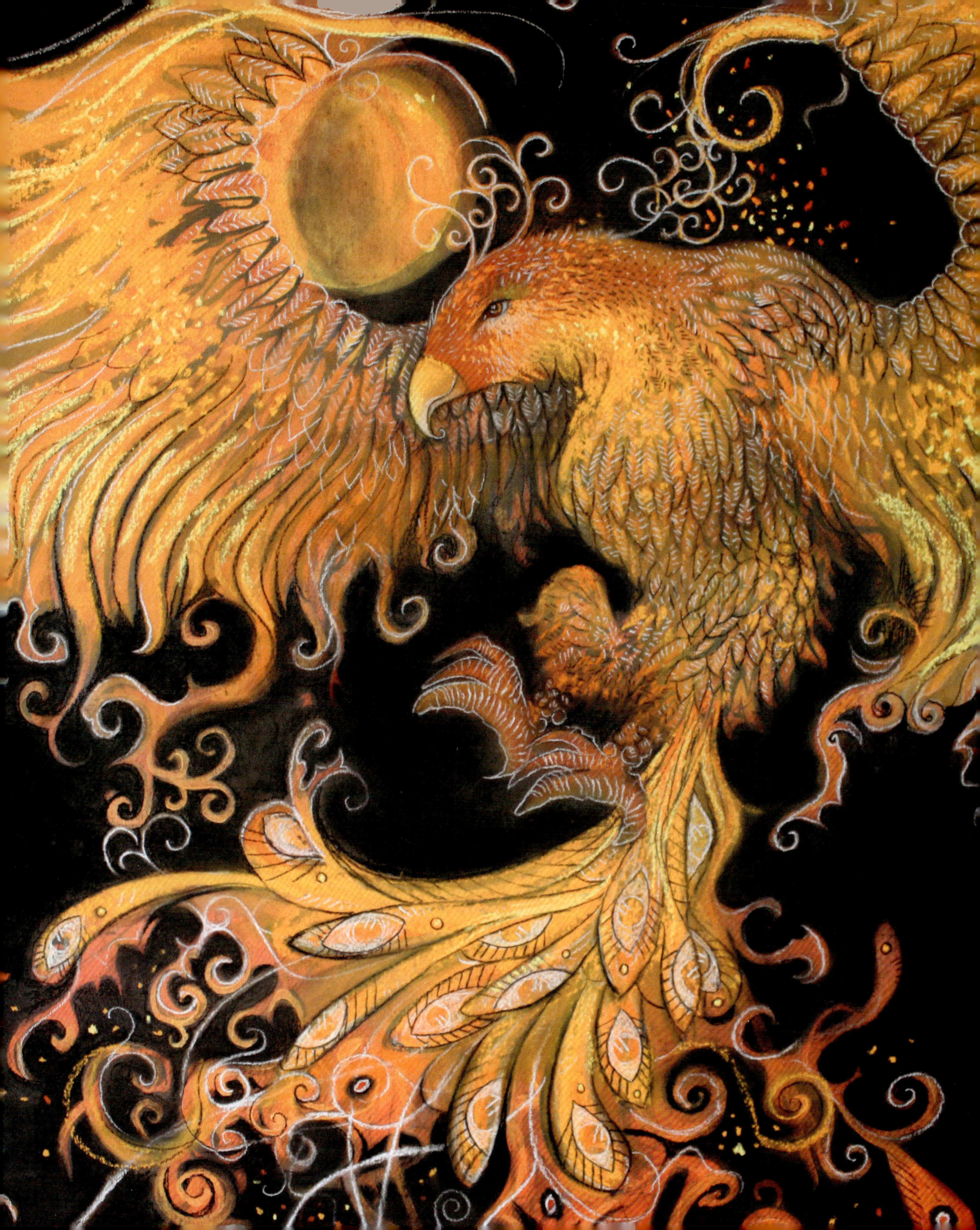

CHAPTER 12

AVOIDING POTENTIAL PITFALLS

MORE AND MORE ACRYLIC MEDIUMS AND ADDITIVES are being manufactured and marketed all the time. This abundance of new materials means more and more possibilities for applications, combinations—and complications! Although manufacturers can anticipate many issues regarding the compatibility of new acrylic mediums with other media, their performance on various substrates, and their suitability for various techniques, they cannot possibly predict all the ways in which artists will try to use them. The trial-and-error process is part of the artistic journey, and painters must bear the onus of understanding their own tools as completely as possible.

New materials mean more possibilities—and more complications!

OPPOSITE In the painted surface opposite, crackle medium was applied to the gesso ground and left to dry, causing the formation of deep crackles and fissures in the subsequent layer of tinted pouring medium.

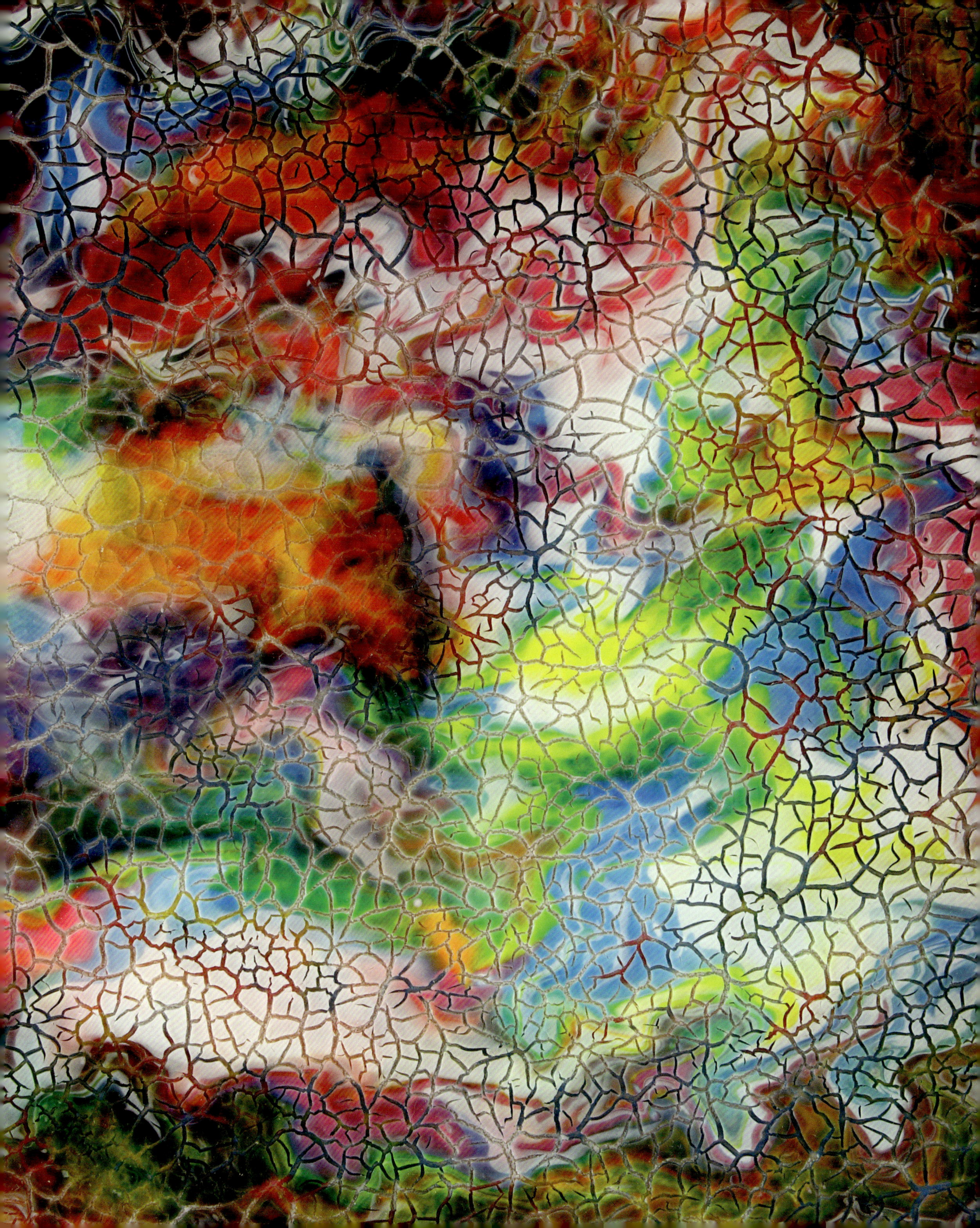

As you work with your materials—not just the new ones, but also those that have been around for a long time—here are some pointers to keep in mind:

ABOVE From left to right: Additional layers of matte polymer medium increasingly opacify and dull the colors beneath.

1. *Not all "clear" mediums actually dry clear.* Gessoes and modeling pastes that are labeled "clear" don't contain titanium white or other pigments. They do, however, contain calcium carbonate or other particulates and usually a quantity of matting agents, as well. These solids give them absorbency and sandability, but they also reduce the clarity of the paint film. Don't use them if you want absolute clarity, but they are ideal for creating colored grounds with the addition of a color.
2. *Matte mediums opacify.* Matte mediums, whether liquid or viscous, contain matting agents. These particles refract light, reducing the clarity of the film and producing a "frosted" appearance. The thicker the application, the more frosted the surface will appear, opacifying what lies beneath.
3. *Beware of bubbles.* Mixing vigorously creates bubbles in a medium, especially a self-leveling medium.

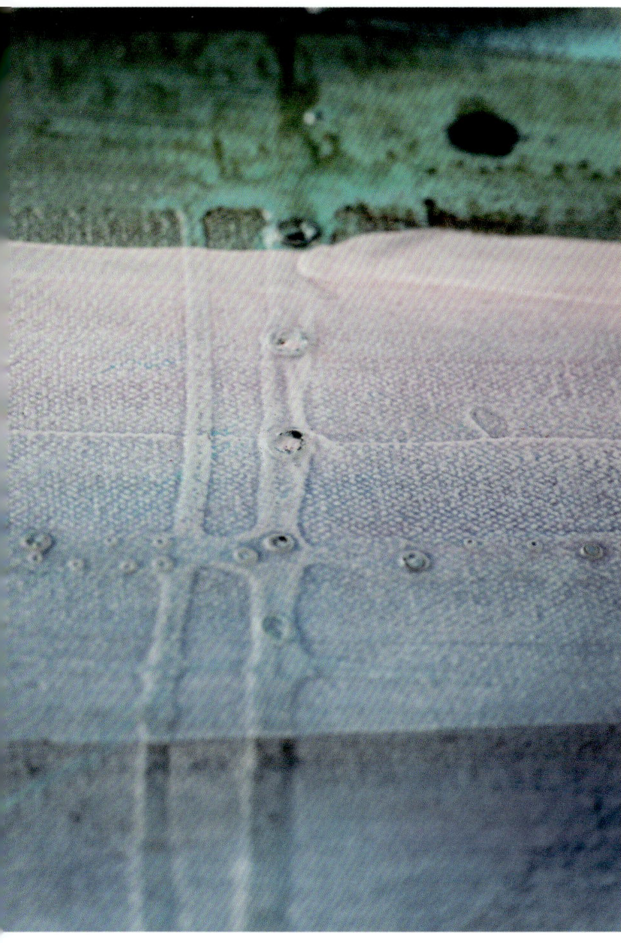

ABOVE When applied quickly, large swirls of self-leveling gel can produce big bubbles. Here, they become part of the landscape.

For fewer bubbles when tinting a medium, allow the mixture to settle for up to a few hours to allow the air to move up and out of it. Self-leveling gels get bubbly when applied with a brush, so spread or pour these mediums instead.

4. *Beware of too much water.* The more water the acrylic contains, the more pathways are formed in the drying paint film to accommodate the water's evaporation. And the more of these small pathways or fissures there are, the more bubbled the film becomes. More bubbles mean less gloss, and as it loses gloss the film becomes more absorbent and more susceptible to damage.
5. *Use a hair dryer with caution.* A hair dryer can be useful if you want to dry your paint more quickly. You should avoid using a hair dryer, however, to dry a thick application of paint or medium. Thick layers need time to cure, and if the curing is accelerated by heat and air, the layer will dry unevenly. A very dry top layer inhibits the flow of volatiles from the sublayer, which may cause them to become trapped, giving the paint film a milky appearance. Even curing is crucial to large structural paint applications; it lets the paint molecules bond and gives the film tensile strength.

A CHECKLIST OF DON'TS

1. Don't layer too quickly. Allow each layer to dry to the touch, at least, before adding another. Allow all layers to fully cure before varnishing.
2. Don't add things that may react with the acrylic paint film (salt, copper or untreated metal, materials containing moisture) unless you are prepared to deal with unexpected or undesirable effects such as separated paint film, mold and bacterial contamination, or oxidation.
3. Don't use acrylic over oil- or wax-based media.
4. Don't forget that mediums contain water. The water will displace water-soluble or loose media.
5. Don't forget that wet mediums will darken dry media such as pastels. Spraying pastels with fixative before adding an acrylic layer will lessen the effect.

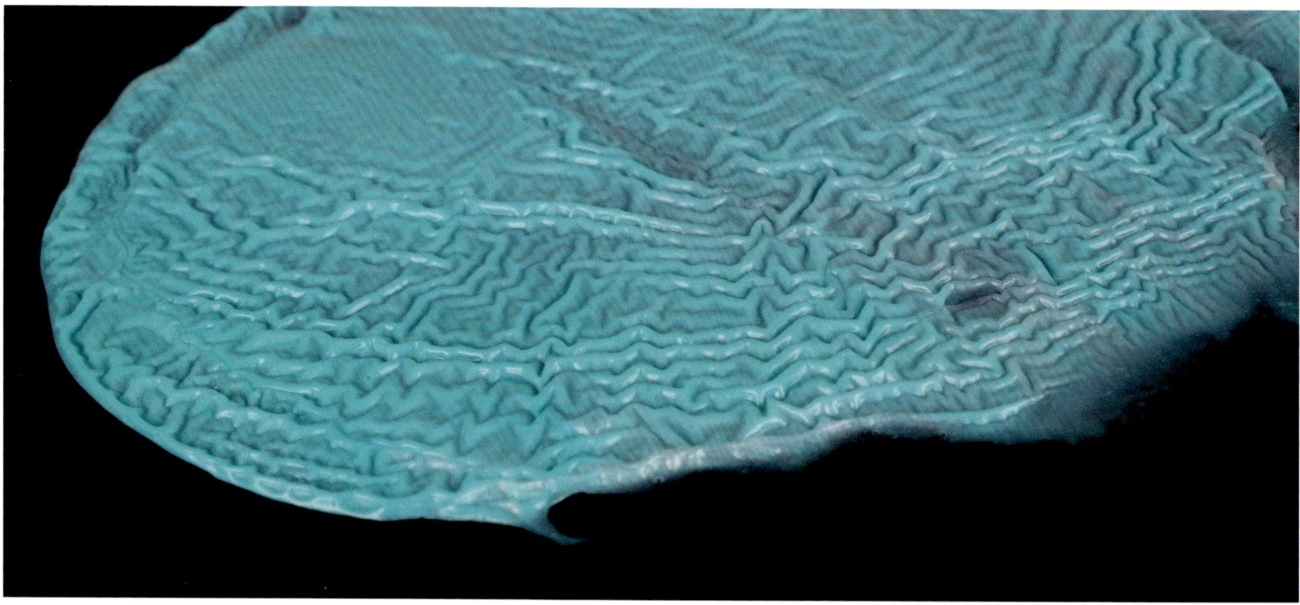

6. *Avoid crazing and cracking.* Sometimes the reason a paint film fails is obvious, but more often than not it's an accident whose cause is hard to decipher. Reasons a paint film can craze or crack include (but are not limited to) the following:
   - Accelerated drying
   - Too much water in the paint film.
   - Layering materials with different levels of flexibility (e.g., a rigid top coat over acrylic on a flexible support)
   - Painting with acrylics over incompatible materials or substrates
   - Expansion and contraction of a substrate covered with a hard or brittle paint film
   - Application of a paint layer to an insufficiently cured primer or first coat of paint
   - Water evaporating from the paint film at varying rates due to thickness of paint (uneven drying)

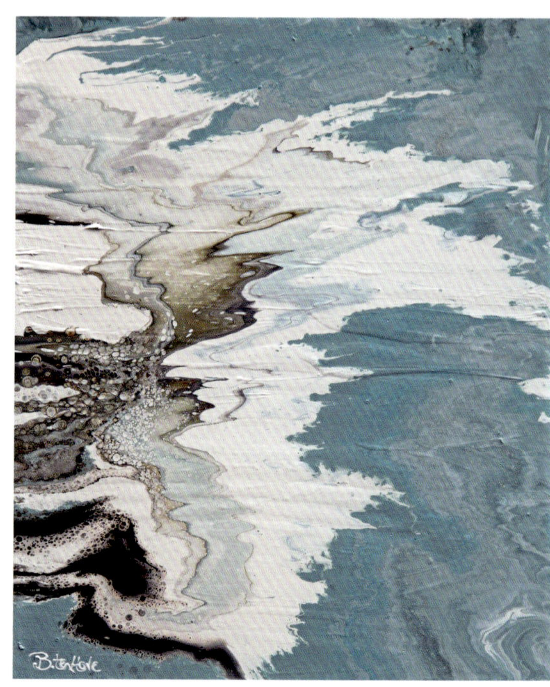

TOP Self-leveling gel and hair dryers are not friends. Even on cool, the forced air of the dryer rapidly seals and shrinks the outer layer of gel, trapping moisture beneath.

BOTTOM **Beth ten Hove, *Healing Waters*, 2017, acrylic on canvas, 20 x 12 inches (51 x 30 cm). Private collection.**

Beth ten Hove's *Healing Waters* was painted using a technique in which all the colors were pre-mixed with water and poured directly onto the surface of the canvas. The paint was then mixed and spread by lifting and rotating the canvas manually. No brushes or other tools were used.

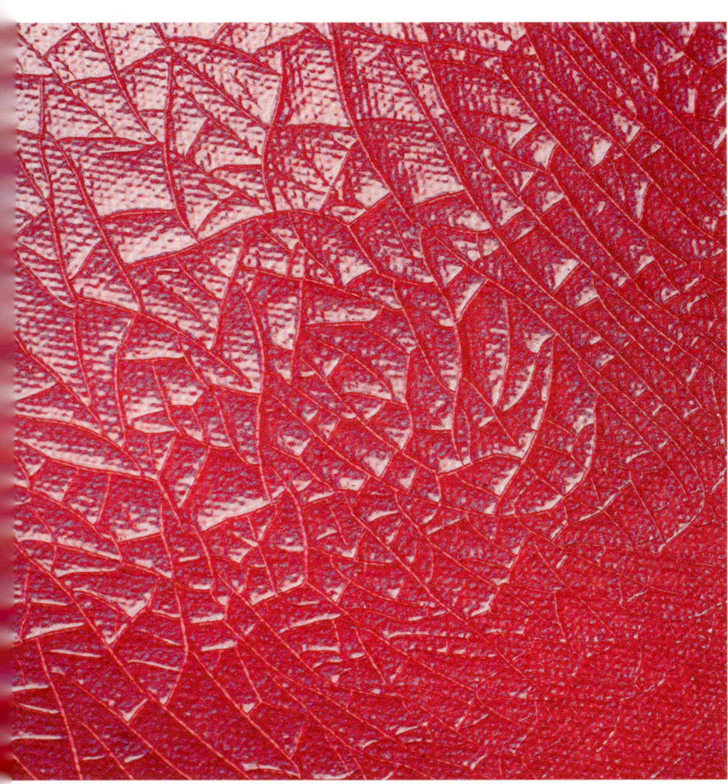

LEFT When overly manipulated and forced to dry too quickly (near an air conditioner with the fan set on high), the surface of this tinted pouring medium shattered.

BELOW Here, a thin layer of tinted liquid medium that was spread over self-leveling gel that was not quite dry. This is what crazing looks like.

BOTTOM A detail of *Healing Waters*. Though not actually accidental, the addition and combination of alcohol and spray paint disrupted the surface of the watered-down acrylic, creating arbitrary patterns. This loss of control over the paint surface's appearance gives it an organic, dynamic feeling.

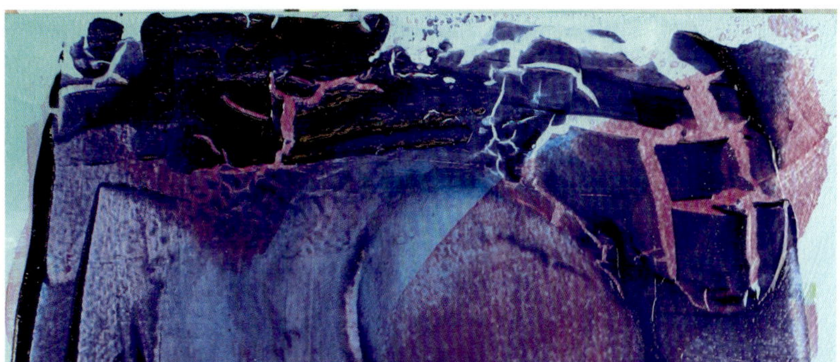

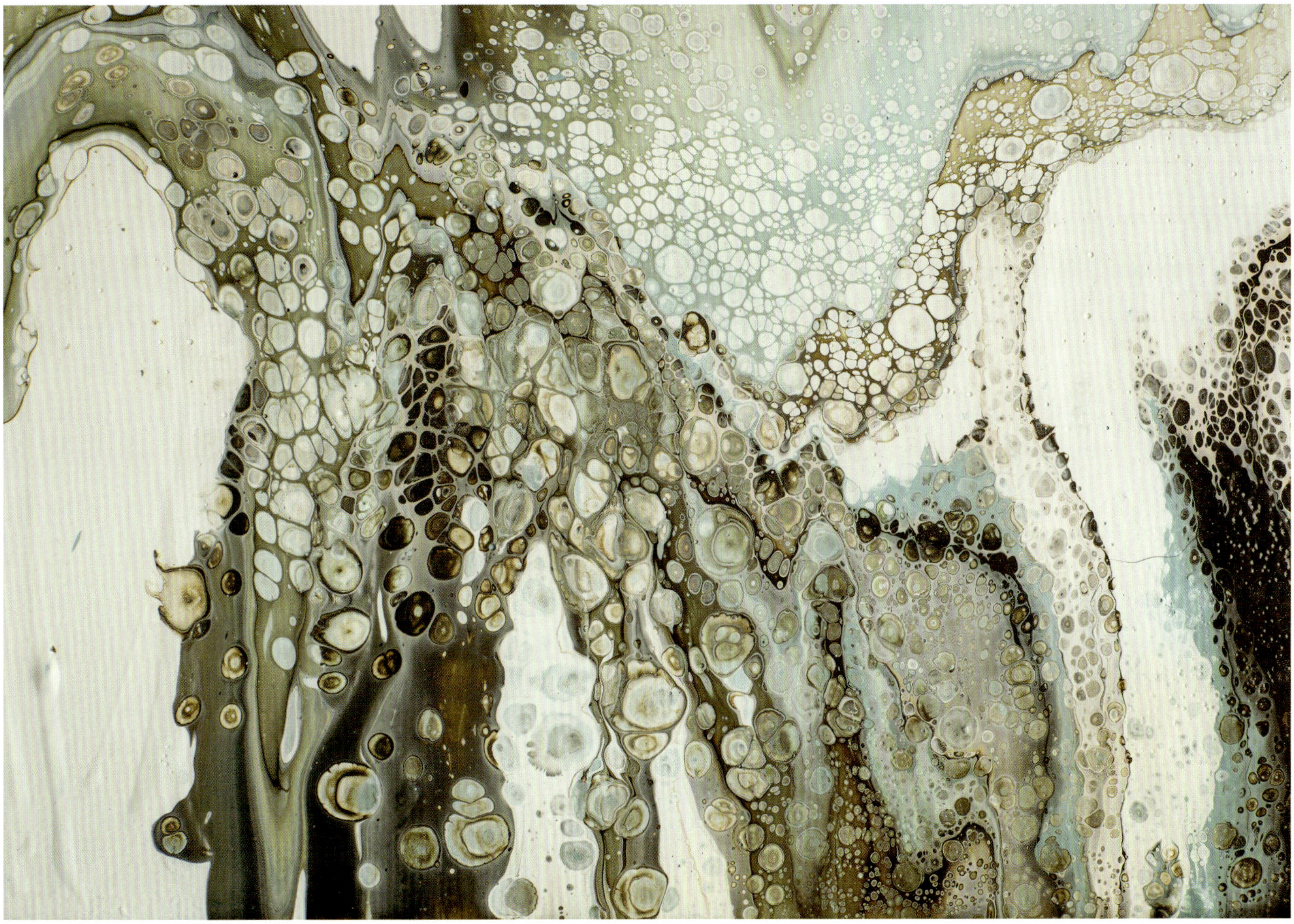

A TROUBLESOME PAINTING

This painting was an exercise in frustration. Making it, I fought with self-leveling gel, water, and gravity. Prior to applying the uppermost layers of self-leveling gel, I liberally diluted paint with water, allowing it to flow on the canvas. The water separated the iridescent colors from the other pigments, disrupting the surface. In an effort to capture the effect, I added a thick layer of self-leveling gel, smoothing out the surface. The painting was almost finished when I added a final (and particularly thick) layer of self-leveling gel to enhance the gloss. Its weight made the canvas sag, revealing the outline of the bracing crossbar and producing a deep cross shape in the middle of the canvas.

To fill in the indentation and level out the painting, I placed a board behind the crossbar, bracing the back of the canvas. New layers of gloss gel followed by self-leveling gel finally got rid of the outline. But then the last layer of self-leveling gel dried irregularly and did not completely self-level. This was probably due to the accumulation of layers and the fact that I did not allow sufficient curing time between them—revealing my impatience with the painting! This labor of several months taught me three important lessons: Allow for sufficient curing time between thick layers. Brace the back of the canvas. And next time use a wood panel instead!

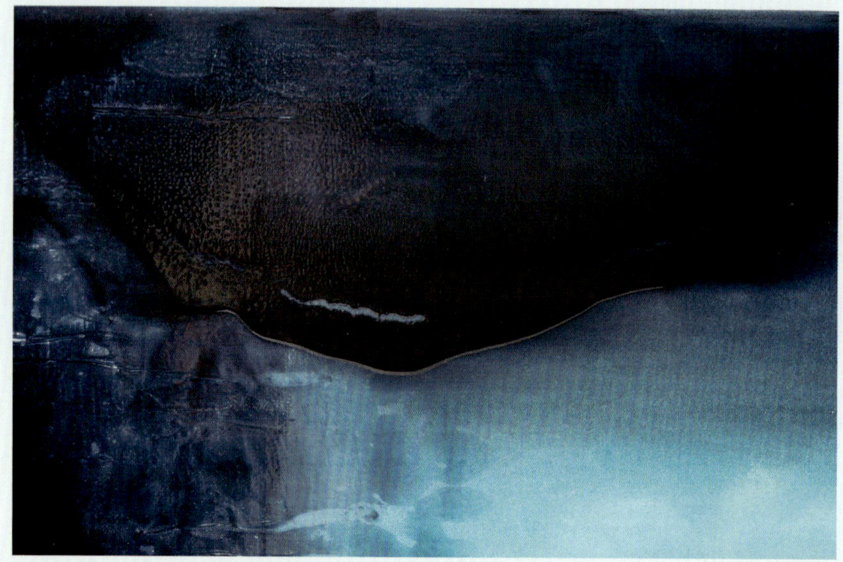

OPPOSITE Rhéni Tauchid, *And the Stars Fell,* 2016, acrylic, charcoal, and metal beads on canvas, 36 x 60 inches (91 x 152 cm). Collection of Tri-Art Mfg. Inc.

LEFT A detail of *And the Stars Fell.* A lip of drying self-leveling gel, sluggishly reaching across the canvas, shows the strain of moving while mostly dry, creating an "orange peel" pattern in the surface. Making sure the painting table was completely level would have prevented this accident.

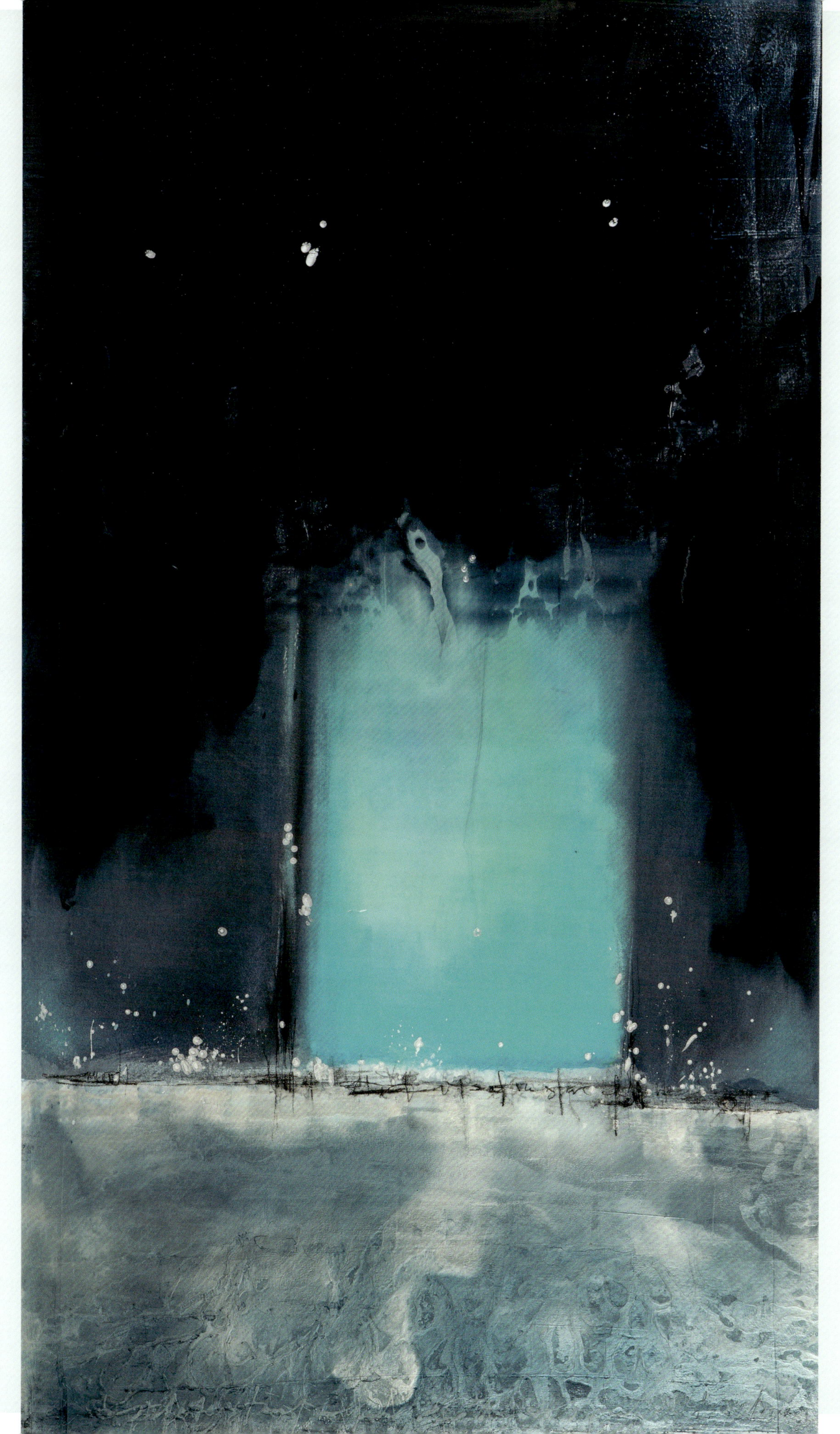

CHAPTER 13

USING FINISHING MEDIUMS

ARE VARNISHES, TOP COATS, OR FINISHING MEDIUMS NECESSARY?

This issue comes up in almost every painting workshop I give. The short answer is "no." Acrylic paintings do not necessarily need to be varnished or sealed. If your painting is done and you are happy with the results and confident that your materials are stable, then there is no need to add a finishing layer. That said, however, there are some good reasons you might want to add a final layer: to protect a delicate surface, create a scratch-resistant surface, unify luster, add depth, or increase UV protection.

Sealing a painting for the purpose of protection is quite different from applying a finishing medium. The purpose of varnishing a painting is to protect it from the accumulation of dirt and dust, water damage and physical damage like scratches, and so on. The acrylic paint surface is actually quite resilient, and unless the uppermost layer is very watered down, it is also water resistant.

There are several good reasons to add a finishing layer.

OPPOSITE Added to a finished painting, a thin, glossy layer of finishing medium (or varnish) unifies the surface, deepens darker colors, and increases the surface luster of the piece.

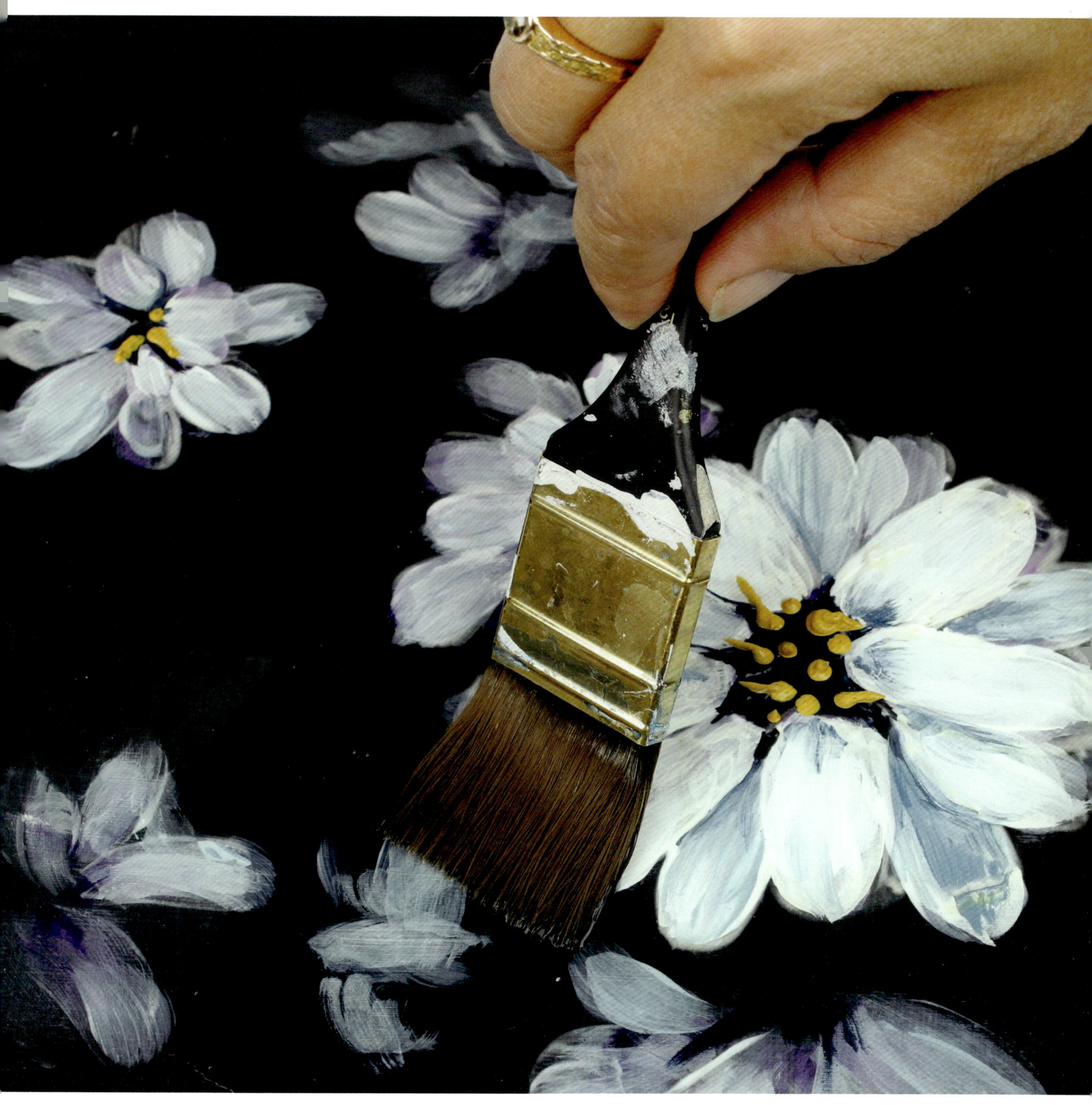

Every brand of acrylics has it own version of a thin finishing medium. These mediums go by various names: final finish, acrylic varnish, polymer varnish. But no matter what it's called, you should test it before using. Read the label and make sure you understand it, then apply the medium to another, nonessential painted surface and let it dry for at least 24 to 48 hours to determine if this is the effect you are looking for, if it is suitable for your purposes, and if it is compatible with the materials you're using. If you have any doubts or questions about it, contact the manufacturer directly. It is imperative that artists do due diligence and inform themselves about the materials they use, especially if they will be exhibiting and selling their work.

CURING

Regardless of how you plan on finishing your painting, the most crucial detail is to be sure the painting is fully cured. There are three stages of "dry" when you're working with acrylics: dry to the touch, dry enough to paint over, and fully cured. Also, be aware that a paint film will not necessarily dry uniformly. How a film dries is dependent on a number of factors, including (but not limited to) the type of paint used; ambient temperature, humidity, and ventilation; the amount of water, retarder, or flow release added; the porosity of the substrate; and the number of layers. Though it may seem dry on the surface, a thicker film will retain moisture for a prolonged period of time as the volatiles in the bottommost layer of the film make their way through the micro-fissures created by the volatiles that have already escaped from the surface layer.

The curing process—from application to the fully cured state—can take anywhere from a few days to several months. A higher-solids acrylic will dry more slowly than one containing more water. Inorganic pigments in a paint

OPPOSITE, TOP TO BOTTOM Top coats and pouring mediums have a smooth, slick consistency. • The top coat levels out smoothly and can be spread over a surface by tipping the painting. • Spreading the top coat with a large palette knife or straightedge is efficient and also discourages the formation of bubbles. • The resulting finish is thick, slick, and extremely glossy, giving the painting increased depth and the appearance of deeper color saturation.

can speed up drying. The addition of mediums increases the acrylic emulsion component, increasing open time. Some mediums, like glazing mediums, contain a larger amount of glycol, and if a retarding additive has been used the film will remain sensitive for a longer period of time. Given all of these scenarios, it is not possible to give a definitive set of parameters for drying time.

Although it is difficult to pinpoint the exact time at which an acrylic paint film is fully cured, it is still something you should strive to predict with some accuracy. This is especially imperative when you're intending to varnish a painting. Some products that are labeled as acrylic varnishes or finishing coats are modified polymer mediums, and these can actually be applied prior to full curing, as they will not inhibit the curing process. That's not true, however, of top coats and mineral spirit–based varnishes. These are tough, create a stronger seal, and in some cases have a tendency to shrink as they harden. If the underlying paint has not cured properly, this shrinkage can create an unstable bond or cause mechanical failure of the top coat.

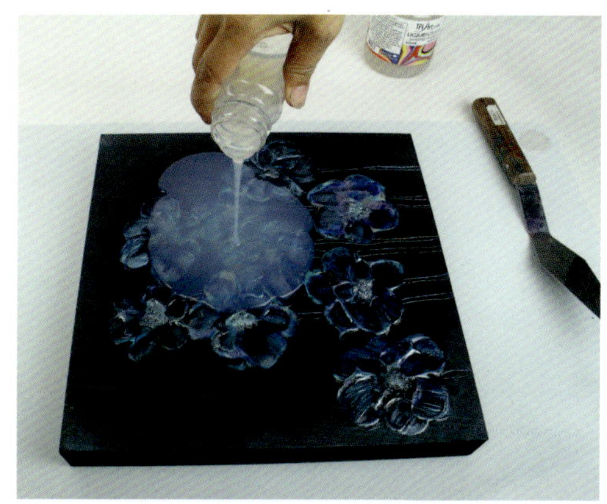

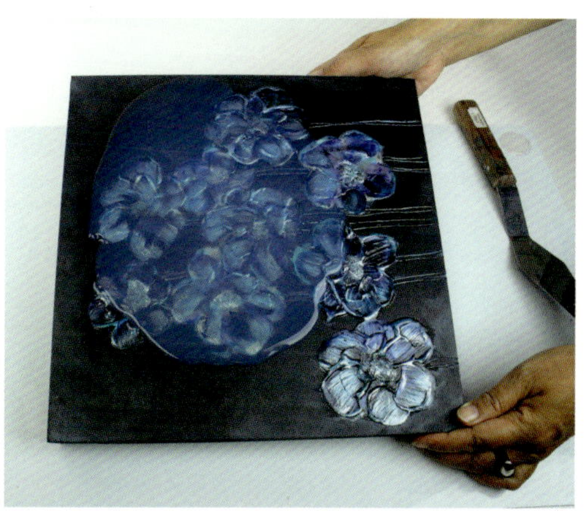

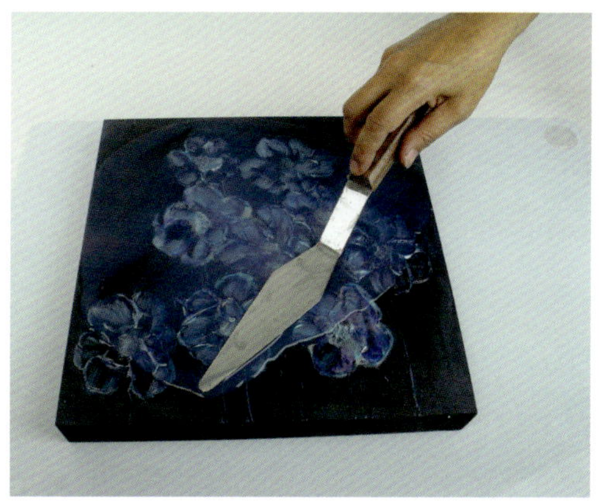

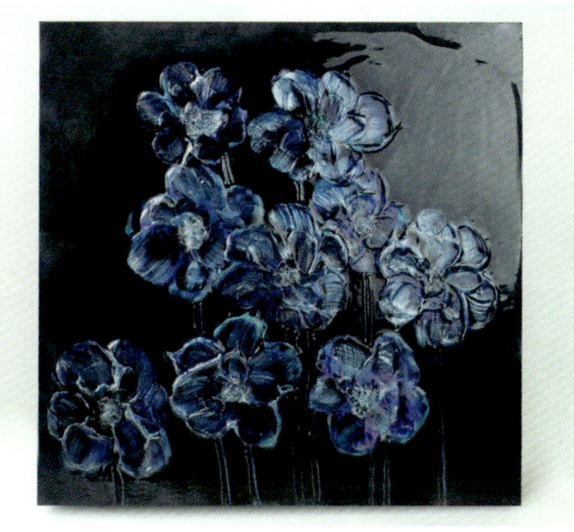

FINISHING FOR AESTHETIC REASONS

A layer of clear medium is often all it takes to give a painting that extra "luxe" appeal. The mediums that are best suited for end applications are those that will not affect the surface texture and that are the most transparent. Choose mediums, finishing mediums, or varnishes that are self-leveling and clear drying. A clean, soft, wide synthetic brush is the ideal tool for very thin mediums. If you are adding a self-leveling or pouring gel as your final medium, spread it with a wide straightedge with as little manipulation as possible. To disrupt the surface of a painting as little as possible while changing its surface luster, use one of the spray acrylic varnishes on the market. Follow the package instructions for best results, and remember to test on another surface first.

FINISHING FOR LIGHTFASTNESS

If you have concerns about the lightfastness of your painting, a finishing layer that contains an ultraviolet light stabilizer (UVLS) will give you an added layer of protection. Keep in mind that this is not a permanent solution to UV exposure and light damage. UVLS is like sunscreen: it protects the surface from some exposure but will not block all UV light, nor will it do so forever. Materials that are susceptible to UV damage include fugitive colors (those with reduced lightfastness, such as fluorescent pigments) and some collage elements (colored decorative papers, color printouts, plant matter). Liquid mediums, spray varnishes, and liquid varnishes and top coats containing UVLS are available.

VARNISHING

If you are going to varnish a work, be sure you know about the product you intend to use. Many products labeled as varnish aren't true varnishes. A true varnish should be removable; if it is not, it should be called a finishing medium or top coat. For example, a polymer medium is not a varnish. It is liquid acrylic without pigment. It will bond permanently with the rest of the acrylic on your support, and any attempt to remove it will almost certainly result in the alteration (or destruction) of the color layer beneath.

Pay heed to the following when varnishing:

1. *Read and understand the application instructions.* Having chosen a varnish, do not fail to read the instructions, as they can often be very specific. If you are tempted to disregard any aspect of the recommended instructions for use, protect yourself from potential disaster by testing the product first. The same holds true for any varnishing or top coat product you have not previously used. Every brand's top coats and varnishes have specific application instructions (either on the product's label or on the manufacturer's website). Take the time to read them!
2. *Apply an isolation coat before applying the varnish.* Once you have decided on the product, set it aside and prepare your surface. Removable varnishes for acrylics will require strong chemicals (acetone or a similar solvent) for removal.

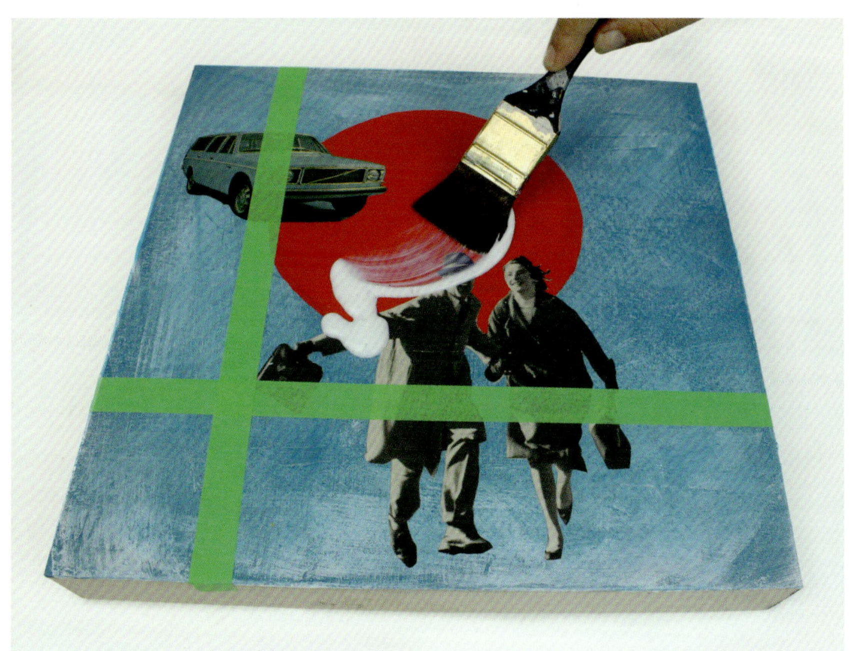

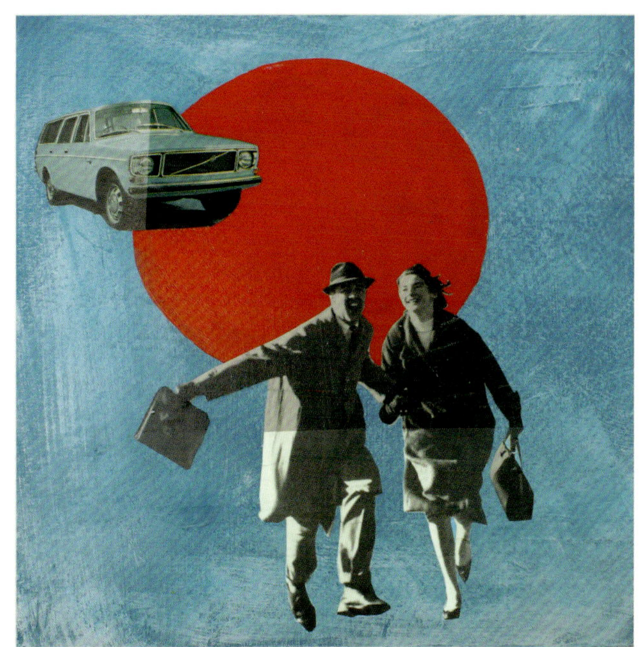

These chemicals will also remove the paint that the varnish is protecting. Hence, the prudent route is to create a shield. Apply a single layer of clear medium to the painted surface—the glossier the better, unless you are rigidly committed to a matte appearance. This will not guarantee the absolute protection of the underpaint, but will create a moderate barrier.

3. *Wait.* Make sure that the acrylic paint film has fully cured before varnishing. This can be very difficult to determine, so you should err on the side of caution and *wait*. A painting that is dry to the touch should be placed with its face against a wall, at an angle. This will prevent dust from accumulating on the surface during curing. Heavy applications should be left to cure for at least a month or two before even considering adding a varnish. Thin paint films can be varnished within one week.
4. *Apply.* Application parameters vary between varnish types and brands. Follow the manufacturer's instructions to the letter. Consider investing in a good varnish brush, one with soft bristles and a strong ferrule. Once your painting is so close to being complete, the last thing you want is a mess of stray bristles on your pristine surface. Cover the painting in a single, uniform layer, being careful not to overwork the varnish as you apply it or it will dry with some brushstrokes still visible. Do not apply a second layer until the first has been left to dry for 24 hours.

ABOVE, LEFT Mediums that contain ultraviolet light stabilizer (UVLS) are of particular importance when incorporating materials that will fade under extended light exposure. This piece has collage elements taken from old magazines. A liberal layer of a liquid medium with UVLS was applied over part of the collage.

ABOVE, RIGHT Both the paper the images were printed on and the ink they were printed with are not lightfast and without protection would fade within a few weeks of direct sun exposure. The background, painted with professional quality acrylic colors, did not suffer any color shift from UV exposure.

CONCLUSION

Mediums—this profusion of toys, these jars of possibility—have transformed the way acrylic painters approach their work. They are the bridges that facilitate connections to other media, the bonding agents for attaching collage elements, and the substances that give color dimension, transparency, and strength. They hold fewer mysteries now that you've gained the vocabulary and the guidelines for pushing your acrylic painting practice in new directions. Still, until you open those jars, play with their contents, and experiment on your own, you will not fully *know* them.

In my first semester at art school, we spent an inordinate amount of time on color theory, making swatches upon swatches. It was a mildly tedious but completely important part of my art education. Learning to paint—and learning about paint—requires hours upon hours of experimenting, the equivalent of rehearsals for a big part. So keep making those color and medium samplers; they will guide you more successfully than any amount of reading or video watching will. They are to a painting education what scales are to a musical one. When it comes to really grasping the sometimes minute differences between mediums, a physical exploration is the best and only way to truly "get" them.

Take the time to try everything. Take a tactile tour of every medium you can get your hands on, make messes, break the rules, get sticky with them. Have fun. It's worth it!

To send you off, another step-by-step demonstration begins on page 247.

OPPOSITE Leftover paint from the production of photos for this book, pulled from my palette.

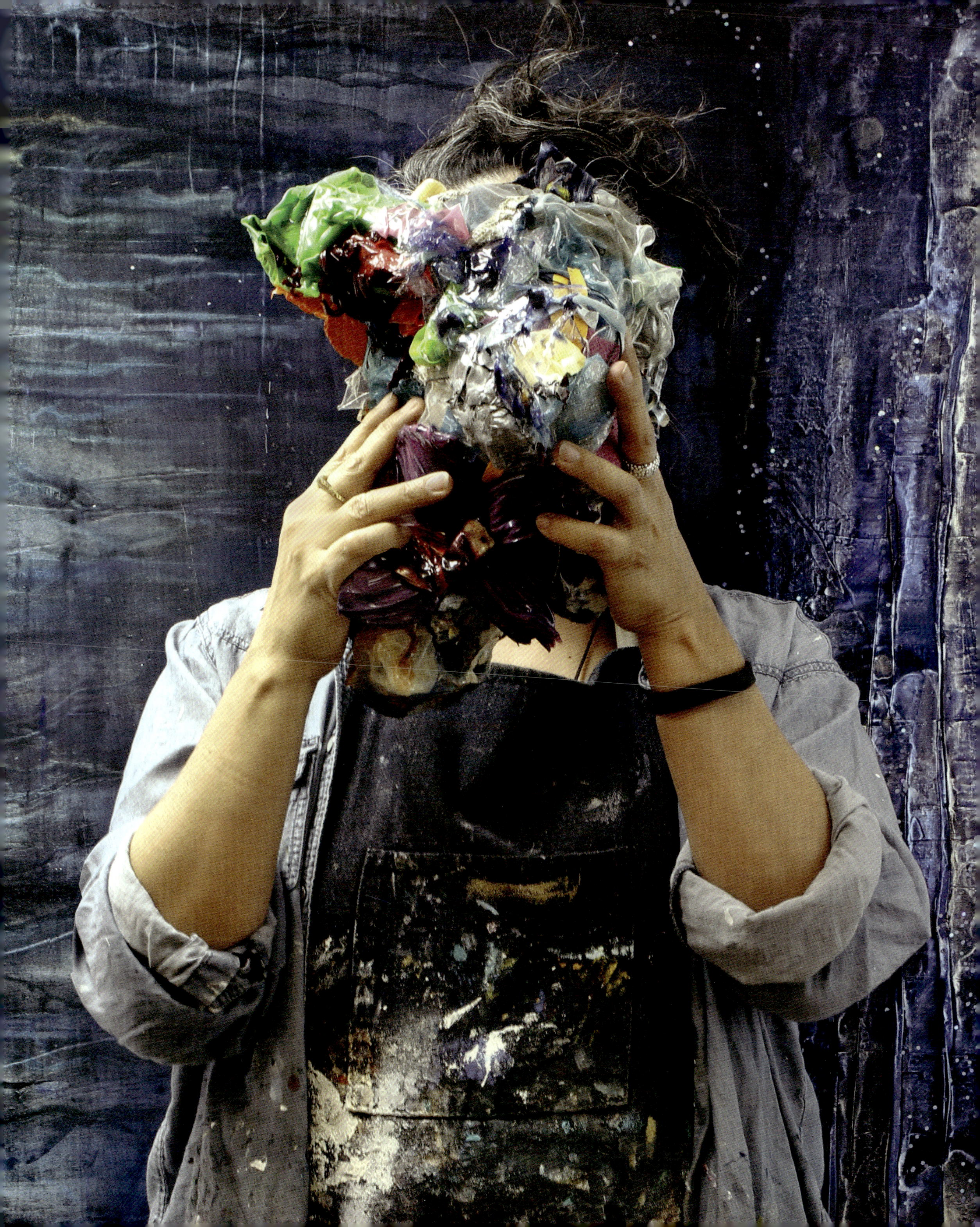

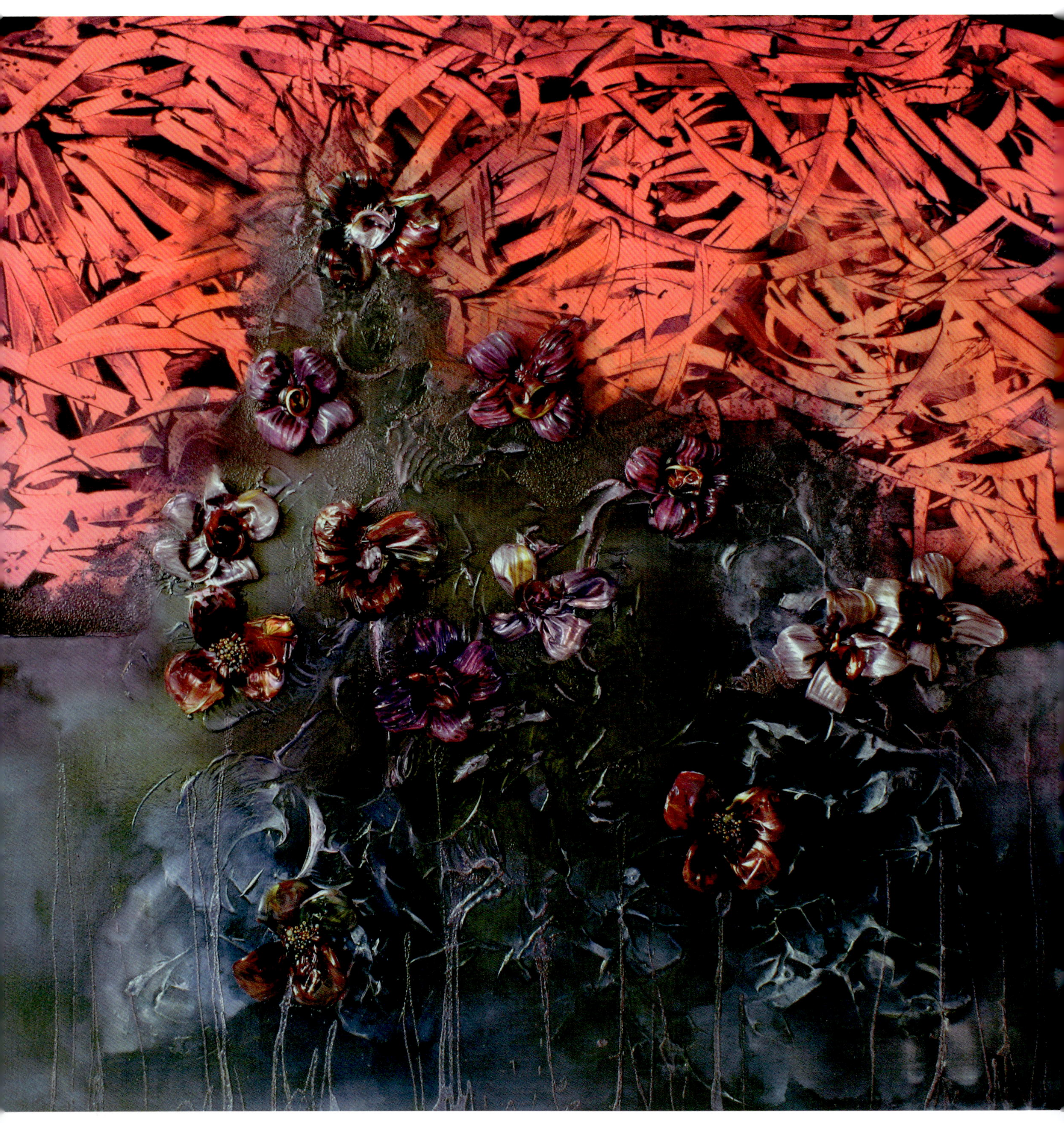

PROCESS

HAVING FUN

The tactile and visual features of acrylic mediums are perhaps best explored when you don't have a definite plan for a painting but instead take more of a "let's see what happens if I do this!" approach. Unconstrained, spontaneous application and mixture of process and materials can reveal fresh aspects of both, and beauty can emerge from this maelstrom of improvisation.

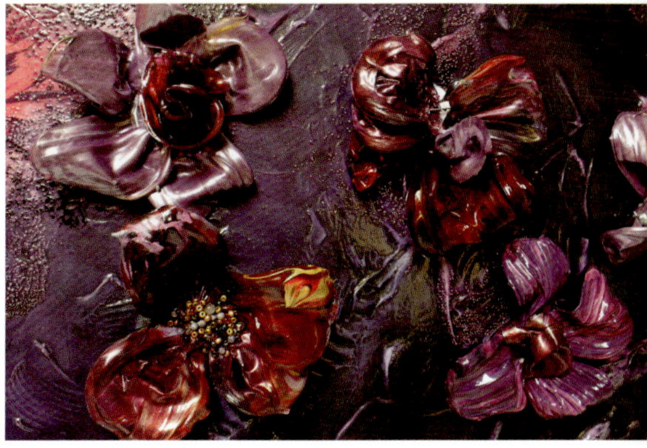

OPPOSITE Rhéni Tauchid, *Wild*, 2018, acrylic, wire, and metal and glass beads on canvas, 48 x 48 inches (122 x 122 cm). Grudging property of the artist.

This painting emerged from the process of producing this book, beginning with a few dry-brush strokes on primed canvas and culminating in a riot of textures, embedded beads, and three-dimensional flowers. Five different types of mediums were used: gloss gel, self-leveling gel, matte polymer medium, clear modeling paste, and pouring medium.

ABOVE A detail of the completed painting *Wild*.

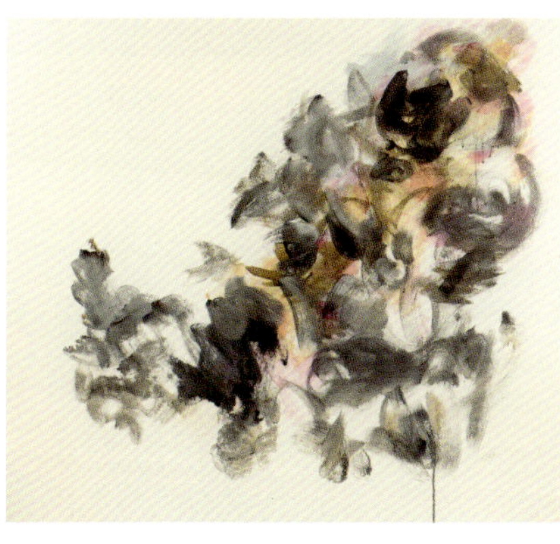

1
I'd been wiping excess paint from my brush onto this canvas while working on another piece. The unintentional marks became the basis of a new work.

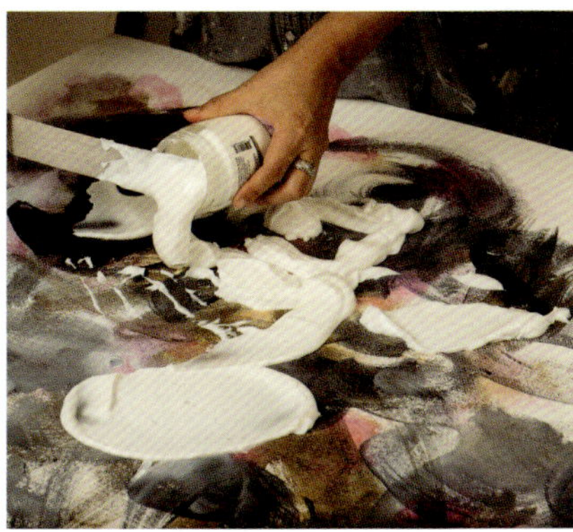

2
Over that loose first layer, I liberally applied gloss gel.

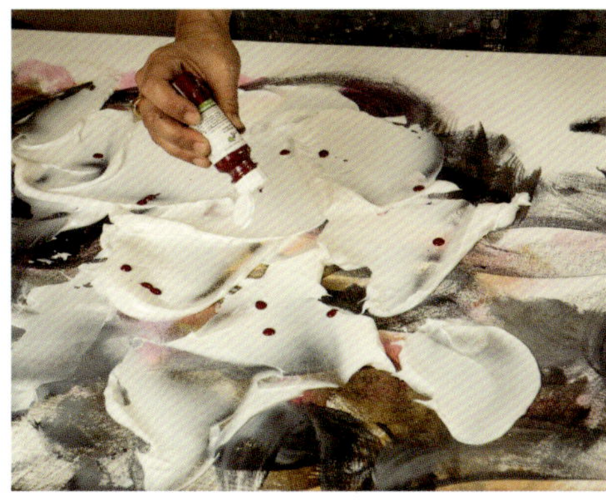

3
These drops of deep quinacridone magenta were grabbed by the gel and diffused into a juicy, transparent pink.

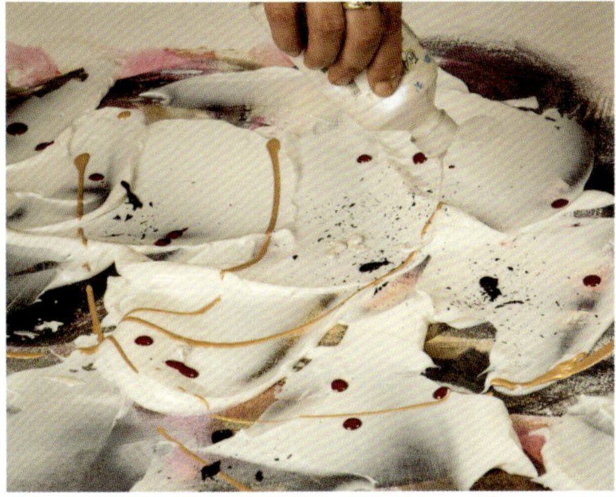

4
Splashes of Payne's grey and a squirt of interference blue rounded out the color mix, echoing the underpainted tones.

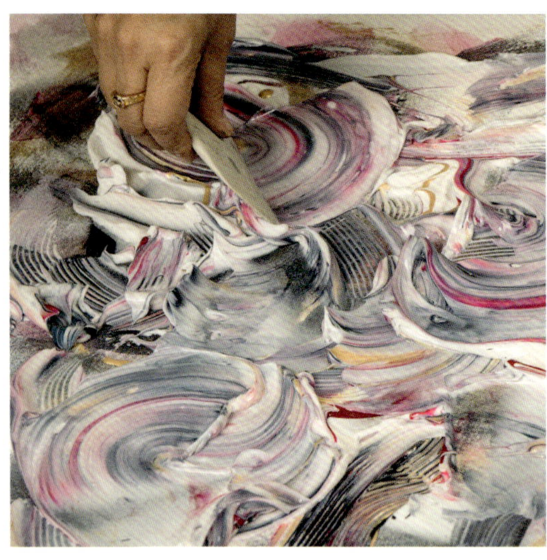

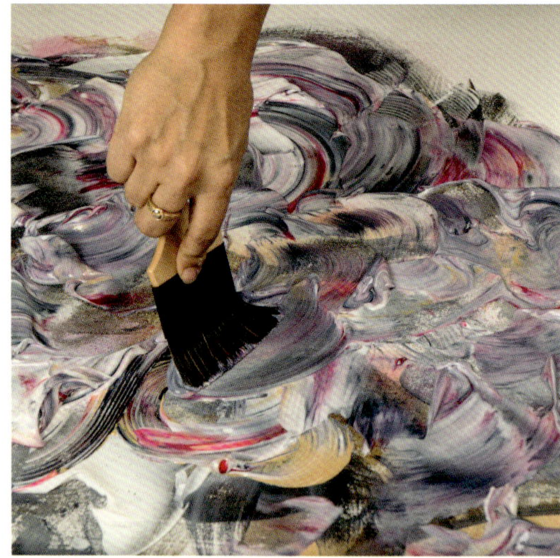

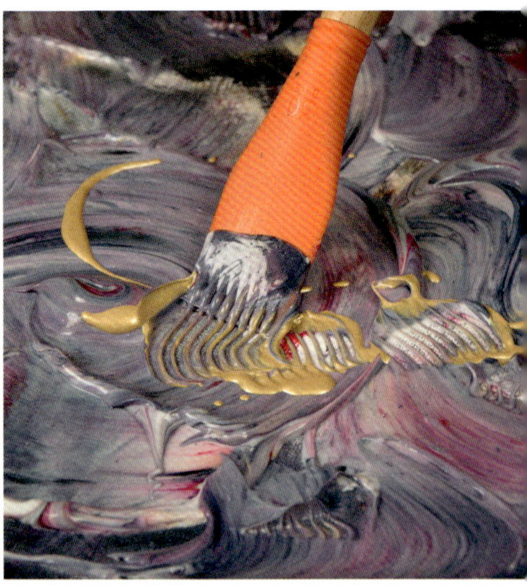

5
Mixing and sweeping through the gel, I broke up the color. Using the smooth edge of a Catalyst flat wedge, I created more circular marks.

6
I then pushed the gel around with the firm but supple hairs of this synthetic bristle brush.

7
A fresh addition of gold bumped up the shine.

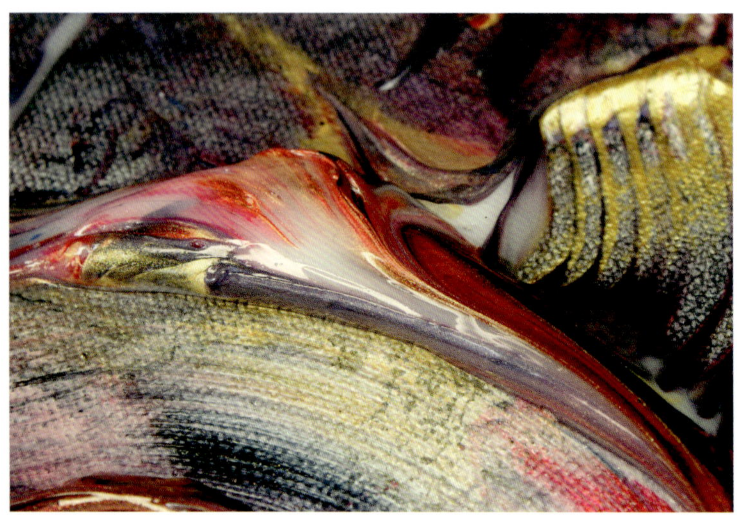

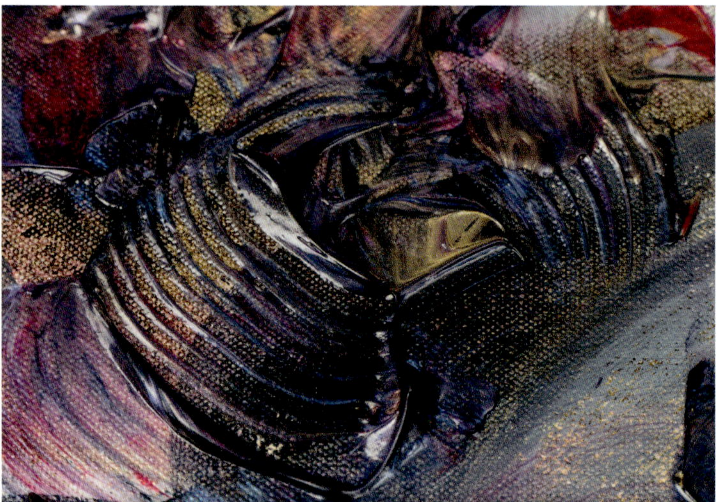

8
A swoosh of partially dried, loosely mixed gloss gel and liquid acrylic colors.

9
As the mediums and colors dried, the white gel clarified and the colors deepened.

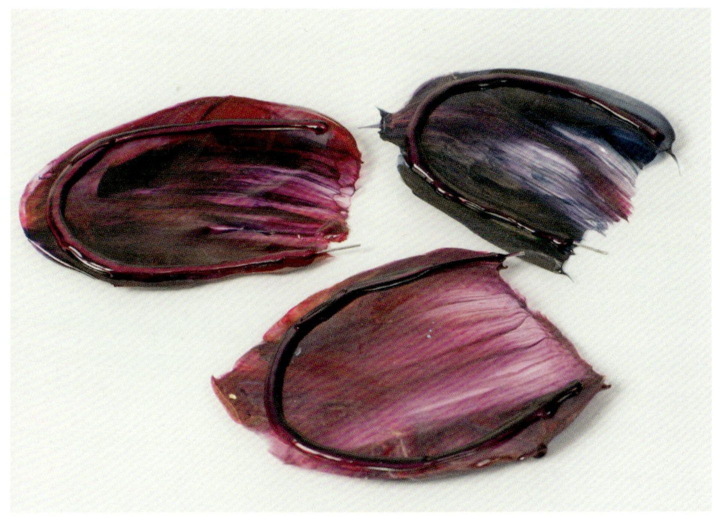

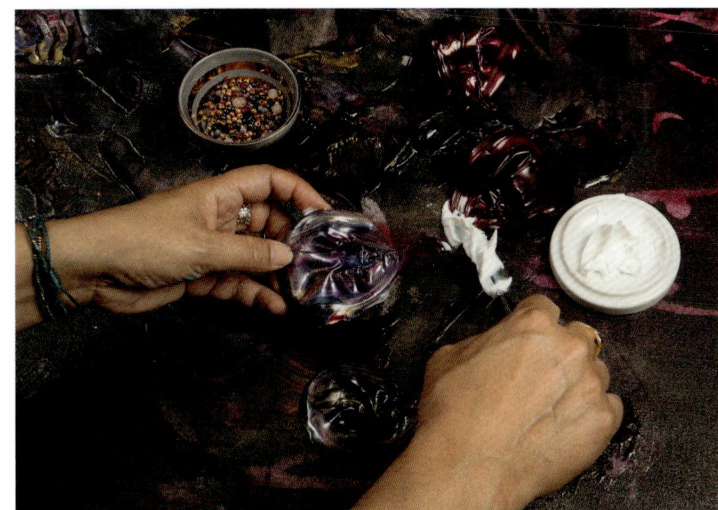

10
Separately, I created palm-size petals of paint and gel. I reinforced the petals with florist's wire and an extruded line of tinted gel.

11
I attached the wire-reinforced petals to the surface of the painting with a generous blob of gel.

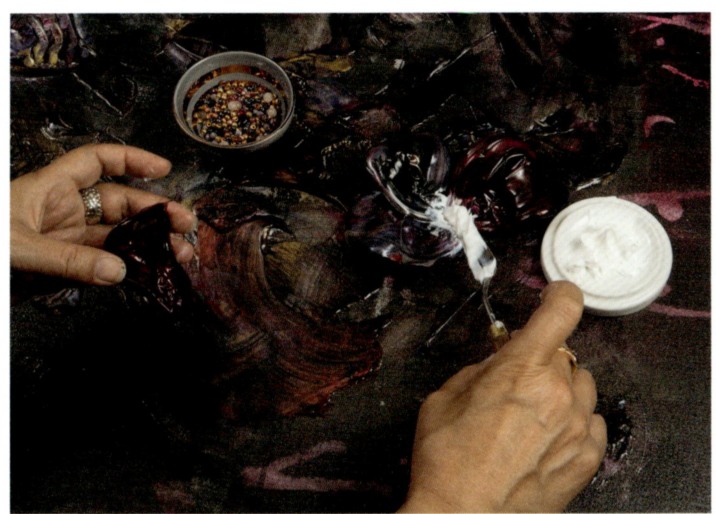

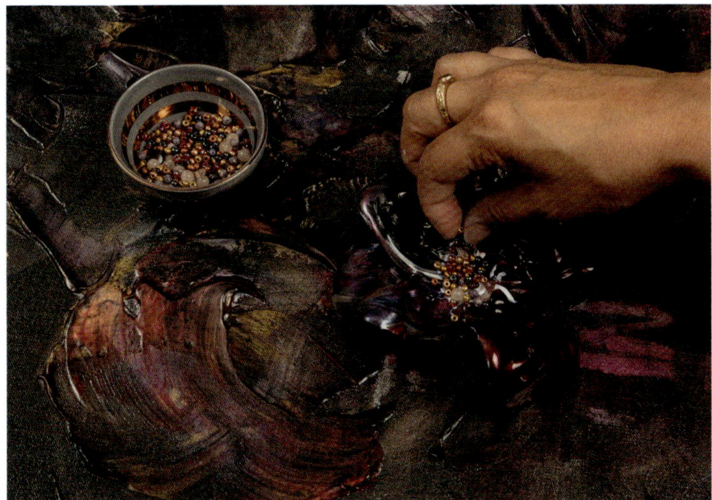

12
Adding some gel to the centers of the flowers helped seal them to the surface.

13
From an assortment of beads, I chose the right ones for a flower's stamen. Dropping beads into the gel finished the flower and added another textural element.

APPENDIX 1

MEDIUMS ROUNDTABLE

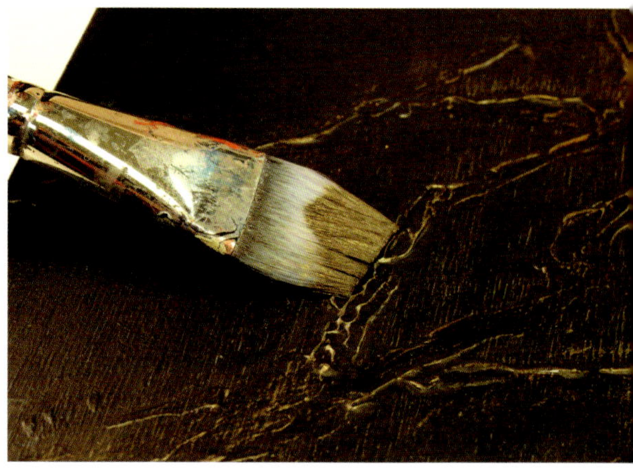

ABOVE Dry-brushing, or scumbling, a light pass of iridescent bronze over light relief brings out even the slightest of textural details.

I hosted a gathering for local acrylic artists and art teachers to discuss and share their experiences with acrylic mediums and experiment with them. My motivation was to find out how other painters use, view, and understand acrylic mediums. The result was a little different from what I had expected: I had anticipated hearing stories of how and when the artists use their favorite mediums, as well as some questions about the newer ones on the market. What I had not expected was the enthusiasm and eagerness, from all of those who attended, to understand more fully the role and scope of the mediums family, to play with them, and to expand their acrylics knowledge.

People were excited to talk about them and almost shy about admitting their limited knowledge of them and how they work. It was a freeing exercise for them to share their experiences with mediums and how they have watched the evolution and expansion of this set of materials.

Most of these people had been painting with acrylic paints for more than ten years. But they weren't necessarily sure even of what I consider some of the most basic characteristics of mediums, the different categories, or any of it, really, and they appreciated being informed about them. The discussions focused on things like the fact that all mediums are water based and will make water-soluble materials (images, text, etc.) "bleed" or smudge. Or on what retarder is, how it works, the proportions to use, and so on. One big topic was identifying differences and similarities between brands. With colors, there's not so wide a gap, as the same

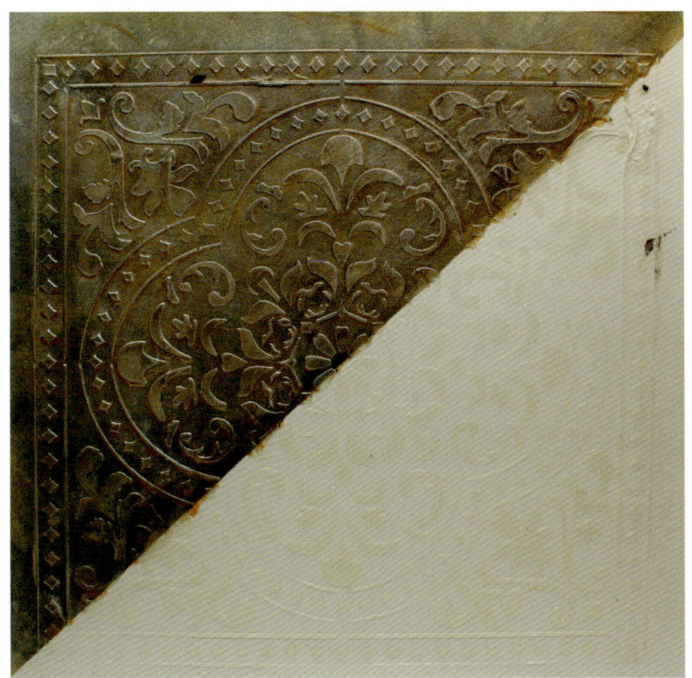

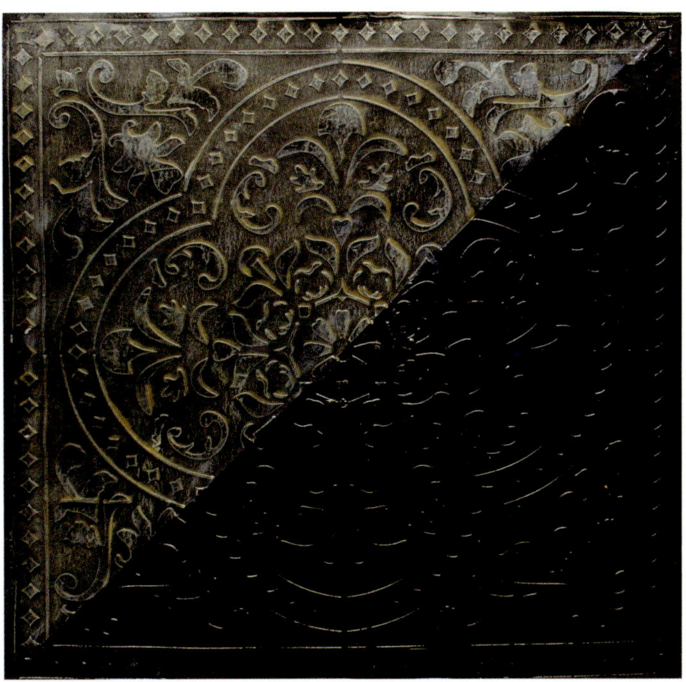

ABOVE Raised stencil textures, produced with gloss gel over white ground (left) and black ground (right). In each case, half the stencil has been accentuated by dry-brushing metallic and interference color over it, as well as by adding a few washes of color to create the appearance of weathered metal.

pigments are fairly universally used. Mediums, on the other hand, do not share the same standard recipes. Everyone agreed that experimentation is obligatory.

I was very interested in learning why some painters were not using mediums. I found that there was a sense of trepidation, a feeling of avoidance, even a noticeable hesitation in people perusing the acrylic-mediums display. Mediums seem to be underused in the acrylic painting world, which, although I personally find it baffling, was not unexpected. Just as it has taken "new" pigments like bismuths and pyrroles a long while to get noticed, so, too, it will take some time for mediums to come into play and become "normalized."

The excitement of discovery from those to whom mediums were relatively new was palpable. The potential of creative possibility firing the imagination, inspiring new directions—this is the big bonus of these types of discussions.

My biggest takeaway from this get-together was experiencing the power of peers sharing information, experiences, accidents, and discoveries. We often forget how solitary the creative time of artists is and how exciting, and rare, it is for artists to come together to learn from one another.

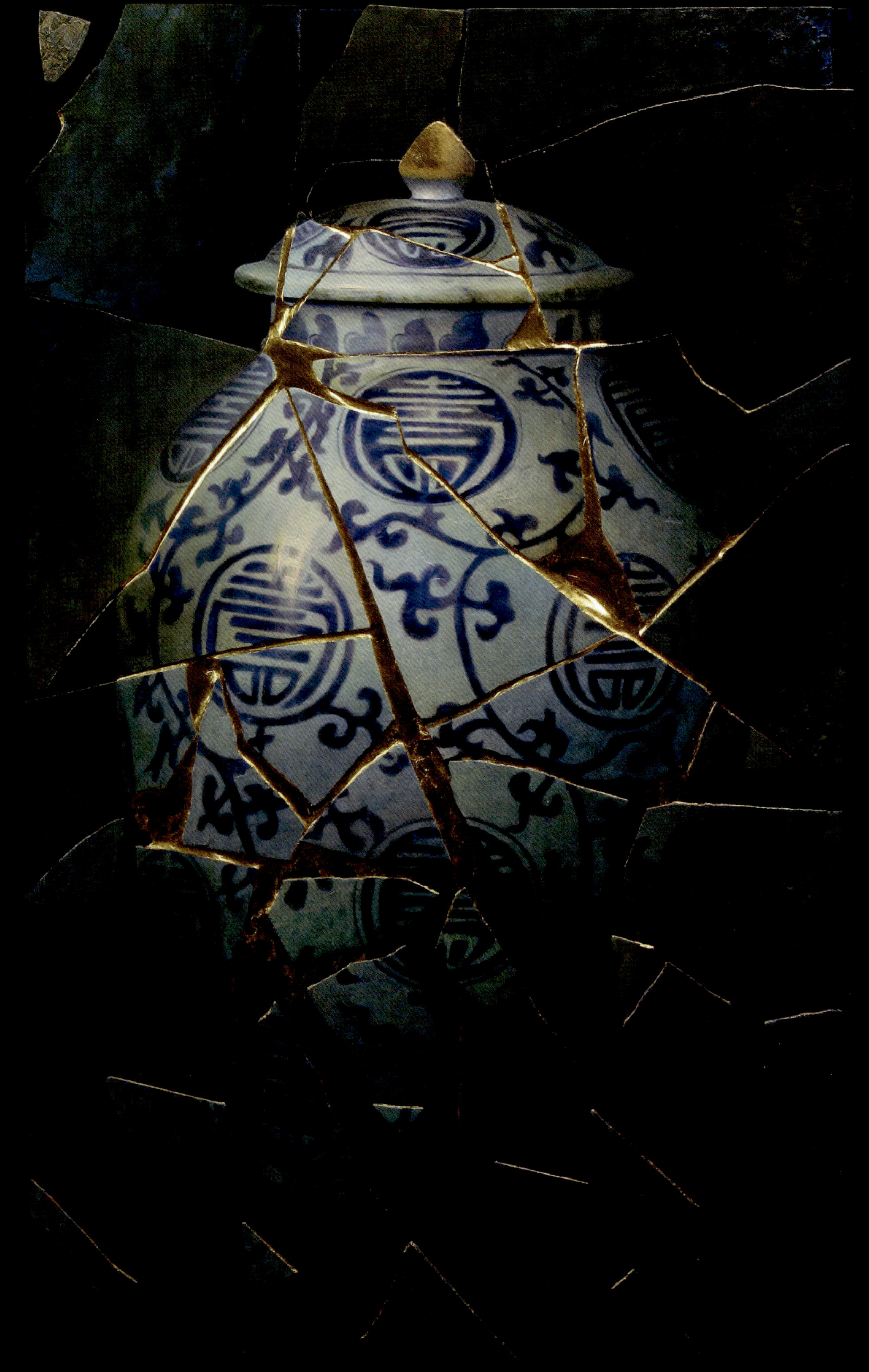

PROCESS

CREATING DRAMATIC EFFECTS

I became aware of Bruno Capolongo's work—and of his thirst for detailed knowledge about his materials—through my capacity as materials consultant at Tri-Art. Through many e mails and telephone calls, we discussed various mediums and troubleshot process, methods, and the intricacies of manipulating the mediums. This type of information gathering and sharing is essential to both artists and the people who produce art materials.

About his work, Capolongo says, "I firmly believe that acrylics are unequaled in versatility and range of effects. The ever-expanding selection of mediums—too long to list—offers breathtaking possibilities to artists of every stripe. It is hard to keep up with the advances in acrylics, both in effects and in quality. Whether one works in the classical realism mode or contemporary and experimental, acrylics provide the artist with every possible medium to realize their vision."

Here's the process Capolongo followed in creating his dramatic work *Fade to Gold*, which imitates the Japanese art of *kintsugi*, in which gold is used to repair broken pottery.

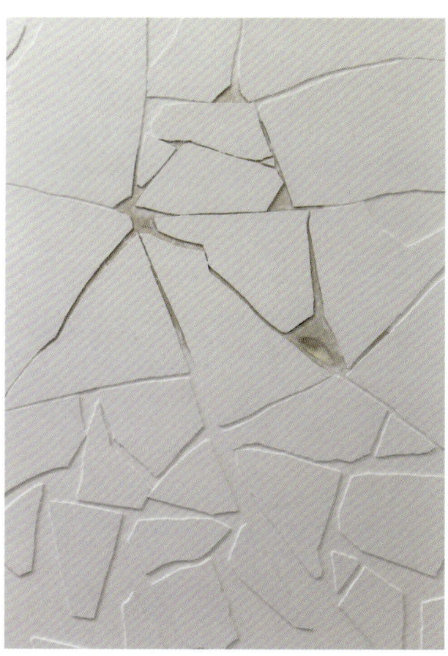

1
Shattered pieces of Magnesiacore panel were arranged and mounted on wood using acrylic medium for glue. Several layers of gesso were applied, creating a bright white surface to encourage later refraction. Exposed wood veins and the bare, deeply carved wood spot to the right of center are ready to receive filler to imitate *kintsugi* repair.

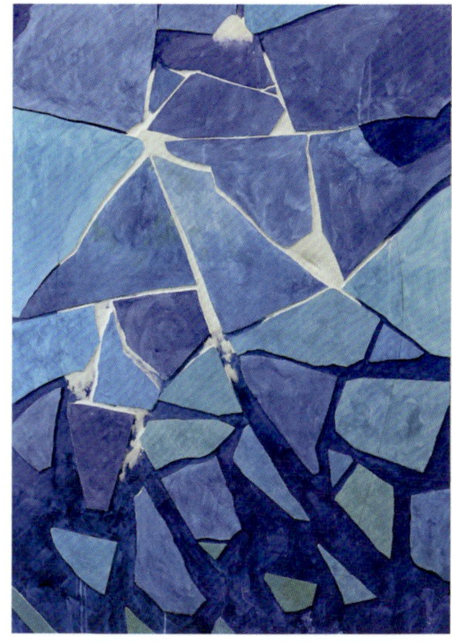

2
Blues were applied with various liquid acrylics for ease of application. Areas of exposed wood were filled with white unprimed filler.

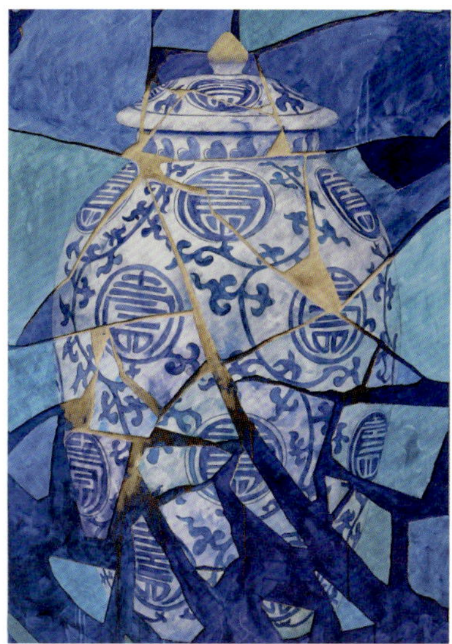

3
The vase pattern and white areas were painted in using a combination of heavy- and soft-body acrylics. Fluid gold acrylic was applied to the *kintsugi* areas and allowed to run and drip. The painting was now ready for a layer of self-leveling gel or pouring medium, which can vary in thickness. For this it was laid flat on a table.

OPPOSITE Bruno Capolongo, *Fade to Gold (Kintsugi Vase in 23kts)*, 2017, acrylics and 23kt gold on mounted broken panels, 60 x 36.25 inches (152 x 92 cm). Collection of the artist.

In the painting's final state, you can see the rich glow and depth resulting from layers of various acrylic mediums and colors.

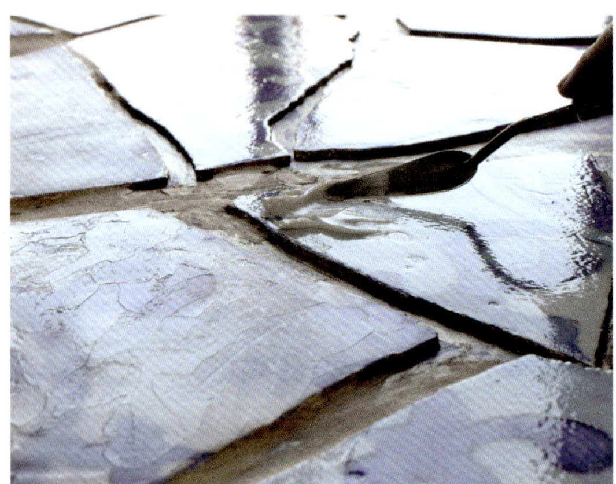

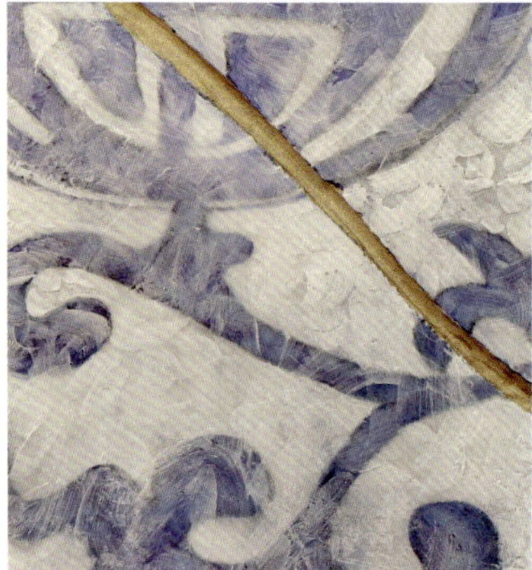

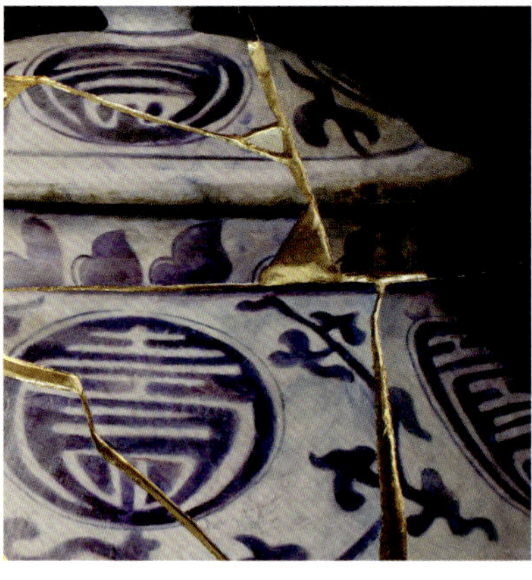

4
This image shows the dried and shiny self-leveling gel on the right side and the heavy matte gel being thickly spread over a "shard" on the left. These clear and translucent layers give tremendous depth and luminosity to the finished painting. The thicker and more numerous these barrier layers, the more depth and refraction.

5
This detail shows strokes of translucent white over the clear barrier coat. More clear gel was then applied over these strokes, and the pattern was then repainted in fluid translucent blues created with fluid and glazing mediums. The gaps in this picture show a rough texture that was smoothed over with either string gel or self-leveling gel before gold was applied.

6
To avoid an overly cool feeling in the painting, Capolongo emphasized warm tones in the vase with fluid acrylic glaze mixtures that included magenta and potter's pink. The subtle complexity of multiple layers has begun to take effect. Highly reflective 23-karat gold was then applied, contrasting with the various satin- and matte-finish vase shards. Little jewel-like fragments—like the triangular piece on the jar's lid, covered in interference turquoise—were used sparingly to intensify the effect.

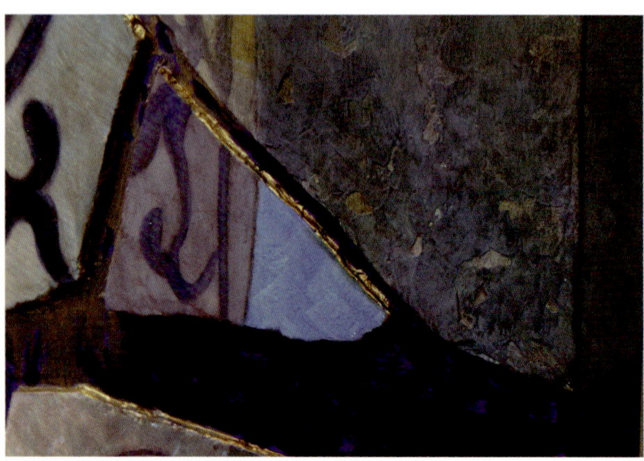

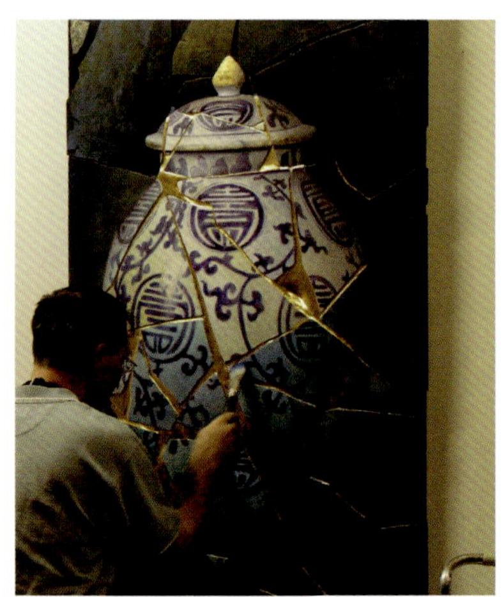

7
This detail shows the eighth-inch thickness of the mounted shards and the varied, highly complex application of materials and colors. At top right, you can see gaps in the surface of barrier layers, exposing layers beneath. Surfaces were stained and wiped with rags and various types of brushes.

8
To create more unity and emphasize the *kintsugi* gold, Capolongo mixed several stains of various blues, including manganese and turquoise. Depending on where they were used, these were further tinted for diversity of effect and shadow. (The term *stain* indicates just how little paint was used.)

APPENDIX 2

ACRYLIC MEDIUMS QUESTIONNAIRE

In addition to my in-person gathering of local painters, I sent out a questionnaire about mediums to other artists around the world. Questions and selected responses follow.

How/when/where did you learn about using acrylic mediums? (More than one answer allowed.)

Personal experimentation: 80%

Workshop: 37.1%

Book: 25.7%

School: 20%

Other: 20% (online course, friend, work)

Which acrylic medium are you most familiar with?

Gel medium came out as the clear leader here, followed closely by polymer (thin) mediums, and close to a third of responders listed three or more mediums.

Which acrylic medium do you use the most?

Even though gels beat out thin mediums as the ones painters are most familiar with, thin mediums are used most often. If I were to speculate as to why, my guess would be that regardless of their boldness in experimentation, the majority of painters will stick to "safer," better-known materials.

Which acrylic medium do you use the least?

It's no surprise that the strangest mediums are the least used. Like that crazy pair of boots that you tried on with no intention of ever buying, let alone wearing, dramatic texturizing mediums, custom mediums, and single-use mediums (opacifying, UV reactive, etc.) are intriguing but rarely go-to choices. Do not dismiss them out of hand, though, as they can be just the thing to transform a floundering painting or give pizzazz to a tired composition. Think of them as special-occasion mediums!

Has your use of mediums increased as the available selection has grown?

It is getting to the point where mediums can no longer be ignored as a viable component of acrylic painting. They used to be marginalized not only in their use but in their placement in art supplies stores.

Yes: 68.6%

No: 31.4%

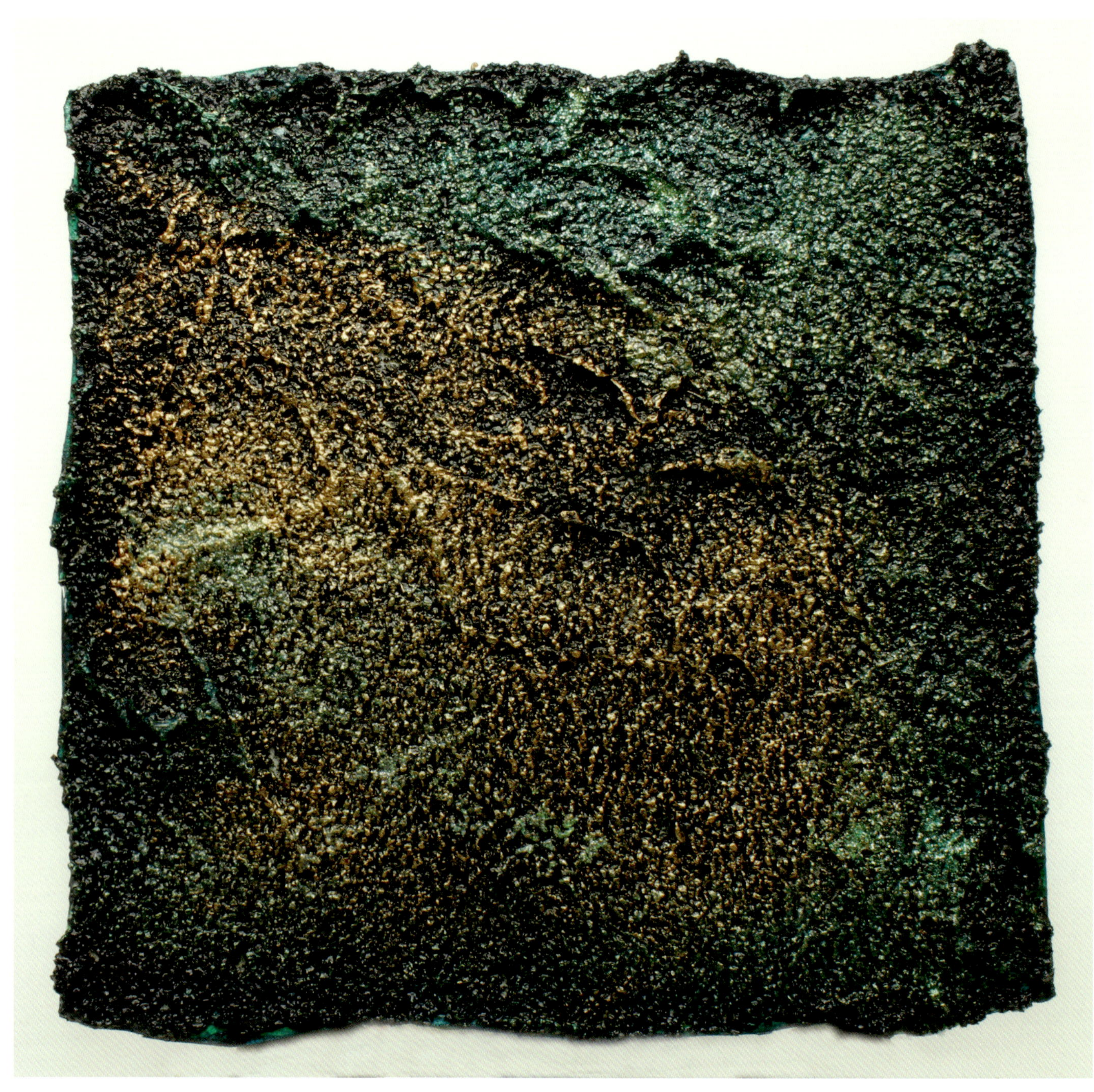

ABOVE A thick slab of copper cinder medium, textured and then lightly painted with reflective colors.

Do you feel that mediums have augmented your painting process? How?

- "Yes! In ALL ways!"
- "They give me depth and texture!"
- "I'm an explorer and will test any medium or product that I can get my hands on!"
- "Surface variety and drying consistency."
- "Using mediums as a glaze has created depth in my paintings and changed my process and approach over time."
- "Yes, the painted surface would look flat and lifeless without gloss medium, and I can achieve texture with modeling gel."
- "Yes. By allowing a wider array of effects through layering."
- "Exceptionally! They allow me to create layers of images and depth."
- "The use of mediums has given more depth to my painting process. This contributed to a stronger approach to paint handling, which has in part become the focus of my visual approach to painting."
- "I play with transparency more. I love embedding things in acrylic."
- "Yes, they've allowed for a lot of breadth and creativity in what I can explore with my students in their art. So many different possibilities!"
- "It was when I switched to using Tri-Art modeling paste as a white that I fell in love with acrylics and decided to stop using oils altogether."
- "Gives me more control over finish."
- "I love to build up semitransparent and transparent layers in my paintings, and I regularly use a textured support layer—granular medium."
- "I do largely abstract and surreal work, so mediums enable me to add depth or stretch the limits of traditional acrylic work."
- "Well, I am experimenting. . . . And I like the results."
- "Absolutely! Nothing can touch the range of mediums now available in acrylics. I was trained at age twelve in oils, and always considered oils 'God's gift to painters.' It is the slow-drying nature of oils that allows lengthy manipulation—especially helpful in tonal gradations and wet-in-wet blending. However, for artists who want to explore more ambitiously, with multiple layers and thick buttery textures—without the lengthy drying time of oils—acrylics are the answer! Even as an encaustic painter I had to concede that the advantage of acrylics is the highly translucent and even perfectly *transparent* mediums now available. More recently I have come to use an increasing amount of fluid acrylics and clear fluid mediums to extend them and increase transparency. Once again, oils have serious limitations in this area, especially if one is keen on keeping the work archival in quality."

Do you use any finishing mediums on your paintings? If so, why?

Thirty percent of painters asked said no. Those who said yes gave these three primary reasons for using a finishing mediums: controlling luster, UV protection, and added vibrancy.

Do you feel that using mediums has increased your overall understanding of how acrylics work?

This question generated both yes and no responses, about half and half. Although mediums have expanded the variety of methods and processes for painters, the overall understanding of those surveyed was that learning about acrylics as a material occurs through the act of painting in a general sense.

Do you have confusion/frustrations around using mediums? Which ones and why?

- "My frustration stems from not having used most of them (other than gel medium) on a regular basis or in combination with one another. I feel I am just scratching the surface of what they can do, but my painting practice began before there were so many options and I am slowly finding ways to incorporate them."
- "I realize I need to experiment and have more knowledge—not so trial-and-error. Happy accidents are all well and good but can be difficult to replicate without an understanding of the underlying process."
- "Yes! Because of the vast selection of mediums, I don't necessarily want to spend money on them if I'm not sure I will use them. Consequently, I end up not experimenting with them, when that is actually the best way to learn about them."
- "I find the texture gets too tacky for my liking."
- "I'm confused but not frustrated. I just need to take more workshops."
- "No, I am game to try anything."
- "No—I find those that I'm aware of useful but know there are many more out there to try and experiment with."
- "Yes, especially the finishing mediums—so many options and confusion about what to use and why to use it."
- "I need to educate myself on all new mediums. Just getting back into painting."
- "Yes, I often feel the directions are not clear, and the labels often lack a clear explanation for their use. My biggest frustration is with the retarders. So far they really don't work as promised."
- "Only when I first started working with them and didn't give myself enough time to master them before incorporating them into a final piece."
- "Mediums don't always perform predictably due to various factors; these can include environmental factors like temperature and humidity but also water and even trapped air content—and, thankfully, rare side-effects from combining different manufacturers' products. "For the most part it is the user's responsibility to test and do due diligence in educating oneself in the use of products."

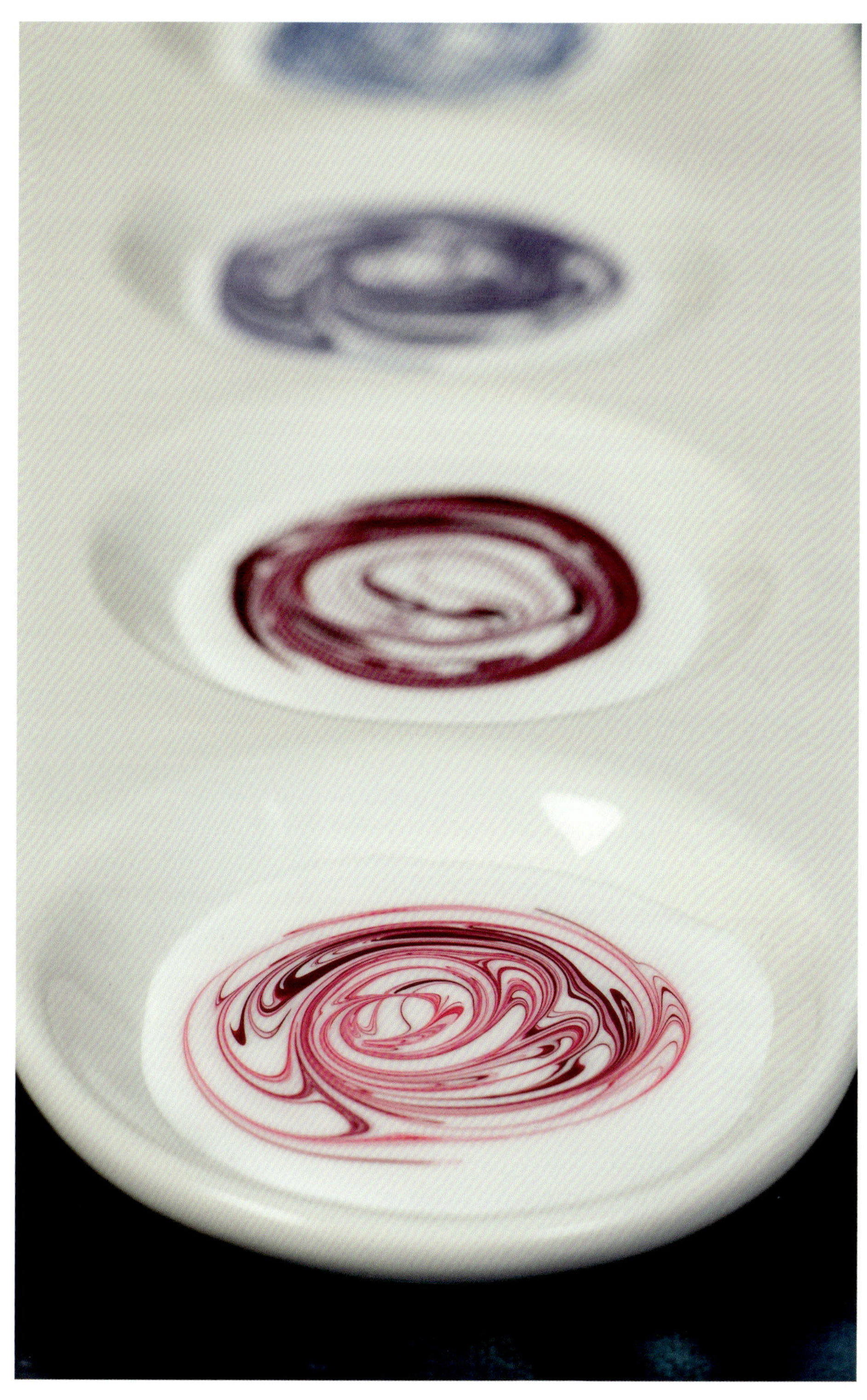

ABOVE Getting ready to glaze: transparent colors swirled into liquid medium.

If you do not use mediums, why not?

The biggest block to using mediums, according to my accumulated data, is unfamiliarity. I have found that once artists become aware of the possibilities offered by mediums, through a tactile exploration and some show and tell, they are much more open to trying them and ultimately using them in their art practice. (Respondents were allowed to choose more than one answer.)

Unfamiliarity: 70%

Aversion: 0%

Confusion: 20%

Indifference: 20%

Unnecessary: 40%

Do you use gesso?

Yes: 48.3%

No: 14.3%

Sometimes: 37.1%

Do you use mediums as an adhesive? If so, to fuse which types of materials?

- "Yes I use a glazing medium to adhere my photo images printed on paper to the canvas. Also, I use it to apply gold leaf."
- "Yes—mostly paper but I also have embedded beads."
- "Yes—usually paper to wood or paper to canvas or to do transfers."
- "Yes, for photo transfers."
- "Paper to canvas, collage, glitter."
- "Papers, cheesecloth, aluminum foil, sometimes heavier objects such as pieces of metal."

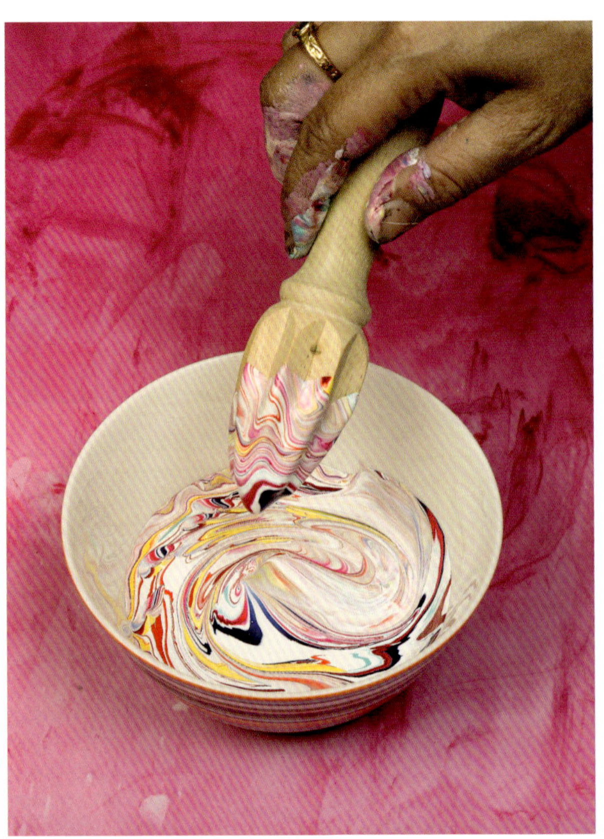

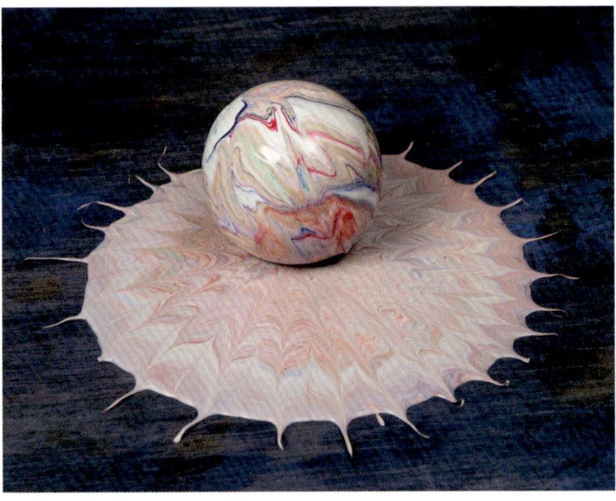

TOP Gesso isn't just for priming. The thick, opaque, white boldness of gesso holds strains of liquid color.

BOTTOM Experimenting with gesso lead to these odd paint skins. Liquid acrylic colors were swirled through white gesso for this photo, after which the mixture was poured into a jar. Pouring the gesso from the jar onto a palette and then "cutting" through it with the sharp side of a palette knife formed the spiky pancake. After the remaining film of gesso had completely dried in the jar, it was gently pulled out in one piece and stretched over a wooden sphere. There's no real point to any of this, but it looks cool!

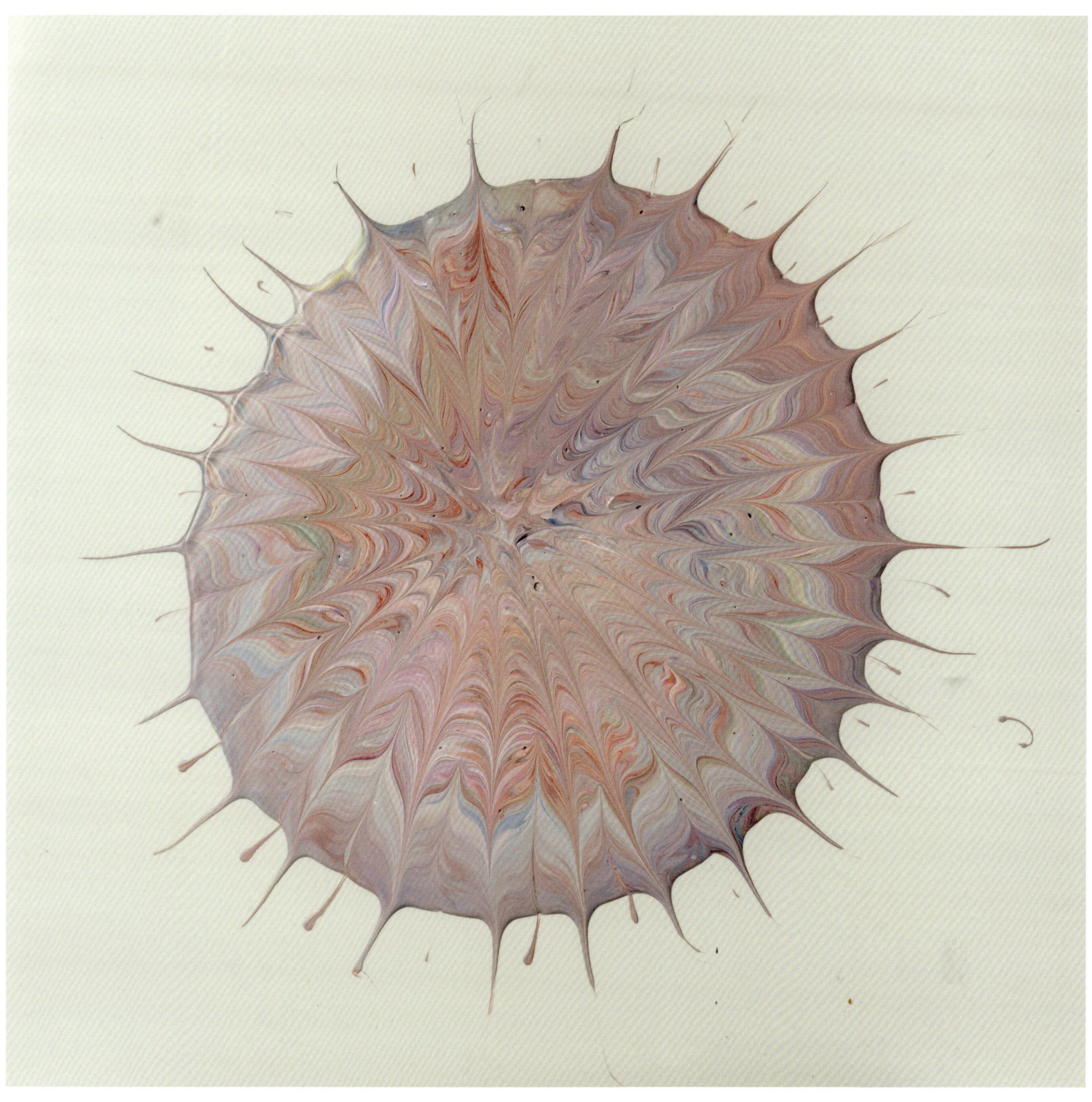

ABOVE A pool of white gesso mixed and marbled through with color. Weird and wonderful.

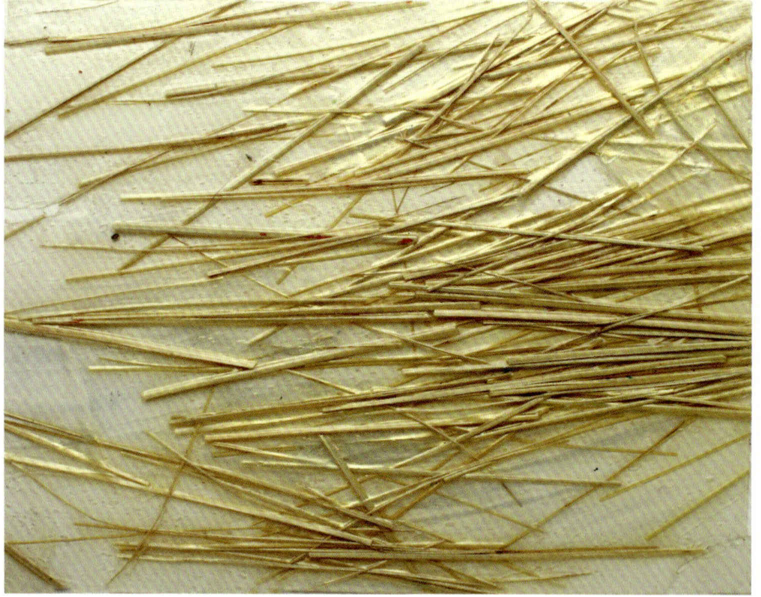

- "I definitely use mediums as an adhesive but mainly to fuse acrylics with acrylic skins."
- "Paper, fabric, metal, objects from nature, anything else I can think of."
- "Fabric, paper, glitter, metallic things, anything loose."
- "I use mediums to adhere thinner fabrics (cheesecloth, panty hose) to the canvas surface, which I can manipulate to build my surface textures."
- "Anything from sticks to buttons."
- "I find that acrylics are often as good as any glue, so I mostly use acrylics themselves as the glue in my artwork. Most notably, I use fluid acrylics to mount pieces of other panels onto my substrate in my *kintsugi* paintings. The bond achieved is better than that of carpenter's glue."

Which techniques do you use mediums for? Tick all that apply.

The more times I asked this question, the more varied the answers became. I take this as a positive sign that acrylic mediums are finding their way into more and more studios. The variety and quantity of techniques they are being used for are growing and will keep growing as more products are developed and more artists experiment and express themselves through their use.

Resist: 23.5%

Create texture: 82.4%

Seal surface: 85.3%

Manipulate surface luster: 82.4%

Glazing: 82.4%

Reduce texture: 35.3%

Extend color: 73.5%

Other: 29.4%

What size of medium do you generally purchase?

Larger sizes of mediums are the logical economical choice.

120ml: 8.8%

250ml: 11.8%

500ml: 32.4%

1000ml: 35.3%

Larger: 11.8%

ABOVE Grasses brushed in iridescent pale gold and embedded in gloss gel.

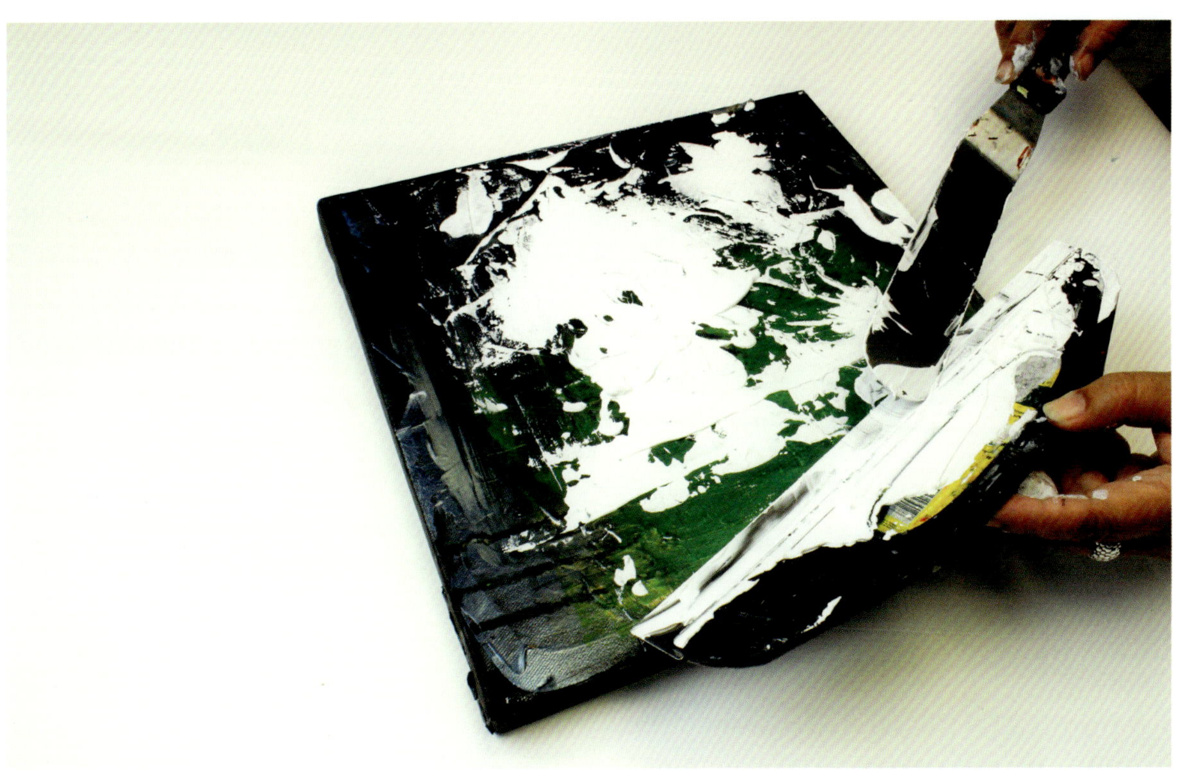

CONTRIBUTING ARTISTS

Nick Bantock
www.nickbantock.com

Diane Black
www.dblackstudio.com

Bruno Capolongo
www.brunocapolongo.com

Pauline Conley
www.paulineconley.ca

Marc Courtemanche
www.marccourtemanche.com

Marie-Claude Delcourt
www.marydelcourt.com

Claire Desjardins
www.clairedesjardins.com

Marion Fischer
no website

Heather Haynes
www.heatherhaynes.com

Lorena Kloosterboer
www.art-lorena.com

Suzy Lamont
www.suzylamont.com

Marie Lannoo
www.marielannoo.com

Connie Morris
www.room6design.ca

Barry Oretsky
www.barryoretsky.com

Lori Richards
www.loririchards.ca

Hester Simpson
www.hestersimpson.com

Ksenia Sizaya
www.kseniasizayaart.com

Alice Teichert
www.aliceteichert.com

Beth ten Hove
www.bethtenhoveart.com

Sharlena Wood
www.sharlenawood.com

Heather Midori Yamada
www.artyamada.com

ABOVE Always try to use up your leftover paints and mediums instead of throwing them out or washing them down the drain.

ABOUT THE PHOTOGRAPHERS

JONATHAN SUGARMAN

www.sugarman.ca

Jonathan is a professor in the Graphic Design program at St. Lawrence College in Kingston, Ontario, where he teaches many photography and design-related courses. He has been an avid photographer since 1978 and an educator since 2002. With over twenty years of professional experience in digital photography and image editing, Jonathan's interests in photography range from events and commercial assignments to portrait and creative work. This publication marks his third collaboration with Rhéni Tauchid.

CONNIE MORRIS

www.room6design.ca

Connie Morris lives and works in Kingston, Ontario. Her career involves digital design, teaching, painting, photography, and illustration work. She has a bachelor of fine art degree from Queen's University, is a graduate of the Queen's Artist in the Community Education Program, and also holds a diploma in new media design from Sheridan College. She has been a part of the Tri-Art team for over ten years, working in the company's education, marketing, and design departments. She works and exhibits predominantly in the medium of two-dimensional painting (using acrylics) but has come to develop a fascination with design and the integration of design, form, and function into everyday spaces and interactions through photography and graphic design. Her works are a marriage and exploration of her visual affinity with fine art and design.

Photo styling: Grace Paulionis

ABOUT ALICE TEICHERT

Born in Paris in 1959, multidisciplinary artist Alice Teichert studied music, philosophy, visual poetry, visual arts, and printmaking in Belgium and France. Since 1989, she has built an international career, with over thirty solo exhibitions in France, Canada, Switzerland, Germany, and Australia. Teichert's paintings are noted for their bold colors and layered transparencies interspersed with script-like line drawings and for their holographic depth. Her works are held in many corporate, private, and public collections worldwide. An English-German monograph on Teichert's work, *In) Formation—On the Philosophy and Art of Alice Teichert*, by Dagmar Täube, has recently been published by Hirmer Publishers, Munich, Germany (hirmerverlag.de/us/titel-3-3/in_formation-1561/).

PRODUCT CREDITS

Tri-Art Manufacturing Inc.
4 Harvey Street
Kingston, ON K7K 5B9
Canada
www.tri-art.ca
(888) 541-0299

Princeton Artist Brush Co.
Princeton, NJ 08540
www.princetonbrush.com
(609) 403-8342

INDEX

Page numbers in *italics* indicate works of artists

G

H

I

K

L

U

ABOUT THE AUTHOR

Rhéni Tauchid (www.rhenitauchid.com) is a writer, painter, and educator. In her professional life, Rhéni is the materials consultant for the Canadian acrylic paint manufacturer Tri-Art Mfg., Inc., a member of the company's product development team, and the founder of the Tri-Art Acrylic Education Program and the Art Noise studio program. She teaches painting workshops and lectures in Canada and abroad.

Rhéni is the author of *The New Acrylics: Complete Guide to the New Generation of Acrylic Paints* (Watson-Guptill, 2005) and its sequel, *New Acrylics Essential Sourcebook: Materials, Techniques, and Contemporary Applications for Today's Artist* (Watson-Guptill, 2009). Her books have been translated into Dutch and German. *Acrylic Painting Mediums & Methods,* her third book, explores the intricacies and possibilities of acrylic mediums. She is based in Kingston, Ontario, Canada.

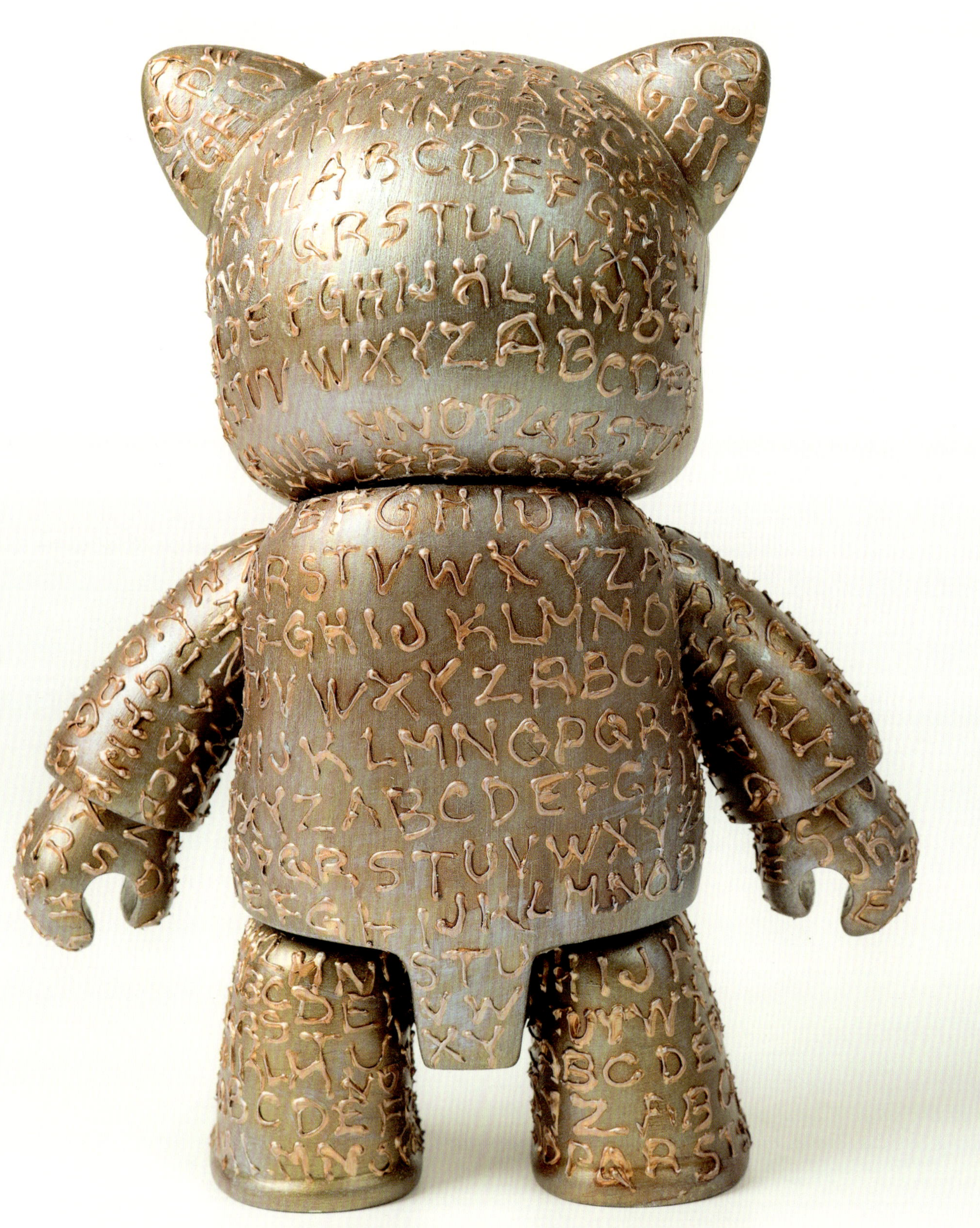

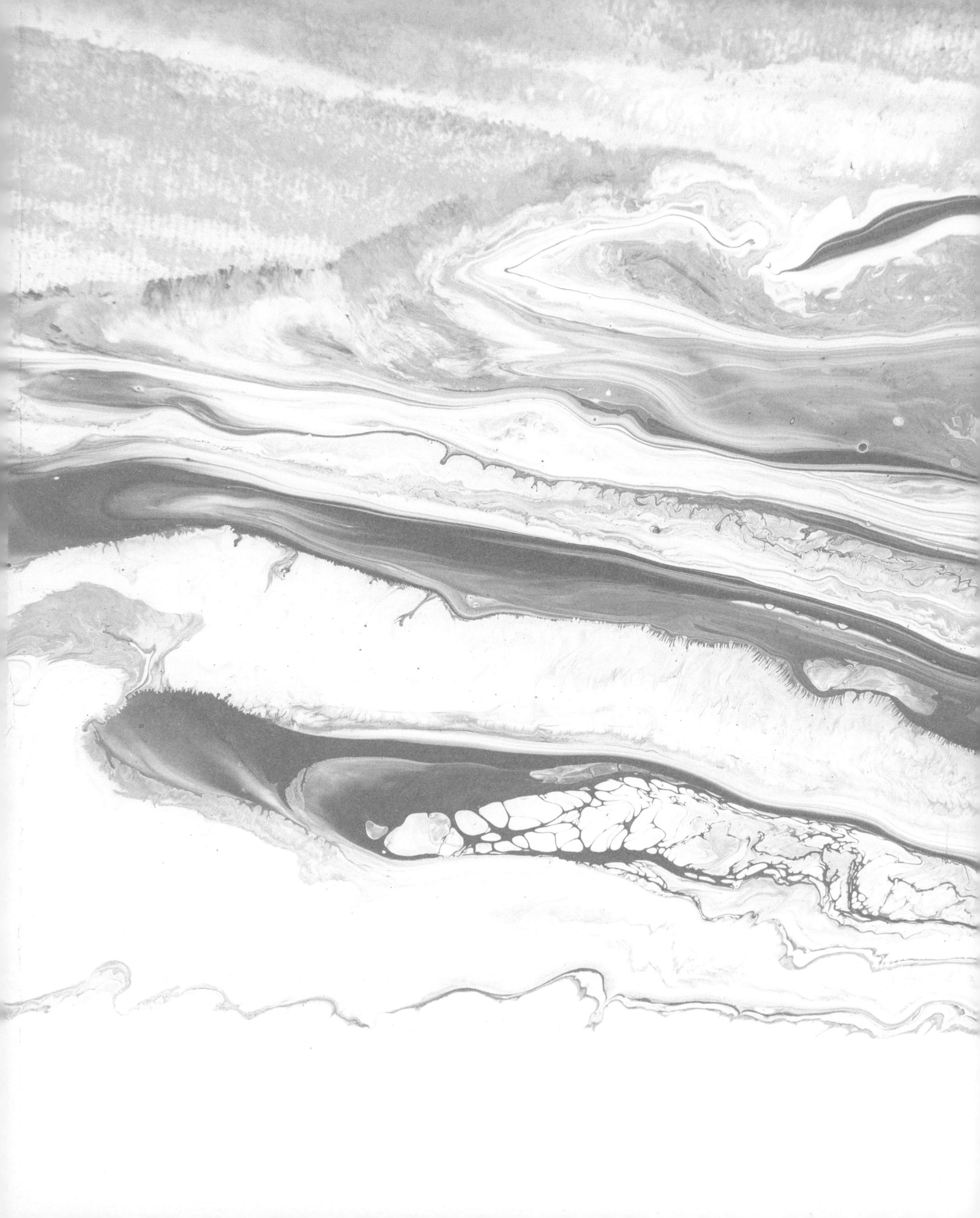

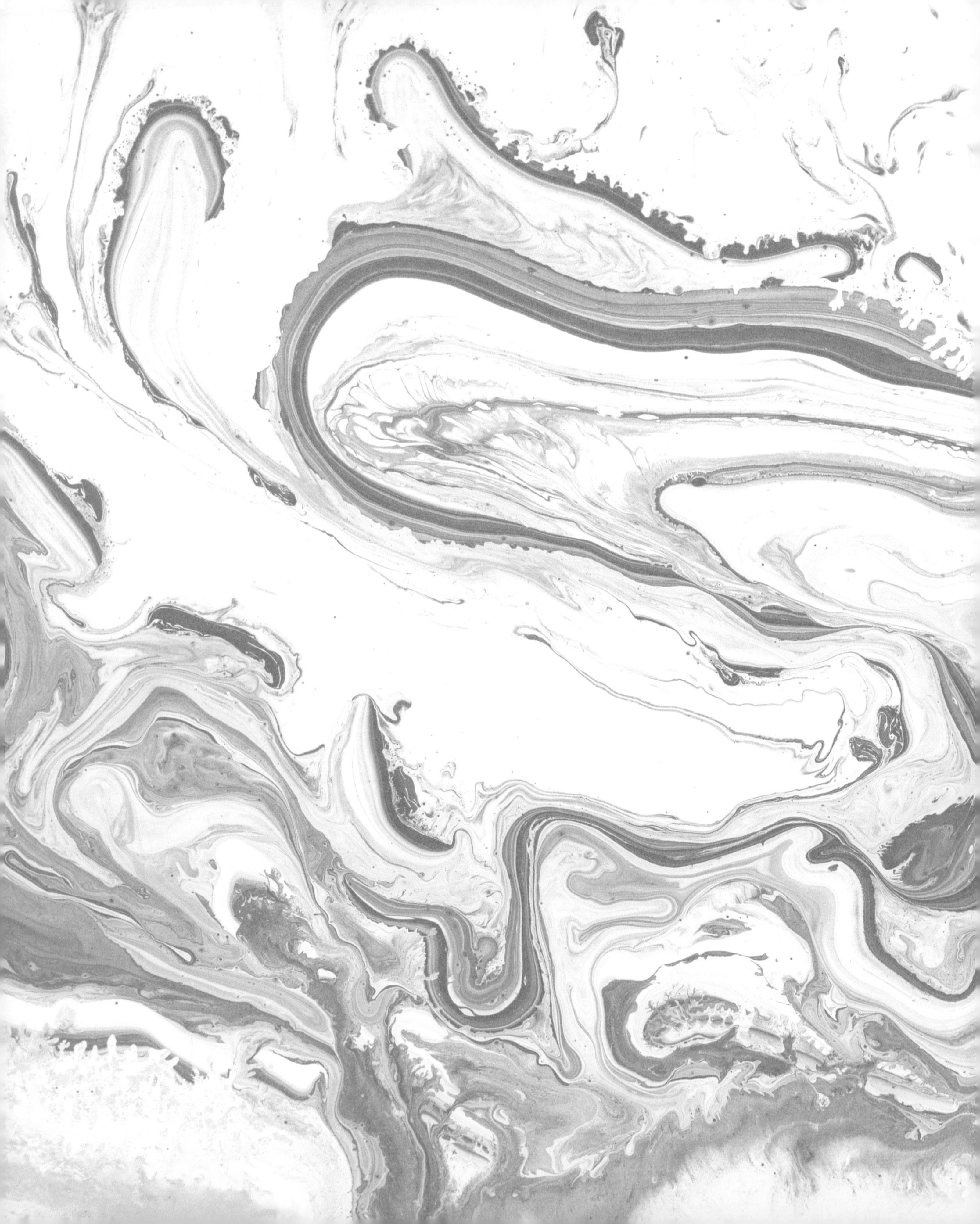